AF581687

Surface Engineering & Coating Technologies for Corrosion and Tribocorrosion Resistance

Surface Engineering & Coating Technologies for Corrosion and Tribocorrosion Resistance

Editor

Yong Sun

Basel • Beijing • Wuhan • Barcelona • Belgrade • Novi Sad • Cluj • Manchester

Editor
Yong Sun
Faculty of Computing,
Engineering and Media
De Montfort University
Leicester
United Kingdom

Editorial Office
MDPI
St. Alban-Anlage 66
4052 Basel, Switzerland

This is a reprint of articles from the Special Issue published online in the open access journal *Materials* (ISSN 1996-1944) (available at: www.mdpi.com/journal/materials/special_issues/Surface_Tribocorrosion).

For citation purposes, cite each article independently as indicated on the article page online and as indicated below:

Lastname, A.A.; Lastname, B.B. Article Title. *Journal Name* **Year**, *Volume Number*, Page Range.

ISBN 978-3-0365-8985-5 (Hbk)
ISBN 978-3-0365-8984-8 (PDF)
doi.org/10.3390/books978-3-0365-8984-8

Contents

About the Editor

Yong Sun

The Editor, Prof. Yong Sun, is a professor of surface engineering in the School of Engineering and Sustainable Development, De Montfort University, Leicester, UK. His research areas include surface engineering and coatings technologies for enhanced tribological properties, as well as the corrosion and tribocorrosion resistance of engineering materials. He has authored and co-authored more than 230 publications with a citation index of more than 8200 and a h-index of 51.

Preface

The corrosion of materials leads to about 3 to 4% economic losses of GDP in an industrial nation and also significantly contributes to greenhouse emissions and climate change, as the production of materials is one of the largest greenhouse emitters. Corroded material replacement leads to increasing demands in materials production. Therefore, from economic and environmental points of view, it is highly necessary and important to enhance the corrosion and tribocorrosion resistance of materials. Since corrosion and tribocorrosion are surface- and subsurface-related material degradation phenomena, surface engineering and coating technologies are the most effective to tackle such problems. This reprint is a collection of the papers published in the Special Issue "Surface engineering & coatings technologies for corrosion and tribocorrosion resistance" in the Materials journal. The Special Issue aims to collate the latest developments in this technologically, economically, and environmentally important area. It provides a forum for researchers to share their original work or insight reviews in this field. This reprint contains an Editorial by the Guest Editor and 12 original research papers and reviews related to this topic contributed by researchers from around the globe. The editor acknowledges the contributions of all the authors and the Editorial team in making this Special Issue a success.

Yong Sun
Editor

materials

MDPI

Editorial

Surface Engineering & Coating Technologies for Corrosion and Tribocorrosion Resistance

Yong Sun

School of Engineering and Sustainable Development, Faculty of Computing, Engineering and Media, De Montfort University, Leicester LE1 9BH, UK; ysun01@dmu.ac.uk

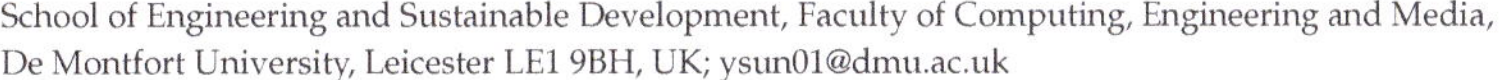

Corrosion of materials not only accounts for about 3 to 4% of economic losses in GDP in an industrial nation, but it also contributes significantly to greenhouse emissions and climate change because material production is one of the largest greenhouse emitters. According to a recent research study [1], steel production will account for 27.5% of carbon emissions worldwide by 2030 and corroded steel replacement will account for up to 9% of these emissions. Therefore, from economic and environmental points of view, it is highly necessary and important to enhance the corrosion and tribocorrosion resistance of materials. Since corrosion and tribocorrosion are surface- and subsurface-related material degradation phenomena, surface engineering and coating technologies are the most effective to tackle the corrosion and tribocorrosion problems of a wide range of engineering materials.

This Special Issue aimed to bring together the latest development in this technologically, economically and environmentally important area, encompassing coating development, corrosion and tribocorrosion characterization and industrial applications. It provides a forum for researchers to share their original work or insight reviews on "Surface engineering and coating technologies for corrosion and tribocorrosion resistance".

This Special Issue contains 12 original research and review papers related to this topic, which have been contributed by researchers from around the globe. A brief description of these published works, which I am honored to edit as a Guest Editor, is given below to highlight the quality and significance of these original research studies.

Coating technologies have been widely used for corrosion protection of automobile bodies made of steel sheets. Steel sheet surfaces are usually treated by phosphating before painting. Advanced high-strength steels (AHSS) that are currently used in automobiles for weight reduction contain a significant amount of Si. A Si oxide film on an AHSS surface can act as a barrier to phosphating. In the work reported in [2], Cho et al., studied the effectiveness of different pickling solutions for removing Si oxide from the surface and improving the phosphatability of AHSS. The results show that the optimal pickling solution was HNO_3-based solution with a HNO_3 concentration higher than 13%. The corrosion resistance of phosphate-treated AHSS was better using the HNO_3-based pickling condition than using the HCl-based pickling condition. This research highlights the importance of pre-treatment in affecting the quality and corrosion performance of the final coating product. In the work reported in [3], Ulbrich et al. studied the tribocorrosion and abrasive wear behaviour of another AHSS steel, i.e., hot-formed 22MnCrB5 steel. The steel was coated with an AlSi coating to prevent oxidation during the hot-forming process. The results show that, compared to the cold-formed state, hot forming followed by appropriate cooling resulted in the formation of a hard and tough martensitic/bainitic structure, which possessed much better corrosion resistance and tribocorrosion resistance in 3.5% NaCl solution.

Surface engineering and coatings are commonly applied to tool steels to improve the performance and service life of cutting and forming tools. In the work reported in [4], Wang et al., prepared Mo coatings on H13 tool steel via the electrospark deposition (ESD) process and studied the mechanical and corrosion properties of the resultant coatings. Under optimal conditions, it was possible to produce a coating of about 35 μm thick

Citation: Sun, Y. Surface Engineering & Coating Technologies for Corrosion and Tribocorrosion Resistance. *Materials* **2023**, *16*, 4863. https://doi.org/10.3390/ma16134863

Received: 3 July 2023
Accepted: 4 July 2023
Published: 6 July 2023

without severe cracks. The coating possessed a high hardness of nearly 1400 HV and a good wear resistance that was seven times higher than that of the substrate. In particular, the corrosion resistance of the coating was much better than that of the substrate due to the excellent corrosion resistance of Mo. The combination improvement in hardness, wear resistance and corrosion resistance would be beneficial in enhancing the working performance of H13 steel dies and molds in practice.

In many applications, composite coatings with multiple elements and phases are used. The electrochemical responses of different elements and phases will affect the corrosion properties of composite coatings. In the work reported in [5], Parchovianská et al., investigated the hydrothermal corrosion behavior of double-layer glass/ceramic composite coatings with passive fillers. The coatings were produced using the polymer-derived ceramic (PDC) synthesis route, and this type of PDC-produced glass/ceramic coatings are suitable for the protection of stainless steel from oxidation at temperatures up to 950 °C [6]. It was found that the composite coating containing polycrystalline Al_2O_3-Y_2O_3-ZrO_2 (AYZ) as an additional filler performed the best. This indicates the potential of PDC coating with AYZ passive filler for applications in harsh environmental conditions.

Galvanizing, or zinc coating, has been commonly used to provide long-term protection against corrosion of steels. However, one of the weaknesses of zinc coating is its relatively low hardness, which affects its sustainable use. Jedrzejczyk and Szatkowska [7] investigated the influence of heat treatment on the hardness and corrosion resistance of hot-dip zinc coating on steel bolts. The Design of Experiments (DOE) approach was used to design the heat treatment experiments. The heat treatment led to the formation of Fe-Zn intermetallic compounds in the coating and, thus, an increased coating hardness from 52 to 204 HV. The authors of [7] demonstrated that through proper selection of the heat treatment conditions, an increase in coating hardness did not lead to a reduction in corrosion resistance, thus maintaining the protective ability of the coating. In many application situations involving the use of galvanized steels, the joining of steels is required and adhesive bonding has become increasingly used. In the work reported in [8], Li et al., studied the interfacial interaction and corrosion resistance between epoxy adhesive and metallic oxide on galvanized steel through a combination of experiments and molecular dynamics (MD) simulation. Their work [8] provided experimental and theoretical evidence regarding atomic bonds at the joint interface.

Surface engineering and coating technologies are also widely used to improve the tribological behavior of titanium alloys, which may also impact corrosion resistance. In the work reported in [9], Zhang et al., studied the corrosion behavior of nitride layer produced on Ti6Al4V alloy through hollow cathodic plasma source nitriding. They found that the nitride layer comprised several sublayers, including a TiN surface top layer, a Ti_2N interlayer, and an interstitial solid-solution α-Ti(N) diffusion zone [9]. Detailed electrochemical corrosion measurements were conducted in Hank's solution. The results showed that the TiN top layer had a better ability for passivation than the untreated alloy, thus leading to a significant reduction in corrosion rate. On the other hand, the Ti_2N interlayer showed deteriorated corrosion resistance when compared to the top TiN layer. This work provides a useful reference for controlled nitriding to achieved optimized corrosion resistance of nitrided titanium alloy. Another category of techniques that has been applied to titanium alloys is laser surface treatments. In the review paper contributed by Wu et al. [10], the authors discussed the fundamentals of the laser shock processing (LSP) technique and its applications to aeronautical materials for improving their wear and corrosion resistance. LSP has many advantages over other mechanical surface-strengthening techniques and can bring about several changes to material surfaces without altering the chemical composition, including grain refinement in the surface region, generation of compressive residual stress to a greater depth, and enhancement of hardness. This review [10] provides a useful reference for researchers to further explore the LSP technique and its application in engineering.

Polymer electrolyte membrane fuel cells (PEMFCs) are one of the promising candidate technologies in vehicles in the fight against greenhouse gas emissions. The bipolar plate is one of the most important components in PENFCs and is exposed to an aggressive environment and corrosive operating conditions. Surface engineering and coating technologies have been used to protect the surface of bipolar plates. In the work reported in [11], Kim et al., deposited a Nb coating on stainless steel, followed by N^+ ion implantation. The authors studied the composition, structure and corrosion performance of the duplex-treated material and found that N+ implantation effectively modified the deposited Nb coating to form an improved protective film with much enhanced corrosion resistance for the metal bipolar plate. This work provides new possibilities for surface modification to improve fuel cell performance and for other applications [11].

There are two contributions relating to corrosion inhibitors in this Special Issue. In the review paper by Wang et al. [12], the authors provided a review on the recent progress of polymer corrosion inhibitors, including natural polymer-based inhibitors and synthetic polymeric inhibitors, with more focus on the structure designs and applications of synthetic polymeric and related hybrid/composite inhibitors. In another review paper contributed by Shapagina and Dushik [13], the application of electrophoretic deposition (EPD) to form an inhibited polymer film on metals for corrosion protection is discussed. The authors highlighted the methods of forming protective films and coatings on metal surfaces from aqueous solutions and demonstrated that EPD is a promising method for modifying metal surfaces to fight against corrosion.

Conflicts of Interest: The author declares no conflict of interest.

References

1. Iannuzzi, M.; Frankel, G.S. The carbon footprint of steel corrosion. *Npj Mater. Degrad.* **2022**, *6*, 101. [CrossRef] [PubMed]
2. Cho, S.; Ko, S.-J.; Yoo, J.-S.; Park, J.-C.; Yoo, Y.-H.; Kim, J.-G. Optimization of Pickling Solution for Improving the Phosphatability of Advanced High-Strength Steels. *Materials* **2021**, *14*, 233. [CrossRef] [PubMed]
3. Ulbrich, D.; Stachowiak, A.; Kowalczyk, J.; Wieczorek, D.; Matysiak, W. Tribocorrosion and abrasive wear test of 20MnCrB5 hot-formed steel. *Materials* **2022**, *15*, 3892. [CrossRef] [PubMed]
4. Wang, W.; Du, M.; Zhang, X.; Luan, C.; Tian, Y. Preparation and Properties of Mo Coating on H13 Steel by Electro Spark Deposition Process. *Materials* **2021**, *14*, 3700. [CrossRef] [PubMed]
5. Parchovianská, I.; Parchovianský, M.; Kaňková, H.; Nowicka, A.; Galusek, D. Hydrothermal Corrosion of Double Layer Glass/Ceramic Coatings Obtained from Preceramic Polymers. *Materials* **2021**, *14*, 7777. [CrossRef] [PubMed]
6. Parchovianský, M.; Parchovianská, I.; Švančárek, P.; Medved', D.; Lenz-Leite, M.; Motz, G.; Galusek, D. High-temperature oxidation resistance of pdc coatings in synthetic air and water vapor atmospheres. *Molecules* **2021**, *26*, 2388. [CrossRef] [PubMed]
7. Jedrzejczyk, D.; Szatkowska, E. The influence of heat treatment on corrosion resistance and microhardness of hot-dip zinc coating deposited on steel bolts. *Materials* **2022**, *15*, 5887. [CrossRef] [PubMed]
8. Li, S.; Zhao, Y.; Wan, H.; Lin, J.; Min, J. Molecular understanding of the interfacial interaction and corrosion resistance between epoxy adhesive and metallic oxides on galvanized steel. *Materials* **2023**, *16*, 3061. [CrossRef] [PubMed]
9. Zhang, L.; Shao, M.; Zhang, Z.; Yi, X.; Yan, J.; Zhou, Z.; Fang, D.; He, Y.; Li, Y. Corrosion behavior of nitride layer of Ti6Al4V titanium alloy by hollow cathodic plasma source nitriding. *Materials* **2023**, *16*, 2961. [CrossRef] [PubMed]
10. Wu, J.; Zhou, Z.; Lin, X.; Qiao, H.; Zhao, J.; Ding, W. Improving the wear and corrosion resistance of aeronautical component material by laser shock processing: A review. *Materials* **2023**, *16*, 4124. [CrossRef] [PubMed]
11. Kim, Y.S.; Choi, J.Y.; Kim, C.H.; Lee, I.S.; Jun, S.; Kim, D.; Cha, B.C.; Kim, D.W. N^+-implantation on Nb coating as protective layer for metal bipolar plate in PEMFCs and their electrochemical characteristics. *Materials* **2022**, *15*, 8612. [CrossRef] [PubMed]
12. Wang, X.; Liu, S.; Yan, J.; Zhang, J.; Zhang, Q.; Yan, Y. Recent progress of polymer corrosion inhibitor: Structure and application. *Materials* **2023**, *16*, 2954. [CrossRef] [PubMed]
13. Shapagina, N.A.; Dushik, V.V. Application of electrophoretic deposition as an advanced technique of inhibited polymer films formation on metals from environmentally safe aqueous solutions inhibited formulations. *Materials* **2022**, *16*, 19. [CrossRef] [PubMed]

Review

Improving the Wear and Corrosion Resistance of Aeronautical Component Material by Laser Shock Processing: A Review

Jiajun Wu [1,*], Zhihu Zhou [1], Xingze Lin [1], Hongchao Qiao [2], Jibin Zhao [2] and Wangwang Ding [3]

[1] College of Engineering, Shantou University, Shantou 515063, China; 21zhzhou@stu.edu.cn (Z.Z.)
[2] State Key Laboratory of Robotics, Shenyang Institute of Automation, Chinese Academy of Sciences, Shenyang 110016, China; hcqiao@sia.cn (H.Q.)
[3] Institute for Advanced Materials and Technology, University of Science & Technology Beijing, Beijing 100083, China
* Correspondence: wujiajun@ustc.edu

Abstract: Since the extreme service conditions, the serious failure problems caused by wear and corrosion are often encountered in the service process for aeronautical components. Laser shock processing (LSP) is a novel surface-strengthening technology to modify microstructures and induce beneficial compressive residual stress on the near-surface layer of metallic materials, thereby enhancing mechanical performances. In this work, the fundamental mechanism of LSP was summarized in detail. Several typical cases of applying LSP treatment to improve aeronautical components' wear and corrosion resistance were introduced. Since the stress effect generated by laser-induced plasma shock waves will lead to the gradient distribution of compressive residual stress, microhardness, and microstruture evolution. Due to the enhancement of microhardness and the introduction of beneficial compressive residual stress by LSP treatment, the wear resistance of aeronautical component materials is evidently improved. In addition, LSP can lead to grain refinement and crystal defect formation, which can increase the hot corrosion resistance of aeronautical component materials. This work will provide significant reference value and guiding significance for researchers to further explore the fundamental mechanism of LSP and the aspects of the aeronautical components' wear and corrosion resistance extension.

Keywords: laser shock processing; aeronautical components; wear resistance; corrosion resistance

Citation: Wu, J.; Zhou, Z.; Lin, X.; Qiao, H.; Zhao, J.; Ding, W. Improving the Wear and Corrosion Resistance of Aeronautical Component Material by Laser Shock Processing: A Review. *Materials* **2023**, *16*, 4124. https://doi.org/10.3390/ma16114124

Academic Editors: Miguel Morales and Amir Mostafaei

Received: 8 March 2023
Revised: 29 May 2023
Accepted: 31 May 2023
Published: 1 June 2023

1. Introduction

The aircraft engine is the power system for modern advanced aircraft [1]. At the same time, as a highly complex and precise power machine, the aircraft engine is also honored as "The crown jewel of modern industry" [2,3]. As for the modern advanced aircraft engine, the aeronautical components such as blades, guide vanes, afterburners, and casings will withstand the centrifugal stresses, thermal stresses, aerodynamic loading, vibration load, wear, and hot corrosion. These extreme service conditions will be harmful to the secure flight of aircraft. With the repaid development of the advanced aircraft, the properties requirements for the aeronautical components are gradually becoming higher and higher, the service conditions will be more extreme such as the increasing working temperatures inside the aircraft engine, more serious centrifugal stresses and various loads, the aeronautical components need to fulfill the requirement of a high thrust-weight ratio for the new generation advanced aero-engines [4,5]. Wear and corrosion are the most severe failure problems often encountered in the service process for aeronautical components [6]. It can easily lead to the cycle slip, plastic accumulation and fatigue cracks for aeronautical components, resulting in the serious aircraft failure or even flight accident [7,8]. Generally, these fatigue failures are caused by wear and corrosion, originating from materials' surface and extending to the inside, resulting in the serious failure for various components [9,10]. According to the existing literature reports, mechanical surface-strengthening technologies have been extensively investigated and applied to adjust the surface integrity parameters such as enhanced microhardness, induced compressive residual stress for the aeronautical components materials without changing any mechanical properties of basal materials, which have essential engineering application significance for reducing the rate of the growth of cracks and even the closing of cracks, and the prevalence of the serious failures of aero-engine components caused by wear and corrosion [2,4,11].

The common mechanical surface-strengthening technologies mainly comprise shot peening (SP) [12], ultrasonic rolling surface strengthening (URSS) [13], ultrasonic shot peening (USP) [14], low plasticity burnishing (LPB) [15], water jet peening (WJP) [16], ultrasonic nano structure modification (UNSM) [17] and so on. The mutual feature of mechanical surface-strengthening technologies is introducing the beneficial compressive residual stress layer, enhancing microhardness, and leading to the microstructure refinement in near-surface layer of targeted material without changing any chemical structures and properties [7]. So these mechanical surface-strengthening technologies can bring about good results and reap some useful benefits for the metallic components. However, the depth of beneficial compressive residual stress layer introduced by these common mechanical surface-strengthening technologies can only be reached to 0.3 mm at most in general, which cannot meet the ever-increasing requirement for the high properties materials and advanced modern equipment [18]. So it's request researchers need to develop and apply more advanced mechanical surface-strengthening technology to tackle the current challenges and problems.

As one of the greatest inventions of natural science in 20th century, the laser has been applied in many engineering fields due to its valuable and exceptional performance. At the same time, many new or novel laser processing technologies utilizing laser are emerged, and these processing technologies have brought great changes to the engineering fields. Laser shock processing (LSP) is a novel mechanical surface-strengthening technology with excellent strengthening effect, controllability, and adaptability [7,8,19,20]. Compared with the common mechanical surface supporting technologies such as SP, URSS, USP and WJP, the adequate beneficial compressive residual stress layer depth induced by LSP treatment can reach over 1 mm. In addition, LSP treatment can obtain more smooth surface morphology relative; that is, LSP can obtain a lower macroscopic surface roughness. For instance, the superalloy FGH97 experimental samples were subjected to LSP treatment and SP treatment; the adequate depth of the compressive residual stress layer of FGH97 caused by SP is about 0.28 mm, while the effective depth of compressive residual stress layer of FGH97 induced by LSP is about 0.84 mm [21]. Gill et al. [22], investigated LSP, CSP, and UNSM on the residual stress of Ni-based superalloy IN718 SPF. And the related experimental results showed that the compressive residual stress layer depth induced by LSP treatment could reach about 0.6 mm, while that caused by the CSP and UNSM treatment are below 0.3 mm. Apart from residual stress, surface topography is also an important parameter to evaluate the qualities of the metallic components and materials treated by mechanical surface-strengthening technologies. Aluminum alloy A356 was modified by LSP treatment and SP treatment. The surface roughness for LSP experimental sample is about 1.1 μm, less than that of SP experimental sample with a value of 5.8 μm [22,23]. So applying LSP treatment to tackle the challenge of the serious failures of aeronautical components can be regarded as a special significance.

LSP treatment has showed evident technical advantages in the fields of surface-strengthening of aeronautical components and related materials. At present, the associated researches mainly focus on the improvements about the fatigue and residual stress of aeronautical components [24–27]. However, the researches, especially the associated summaries about wear and corrosion resistance, are rare in relative. So it is necessary to summary the related researches about the wear and corrosion resistance improvements of aeronautical components by LSP. This work summarized the fundamental mechanism of LSP in detail, and introduced several typical cases about apply LSP treatment to improve the wear and corrosion resistance of aeronautical components materials. This work will provide fundamental reference values and guiding significance for researchers to further explore the primary mechanism of LSP and the aspects of the aeronautical components' wear and corrosion resistance extension.

2. Fundamental Mechanism of Laser Shock Processing

As the latest peening technology was initially introduced in the engineering fields, LSP aims to improve the mechanical performances of aeronautical components [28]. Figure 1 shows the common fundamental mechanism schematic of LSP [29]. Based on the physical interaction process between the pulsed laser and metallic material, LSP utilizes the stress effect generated by high-pressure laser-induced plasma shock waves to treat the materials' surface [30]. When a nanosecond pulsed laser beam irradiates the surface of metallic material with the power density of GW/cm^2 level, the surface layer of the metallic target will absorb the pulsed laser energy and cause explosive vaporization [23]. The vaporized particles will continue to absorb the pulsed laser energy and induce the plasma plume formation at high temperatures (over 10,000 K) nearly simultaneously [30]. Since the confined ablation mode in the LSP process, the laser-induced plasma will expand rapidly and form super-high-pressure shock waves (GPa level) acting on the metal target's surface and propagating to the inside material [2,30–32].

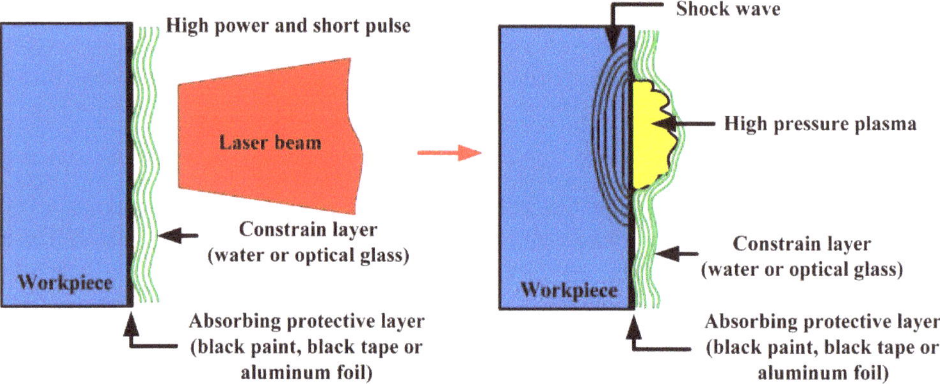

Figure 1. Fundamental mechanism schematic of LSP [29].

As shown in Figure 1, the LSP treatment process adopts the confined ablation mode with a transparent constrained layer and an absorbing protective layer in general. The shocked surface must be coated with absorbing protective layer such as black paint, black tape, or aluminum foil, and then covered with a thin transparent constrain layer such as water or optical glass [30]. It is well known that, the temperature for lasers in engineering fields is very high. So the interaction between laser and material will inevitably lead to the laser thermal effect, which harms the properties improvement of metallic materials. So applying the absorbing protective layer can protect the metal target from laser thermal ablation and create a pure mechanical stress effect in the metallic material [2,17]. The plasma plume generated by the interaction of the laser and material contains enormous energy, which will further induce the formation of shock waves. To improve the pressure of laser-induced plasma shock waves, Clauer et al. [33], proposed the confined ablation mode, the black coating layer coated on the shocked surface of the metal target and then covered with a transparent constrain layer. The transparent constrain layer can be considered a confined space, preventing the plasma from expanding extensively [23]. As a result, the pressure of laser-induced plasma shock wave can be increased by up to two orders of magnitude compared to the plasma generated in the vacuum (without constraint layer) [2,34]. Apart from increase of laser-induced plasma shock wave pressure, the action time of the shock wave can also be increased due to the restrictive effect of the constraint layer [35]. According to Fabbro's theory [32], the action time for the laser-induced plasma shock wave is about 2–3 times of the pulsed width in general. Since the pulsed width for the laser generally exceeds 30 ns, the laser-induced plasma shock waves' action time is below 100 ns. Hence, the strain rates for the LSP process can be reached to the magnitude of 10^7 s^{-1}. And the interaction mechanism of laser-induced plasma shock wave and target material can be referred to the shock wave dynamics. Only when the pressure of the induced shock wave exceeds the dynamic yield strength (${\sigma_Y}^{dyn}$) in a one-dimensional stress state, the severe plastic deformation (SPD) of the metallic target can occur. When the internal stress over the Hugoniot elastic limit (*HEL*), the one-dimensional strain state will be occurred [36]. According to the theory of Johnson and Rhode, the dynamic yield strength and *HEL* can meet the following equations [37].

$$HEL = \sigma_Y^{dyn} \frac{1-v}{1-2v} \tag{1}$$

where ν is the Poisson's ration. The dynamic yield strength can be computed by the Cowper-Symond constitutive equation.

$$\sigma_Y^{dyn} = \sigma_Y^{sta} \left[1 + \left(\frac{\varepsilon'}{D} \right)^{\frac{1}{q}} \right] \tag{2}$$

where ${\sigma_Y}^{sta}$ is the static yield strength, D and q are the constants of material, and ε' is strain rates, which can be expressed as follows.

$$\varepsilon' = \frac{1}{t_0} \tag{3}$$

where t_0 is the action time of the stress wave.

During the LSP process, the deformation induced by LSP generated stress effect can be considered to exist in a one-dimensional stress state, and the surface of the metallic target exists in a one-dimensional strain state [36]. Since the peak pressure of the laser-induced plasma shock waves can be reached to GPa magnitude, which is higher than the dynamic yield strength, SPD will arise in the near-surface layer of metal targets and accompany by the stress effect [38]. The stress effect induced by LSP treatment can lead to cold working in the microstructure refinement evolution and the introduction of beneficial compressive residual stress [2]. Compared with other laser processing technologies, LSP treatment is characterized by using the laser-generated stress effect rather than laser thermal effect; that is LSP utilizes the stress effect of laser-induced plasma shock waves to form a gradient distribution of compressive residual stress, microhardness, and microstructure in the near-surface layer of the metallic target, thereby improving the fatigue performance, wear resistance, corrosion resistance, and tensile property, etc. [7,8].

The LSP process parameters mainly consists of laser parameters (pulsed laser energy, wavelength, laser pulse duration, laser beam spot diameter, and overlap rate), absorbing protective layer, and constrain layer. Generally, the absorbing protective layer and constrain layer is determined in most application. So the researchers usually adjust the laser parameters to obtain the desired LSP effects or results. The laser power density can be expressed as follows.

$$I = \frac{4E}{\pi d^2 \tau} \tag{4}$$

where I is laser power density, E is pulsed laser energy, τ is laser pulse duration. So it can be indicated that laser power density is the comprehensive parameter for laser parameters.

According to laser power density, the peak pressure of laser-induced plasma shock wave can be expressed as follows.

$$P = 0.01\sqrt{\frac{\xi}{3+2\xi}}\sqrt{Z}\sqrt{I} \tag{5}$$

where ξ is fraction of absorbed energy in the range between 0.1 and 0.5, Z is the reduced acoustic impedance, which can be expressed as follows.

$$\frac{2}{Z} = \frac{1}{Z_1} + \frac{1}{Z_2} \tag{6}$$

where Z_1 and Z_2 are acoustic impedance for targeted material and constrained layer, respectively. The acoustic impedance can be calculated by the following equation.

$$Z_i = \rho_i D_i \tag{7}$$

where i = 1 or 2, ρ_i and D_i denotes mass density and velocity of laser-induced plasma shock wave, respectively.

According to the above equations, the peak pressure of laser-induced plasma shock wave is proportional to laser power density. With the higher laser power density, which will lead to higher peak pressure of laser-induced plasma shock waves and more severe plastic deformation. As a result, the compressive residual stress, microhrdness, and the degree of grain refinement will be increased, which means that the better LSP effect. However, the excessive laser power density always lead to laser ablation phenomenon for absorbing protective layer, which will reduce the LSP effect and lead to the surface thermal ablation. So it should be select suitable laser parameters to obtain the optimal LSP effect.

Apart from laser power density, the overlap is an import parameter to affect the LSP effect. The schematic of overlap rate is shown in Figure 2. And the overlap rate can be expressed as follows:

$$\eta = \frac{\delta}{2R} \times 100\% = \frac{\delta}{d} \times 100\% \tag{8}$$

where η is the overlap rate ($0\% \leq \eta < 100\%$), δ is the overlap distance between two adjacent laser spots, R is the radius of laser beam spot, d = 2R.

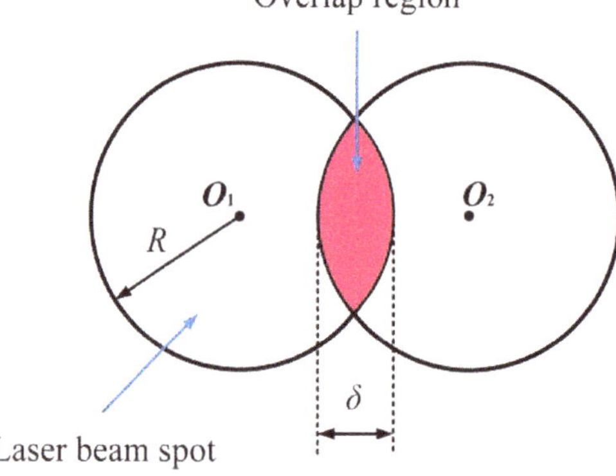

Figure 2. Schematic of overlap rate.

Generally, the higher overlap rate can be regarded as the more shocked times in the same LSP region. So the LSP effect such as induced compressive residual stress, microardness will be increased with overlap rate. Apart from the enhanced mechanical properties, the distribution of surface mechanical properties will be become more uniform. For instance, the effect of overlap rate on the residual stress distribution was investigated by Hu et al. [39], and the related results is shown in Figure 3. The experimental results showed that the increasing the overlap rate can improve the residual stress and also make it more uniform. So select a higher overlap rate can be regarded as a good means to obtain a better LSP effect.

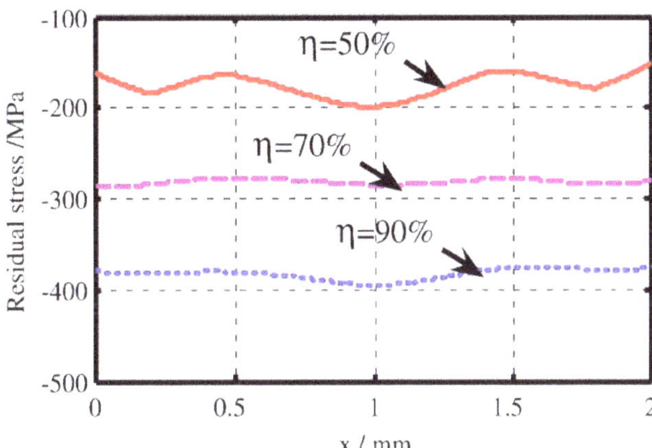

Figure 3. Residual stress distribution on the top surface for different overlap rates [39].

3. LSP Treatment for Wear and Corrosion Resistance Improvement

The serious failures of aeronautical components due to wear and corrosion are most common phenomenon in aerospace industry. The common feature for these failures are originating from the surface and subsurface of materials. As an advanced mechanical surface-strengthening technology, LSP treatment shows tremendous advantages in improving mechanical performances for metallic materials and components. So applying LSP treatment to enhance wear and corrosion resistance can significantly solve failure problems for the aeronautical details.

3.1. Wear Resistance Improvement for Aeronautical Components Material

Since the operation speed of an aero-engine is very high, so the wear problems among the aeronautical components are inevitable [40]. Therefore, improving the wear resistance, especially the impact wear resistance for aeronautical components and related materials by post-treatment is very important.

Since the excellent performance, such as low mass density, high specific strength and excellent wear resistance, Ti-6Al-4V titanium alloy is widely used in the aerospace industry [41]. In the aerospace industry, the blades of aircraft engines are constantly subjected to heavy wear conditions caused by dusts, sands, and other particles entrained by airflow [42]. And there are many investigations have studied the wear mechanism and response of Ti-6Al-4V alloy by different surface-strengthening technologies. In the aspect of LSP, the impact wear behavior of Ti-6Al-4V alloy subjected to LSP treatment was investigated by Yin et al. [41]. In this LSP experiment, the laser parameters used were 5 J and 7 J for pulsed laser energy, 1064 nm for wavelength, 3 mm for laser spot diameter, 10 ns for laser pulse duration, and 33% for overlap rate. According to given laser parameters and Equation (4), the laser power density for 5 J LSP treatment and 7 J LSP treatment are 7.08 GW/cm^2 and 9.91 GW/cm^2, respectively. After the LSP treatment, the impact wear test with initial impact velocities of 60 mm/s and 150 mm/s was conducted for these experimental samples (before and after LSP treatment). The 3D-profile micrographs of the worn scars of experimental samples in different impact wear test conditions are shown in Figure 4. The width and maximum wear depth of the worn scars for the experiment samples in other impact wear test conditions are presented in Figure 5.

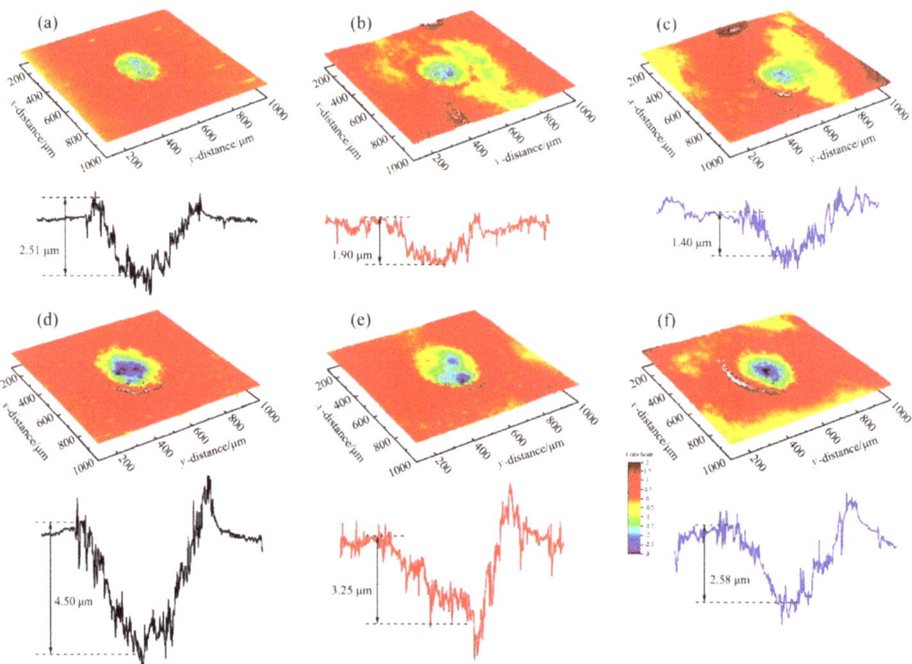

Figure 4. 3D-profile micrographs of the worn scars of experimental samples in different impact wear test conditions: (**a**) prior LSP treatment with initial impact velocities of 60 mm/s; (**b**) 5 J LSP treatment with initial impact velocities of 60 mm/s; (**c**) 7 J LSP treatment with initial impact velocities of 60 mm/s; (**d**) prior LSP treatment with initial impact velocities of 150 mm/s; (**e**) 5 J LSP treatment with initial impact velocities of 150 mm/s; (**f**) 7 J LSP treatment with initial impact velocities of 150 mm/s [41].

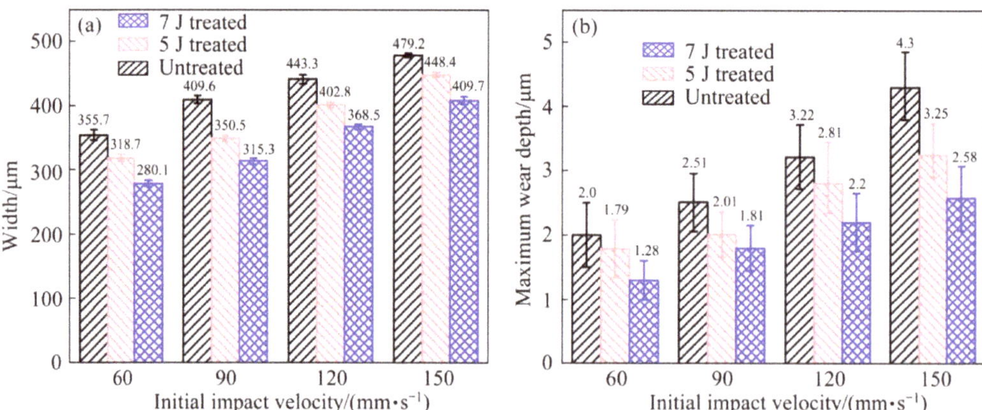

Figure 5. Width and maximum wear depth of wear scars for the experimental samples at different initial impact velocities: (**a**) width; (**b**) maximum wear depth [41].

At the determined initial impact velocities, the width and maximum wear depth of worn scars have similar change laws. The lower width and maximum wear depth of worn scars means a better wear resistance. With the initial impact velocity of 60 mm/s, the width of worn scars for the experiment samples before LSP treatment (untreated), and with 5 J LSP treatment, and 7 J LSP treatment are 335.7 μm, 318.7 μm, and 280.1 μm, respectively. While the maximum wear depth of worn scars for the experiment samples with prior LSP treatment, 5 J LSP treatment, and 7 J LSP treatment are 2.0 μm, 1.79 μm, and 280.1 μm, respectively. When with the initial impact velocity was increased to 150 mm/s, both the width and the maximum wear depth of worn scars were increased. The width of worn scars for the experiment samples with prior LSP treatment, 5 J LSP treatment, and 7 J LSP treatment are increased to 479.2 μm, 448.4 μm and 409.7 μm, respectively. While the maximum wear depth of worn scars for the experiment samples with prior LSP treatment, 5 J LSP treatment and 7 J LSP treatment are increased to 4.3 μm, 3.25 μm and 2.58 μm, respectively. No matter the different impact wear test conditions, it can be indicated that the wear resistance of Ti-6Al-4V titanium alloy is improved after the LSP treatment, which is mainly caused by the microhardness improvement and the introduction of beneficial compressive residual stress [6,43]. And the microhardness and residual stress distribution of experimental samples are shown in Figure 6. In this work, the residual stresses was measured by the Proto-LXRD X-ray diffractometer with $\sin^2 \psi$ method, the microhardness was measured by a standard Vickers indenter with an indentation load of 500 g and the dwell time of 10 s.

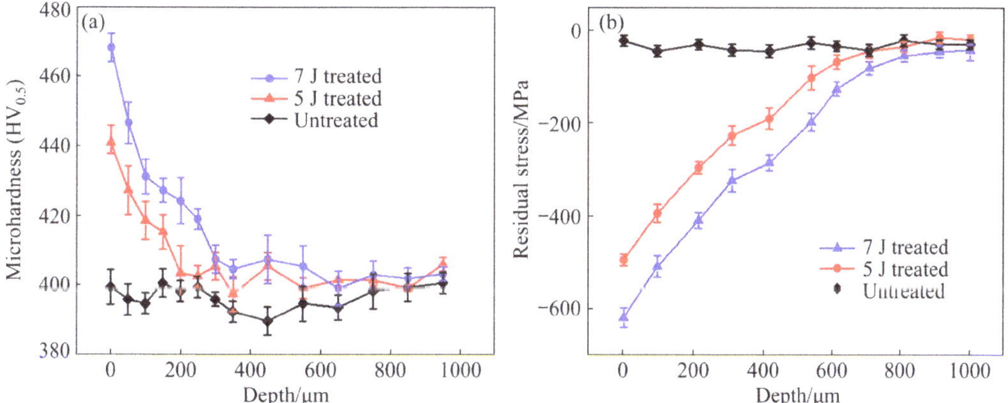

Figure 6. Microhardness and residual stress distribution of experimental samples before and after LSP treatment: (**a**) microhardness; (**b**) Residual stress [41].

It is well known that wear resistance of metallic materials is well intimated with the microhardness, which can be expressed as follows [44].

$$V = K\frac{PL}{H_V} \tag{9}$$

where V is the worn volume, K is the wear factor, P is the positive load during impact wear, L is the impact wear distance, and Hv is the microhardness.

According to Figure 6, the microhardness of Ti-6Al-4V titanium alloy is increased after LSP treatment. So it can be indicated that the experimental sample's worn volume will decrease after LSP treatment. Therefore, the wear resistance will be increased since the shock wave is induced in the LSP process, which will introduce beneficial compressive residual stress. As for metallic components, the compressive residual stress can release and hinder more severe plastic deformation and delay microcracks initiation and propagation, weakening the wearing effect [45].

Of course, the wear resistance improvement level is different in different LSP parameters. Interestingly, under the same initial impact velocity, the wear scars for the experimental sample with 7 J have the shortest width and the shallowest maximum depth. The 5 J LSP treated sample, and finally, the untreated sample. So the wear resistance of the experimental sample with 7 J is better than that with 5 J. The main reason is that the microhardness and induced compressive residual stress are increased through the LSP treatment. Higher laser pulsed energy can lead to higher pressure laser-induced plasma shock waves and produce a more obvious strengthening effect. As shown in Figure 4, the maximum microhardness and maximum compressive residual stress of all experimental samples are presented at the material's surface (e.g., surface microhardness or surface compressive residual stress). In terms of microhardness, the surface microhardness of the Ti-6Al-4V titanium alloy experimental sample prior LSP treatment is about 390 HV, when with the LSP parameter of 5 J, the microhardness of the Ti-6Al-4V titanium alloy experimental sample is increased to about 440 HV, which is about 12.8% higher than the untreated sample. When the laser pulse energy is increased to 7 J, the microhardness of the Ti-6Al-4V titanium alloy experimental sample is increased to about 470 HV, which is about 6.82% higher than the 5 J LSP treated experimental sample. Regarding residual stress, its change law is similar to the microhardness. The surface compressive residual stress for the untreated experimental sample, 5 J LSP treated experimental sample, and 7 J LSP treated experimental sample are about 30 MPa, 495 MPa, and 625 MPa, respectively. All in all, the wear resistance improvement of the Ti-6Al-4V titanium alloy after the LSP was attributed to the enhancement of surface microhardness and beneficial compressive residual stress.

3.2. Corrosion Resistance Improvement for Aeronautical Components Material

With the rapid development of the aerospace industry, aeronautical components are usually used in extreme conditions such as increasing working temperature, growing working pressure, super-high-speed rotation, and beast corrosion environment (containing sodium, sulfur and chloride) [2,46]. Since the high-temperature strength, great anti-oxidation, and excellent hot-corrosion resistance, the Ni-based superalloy has become a super-excellent material to manufacture aeronautical components such as blades, guide vanes, afterburners, turbine, and so on [47]. Since the complex working condition, the hot-corrosion and high-temperature oxidation inevitably occur in the aeronautical components, especially the hot-section components [48]. Hot corrosion and high-temperature oxidation can degrade the material's performance and reduce the service life of these components [49]. So the hot corrosion resistance and high-temperature oxidation resistance of superalloy must be enhanced.

To enhance the hot-corrosion resistance of the aeronautical components' material, Cao et al. [50], selected Ni-based superalloy GH202 as experimental material and investigated the hot corrosion behavior of GH202 treated by LSP. In this work, the LSP experiment was performed in a Q-switched Nd3+:YAG laser with a repetition rate of 0.5 Hz and wavelength of 1.6 μm, and the hot corrosion environment was selected as 800 °C and 900 °C at molten salt. The corrosion kinetics curve of GH202 experimental before and after LSP treatment is displayed in Figure 7 [50]. Since oxidation films were formed on the material's surface at high temperatures, the mass of experimental samples before and after LSP were increased. After the hot corrosion time was over 1h, all experimental samples' mass were decreased. As observed from Figure 7, when the hot corrosion time is around 1 h, the mass gain of the experimental sample after the LSP treatment is less than that of the experiment sample before the LSP treatment. So it can be indicated that the mass loss of the experimental sample after the LSP treatment is less than that of the observed sample before the LSP treatment. So LSP treatment can significantly improve the hot corrosion resistance of GH202 superalloy.

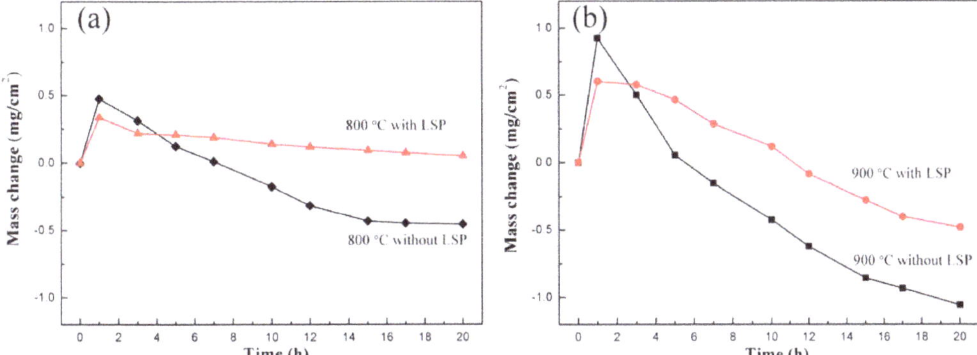

Figure 7. The corrosion kinetics curve of GH202 experimental before and after the LSP treatment: (**a**) at 800 °C; (**b**) at 900 °C [50].

Apart from superalloy GH202, the hot corrosion resistance of the Ni-based single-crystal superalloy was also increased. The hot corrosion behavior of Ni-based single-crystal superalloy treated by LSP was investigated by Geng et al. [46], and its hot corrosion kinetics is shown in Figure 8. In this experiment, the laser parameters were 7 J for laser pulse energy, 1064 nm for wavelength, 20 ns for pulse width, 3mm for spot diameter, and 50% for overlap rate. According to the given laser parameters and the Equation (4), the laser power density in this experiment is 4.95 GW/cm^2. Figure 8 shows that the mass of the non-LSP and LSP experimental samples are increased, reflecting the negative effect of harsh conditions on the material. Interestingly, after two cycles, the mass gains of the LSP-treated experimental sample were more stable during the subsequent cycles. In contrast, the mass gains of non-LSP-treated experimental samples continue to increase.

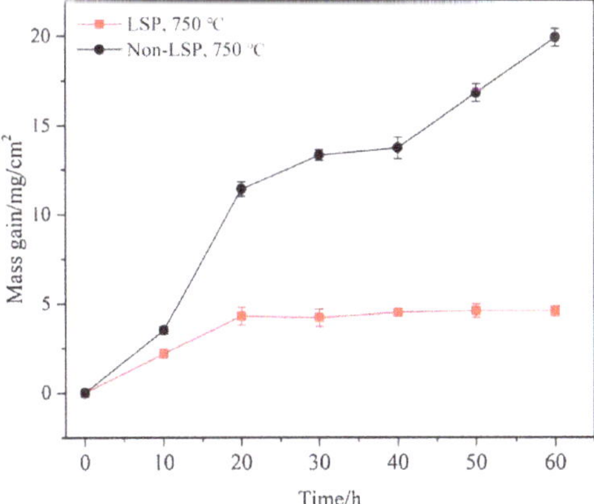

Figure 8. Mass changes of the Ni-based single-crystal superalloy experimental sample before and after the LSP treatment [46].

Compared with Figures 7 and 8, it can be seen that the phenomenon of superalloy GH202 represented weight loss, while the phenomenon of single-crystal superalloy represented weight addition. But the two phenomenons all reflect the improvement of the experimental sample's hot corrosion resistance by the standard LSP treatment.

It is well known that the mechanical properties of near-surface layer can affect the corrosion behaviour of metallic materials components strongly [51]. The hot corrosion resistance improvement of the aeronautical components treated by LSP mainly result from the combine influences of the beneficial compressive residual stress, working hardening, grain refinement, and crystal defects (eg., high-density dislocations) in the near-surface layer [52–54]. The cross-section microstructures of superalloy GH202 experimental specimens before and after the LSP treatment characterized by EBSD is presented in Figure 9 [50]; the different colors represents different orientations. From the Figure 9a, it can be seen that the grains of the superalloy GH202 experimental sample before LSP treatment were distributed unevenly, and some annealing twins can be seen trough the grains. Since the high-pressure plasma shock waves induced by LSP treatment, an arc distribution of the grains which was close to the near-surface layer of superalloy GH202 experimental sample can be observed. Compared with the superalloy GH202 experimental sample before LSP treatment, the grains of superalloy GH202 experimental sample after the LSP treatment were refined significantly. In addition, the twins density for experimental sample after the LSP treatment was also increased significantly.

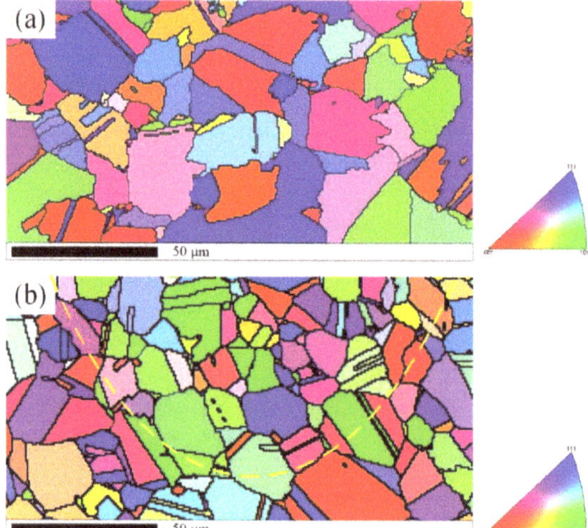

Figure 9. The cross-section microstructures of superalloy GH202 experimental specimens characterized by EBSD before and after the LSP treatment: (**a**) before the LSP treatment; (**b**) after the LSP treatment [50].

It is well known that the grain refinement of microstructures can lead to the increase of dislocation density, which will contribute to the improvement of microhardness. According to the Hall-Petch theory, the microhardness caused by grain refinement can be expressed as follows [55].

$$H_V = H_0 + \alpha G b \rho \tag{10}$$

where H_V is the microhardness, H_0 is the initial microhardness, G is the shear modulus, α is the material's constants, b is the Burgers vector, and ρ is the dislocation density. Hence, LSP induce the dislocation density enhancement will contribute to enhance the microhardness in near-surface layer. In addition, the dislocation density enhancement will facilitate the formation of the homogeneous oxidation film and lead to the microstructure evolution, thereby providing the diffusion paths for the elements such as Cr, Al, and Ti, further producing a protective oxidation film. As a result, the hot corrosion resistance of metallic materials can be improved evidently after the LSP treatment [49,50,52].

In addition, some studies [54,55] have proved that the dislocations in near-surface layer plays an important role in the effect of grain refinement. The movement and accumulation for the dislocations can lead to the grain refinement in the near-surface layer of LSP treated samples, which also help to enhance the corrosion resistance. To reveal the fine structures of microstructures for the superalloy GH202 after the LSP treatment, the representative top surface of experimental sample after the LSP treatment was characterized by transmission electron microscopy (TEM) observation. The TEM image of top surface for superalloy GH202 experimental sample before and after the LSP treatment are shown in Figures 10 and 11, respectively.

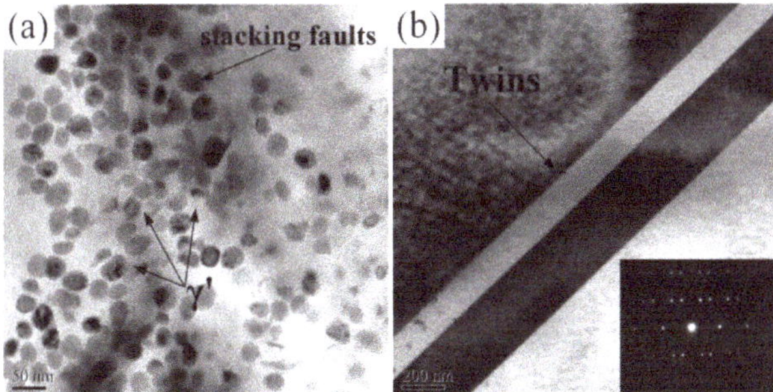

Figure 10. TEM image of top surface for superalloy GH202 experimental sample before the LSP treatment: (**a**) γ' phase and stacking faults; (**b**) annealing twins [50].

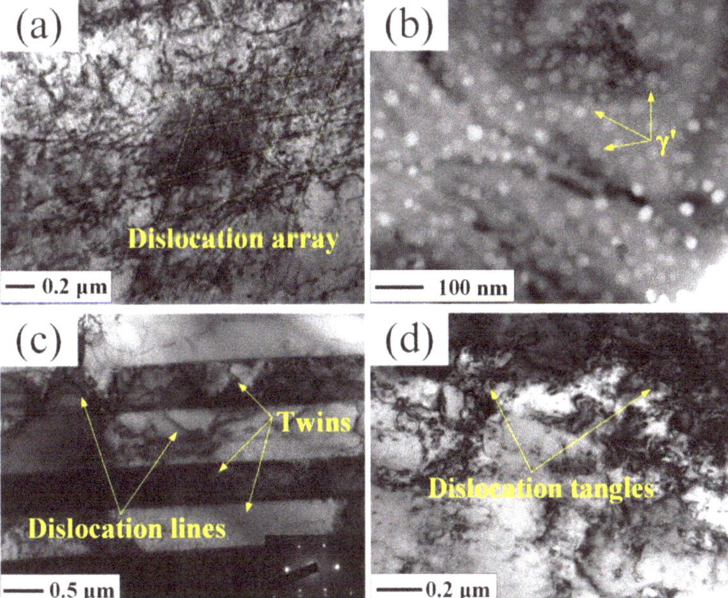

Figure 11. TEM image of top surface for superalloy GH202 experimental sample after the LSP treatment: (**a**) Dislocation array; (**b**) γ' phases; (**c**) Twins and dislocation lines; (**d**) Dislocation tangles in the grain boundaries [50].

Compared with the superalloy GH202 experimental sample before the LSP treatment (Figure 10a), the dislocation density of the superalloy GH202 experimental samples after the LSP treatment was increased significantly (Figure 11a). In addition, the dislocation array can be seen in Figure 11a, which is caused by the reaction and motion of dislocations activated by the laser-induced plasma shock waves. The γ' phases were dispersed evenly after the LSP treatment. At the same time, many twins were present on the surface of superalloy GH202 experimental sample after the LSP treatment, and a few dislocation lines through the twins were observed. And there are several dislocation tangles can be seen in the grain boundaries, which will lead to the limitation of the dislocation motion and the corrosion resistance improvement for the metallic materials or components [56,57].

4. Conclusions and Outlook

LSP is a novel surface-strengthening technology to modify microstructures and induce compressive residual stress on the near-surface layer of materials, thereby enhancing mechanical performance. The main LSP parameters mainly consists of laser parameters (pulsed laser energy, wavelength, laser pulse duration, laser beam spot diameter, and overlap rate), absorbing protective layer, and constrain layer. Select the suitable LSP parameters can obtain a better LSP effect. As an advanced mechanical surface-strengthening technology, LSP treatment shows tremendous advantages in improving mechanical performance for aeronautical components. This work summary the fundamental mechanism of LSP in detail and introduce several typical cases of applying LSP treatment to enhance the wear and corrosion resistance of aeronautical components materials, which will provide significant reference value and guiding significance for researchers to explore further the fundamental mechanism of LSP and the aspects of the wear and corrosion resistance extension of the aeronautical components.

(1) LSP utilizes the stress effect of laser-induced plasma shock waves to form a gradient distribution of compressive residual stress, microhardness, and microstructure in the near-surface layer of the metallic target, thereby improving the mechanical properties.
(2) The wear resistance improvement of metallic material treated with the LSP treatment is attributed to the enhancement of microhardness and beneficial compressive residual stress.
(3) The increased hot corrosion resistance of metallic materials treated by LSP is mainly attributed to the introduction of beneficial compressive residual stress, grain refinement, and crystal defects.

Author Contributions: Conceptualization, J.W. and W.D.; methodology, J.W.; validation, Z.Z., X.L. and H.Q.; formal analysis, J.Z.; investigation, J.W.; resources, J.W., H.Q. and J.Z.; data curation, J.W. and W.D.; writing—original draft preparation, J.W. and Z.Z.; writing—review and editing, J.W., Z.Z., H.Q., J.Z. and W.D.; visualization, J.W.; supervision, W.D.; project administration, J.W.; funding acquisition, J.W., H.Q. and J.Z.; All authors have read and agreed to the published version of the manuscript.

Funding: This research was funded by "Scientific Research Foundation of Shantou University (NTF22001)" and "National Natural Science Foundation of China (51875558)".

Institutional Review Board Statement: Not applicable.

Informed Consent Statement: Not applicable.

Data Availability Statement: All data supporting the conclusions of this manuscript are included within the manuscript.

Conflicts of Interest: The authors declare no conflict of interest.

References

1. Nie, X.F.; Li, Y.H.; He, W.F.; Luo, S.H.; Zhou, L.C. Research Progress and Prospect of Laser Shock Peening Technology in Aero-engine Components. *J. Mech. Eng.* **2021**, *57*, 293–305.
2. Zhou, R.; Zhang, Z.; Hong, M.H. The art of laser ablation in aeroengine: The crown jewel of modern industry. *J. Appl. Phys.* **2020**, *127*, 080902.
3. Shi, P.Y.; Liu, X.H.; Wang, T.; Wang, K.X.; Li, Y.; Zhang, F.S.; He, W.F.; Li, Y.H. Tensile and High Cycle Fatigue Properties and Fracture Mechanism of Near β Titanium Alloy Strengthened by Laser Shock Peening. *Surf. Technol.* **2022**, *51*, 58–65, 166.
4. Lu, G.; Liu, H.; Lin, C.; Zhang, Z.; Shukla, P.; Zhang, Y.; Yao, J. Improving the fretting performance of aero-engine tenon joint materials using surface strengthening. *Mater. Sci. Technol. Lond.* **2019**, *35*, 1781–1788. [CrossRef]

5. Wu, J.J.; Huang, Z.; Qiao, H.C.; Wei, B.X.; Zhao, Y.J.; Li, J.F.; Zhao, J.B. Prediction about residual stress and microhardness of material subjected to multiple overlap laser shock processing using artificial neural network. *J. Cent. South Univ.* **2022**, *29*, 3346–3360. [CrossRef]
6. Yan, L.; Wu, H.N.; Tian, L.; Xu, P.W.; Deng, T.; Nie, X.F. Process Analysis and Friction Wear Performance Study of Laser Impact Strengthened MB8 Magnesium Alloy. *Surf. Technol.* **2022**, *51*, 45–57.
7. Zhou, L.C.; He, W.F. *Gradient Microstructure in Laser Shock Peened Materials: Fundamentals and Applications*; Zhejiang University Press: Hangzhou, China, 2021; ISBN 978-981-16-1747-8.
8. Li, Y.H. *The Theory and Technology of Laser Shock Processing*; Science Press: Beijing, China, 2013; ISBN 978-7-03-037459-2.
9. Guo, W.; Sun, R.; Song, B.; Zhu, Y.; Li, F.; Che, Z.; Li, B.; Guo, C.; Liu, L.; Peng, P. Laser shock peening of laser additive manufactured Ti_6Al_4V titanium alloy. *Surf. Coat. Technol.* **2018**, *349*, 503–510. [CrossRef]
10. Zhang, Y.K.; Lu, J.Z.; Luo, K.Y. *Laser Shock Processing of FCC Metals*; Springer: Berlin/Heidelberg, Germary, 2013; ISBN 978-3-642-35673-5.
11. Yu, Y.Q.; Zhou, L.C.; Gong, J.E.; Fang, X.Y.; Zhou, J.; Cai, Z.B. Microstructure and Fretting Fatigue Behaviour of GH4169 Superalloy after Laser Shock Peening. *Surf. Technol.* **2022**, *51*, 38–48.
12. Li, M.R.; Wang, R.Q.; Tian, T.Y.; Mao, J.X.; Hu, D.Y. Low Cycle Fatigue Life Prediction of DD6 Single Crystal Superalloy by Shot Peening. *Surf. Technol.* **2022**, *51*, 1–9.
13. Meng, C.; Zhao, Y.C.; Zhang, X.Y.; Wang, Y.; He, Y.; Zhang, J. Research and Application of Ultrasonic Rolling Surface Strengthening Technology. *Surf. Technol.* **2022**, *51*, 179–202.
14. Cai, J.; Kiplagat, C.C.; Li, W.; Shi, D.J.; Lin, S. Surface Roughness Numerical and Test Evaluation of FGH97 Powder Superalloy by Ultrasonic Shot Peening. *Surf. Technol.* **2021**, *50*, 250–257.
15. Prevey, P.S.; Cammett, J.T. The influence of surface enhancement by low plasticity burnishing on the corrosion fatigue performance of AA7075-T6. *Int. J. Fatigue* **2004**, *26*, 975–982. [CrossRef]
16. Stolárik, G.; Nag, A.; Petrů, J.; Svobodová, J.; Hloch, S. Ultrasonic Pulsating Water Jet Peening: Influence of Pressure and Pattern Strategy. *Materials* **2021**, *14*, 6019. [CrossRef] [PubMed]
17. Sembiring, J.P.B.A.; Amanov, A.; Pyun, Y.S. Artificial neural network-based prediction model of residual stress and hardness of nickel-based alloys for UNSM parameters optimization. *Mater. Today Commun.* **2020**, *25*, 101391. [CrossRef]
18. Wu, J.J.; Lin, X.Z.; Qiao, H.C.; Zhao, J.; Ding, W.W.; Zhu, R. Microstructural Evolution and Surface Mechanical Properties of the Titanium Alloy Ti-13Nb-13Zr Subjected to Laser Shock Processing. *Materials* **2023**, *16*, 238. [CrossRef]
19. Li, X.; Zhou, L.; Zhao, T.; Pan, X.; Liu, P. Research on Wear Resistance of AISI 9310 Steel with Micro-Laser Shock Peening. *Metals* **2022**, *12*, 2157. [CrossRef]
20. Ouyang, P.; Luo, X.; Dong, Z.; Zhang, S. Numerical Prediction of the Effect of Laser Shock Peening on Residual Stress and Fatigue Life of Ti-6Al-4V Titanium Alloy. *Materials* **2022**, *15*, 5503. [CrossRef] [PubMed]
21. Gao, Y.K.; Zhong, Z.; Lei, L.M. Influence of Laser Peening and Shot Peening on Fatigue Properties of FGH97 Superalloy. *Rare Metal. Mater. Eng.* **2016**, *45*, 1230–1235.
22. Gill, A.; Telang, A.; Mannava, S.R.; Qian, D.; Pyoun, Y.S.; Soyama, H.; Vasudevan, V.K. Comparison of mechanisms of advanced mechanical surface treatments in nickel-based superalloy. *Mater. Sci. Eng. A* **2013**, *576*, 346–355. [CrossRef]
23. Lu, G.; Sokol, D.W.; Zhang, Y.; Dulaney, J.L. Nanosecond pulsed laser-generated stress effect inducing macro-micro-nano structures and surface topography evolution. *Appl. Mater. Today* **2019**, *15*, 171–184. [CrossRef]
24. Gu, H.Q.; Yan, P.; Jiao, L.; Chen, S.Q.; Song, Y.F.; Zou, S.K.; Wang, X.B. Effect of laser shock peening on boring hole surface integrity and conformal contact fretting fatigue life of Ti-6Al-4V alloy. *Int. J. Fatigue* **2023**, *166*, 107241. [CrossRef]
25. Wang, B.H.; Cheng, L.; Li, D.C. Study on very high cycle fatigue properties of forged TC4 titanium alloy treated by laser shock peening under three-point bending. *Int. J. Fatigue* **2022**, *156*, 106668. [CrossRef]
26. Zhang, H.P.; Cai, Z.Y.; Wan, Z.D.; Peng, P.; Zhang, H.Q.; Sun, R.J.; Che, Z.G.; Guo, C.; Li, B.; Guo, W. Microstructure and Mechanical properties of laser shock peened 38CrSi steel. *Mater. Sci. Eng. A* **2020**, *788*, 139486. [CrossRef]
27. An, Z.; He, W.; Zhou, X.; Zhou, L.; Nie, X. On the Microstructure, Residual Stress and Fatigue Performance of Laser Metal Deposited TC17 Alloy Subjected to Laser Shock Peening. *Materials* **2022**, *15*, 6501. [CrossRef] [PubMed]
28. Azhari, A.; Sulaiman, S.; Rao, A.K.P. A review on the application of peening processes for surface treatment. *IOP Conf. Ser. Mater. Sci. Technol.-Lond.* **2016**, *114*, 012002. [CrossRef]
29. Wu, J.J.; Liu, X.J.; Zhao, J.B.; Qiao, H.C.; Zhang, Y.N.; Zhang, H.Y. The online monitoring method research of laser shock processing based on plasma acoustic wave signal energy. *Optik* **2019**, *183*, 1151–1159. [CrossRef]
30. Hu, Y.X. Research on the Numerical Simulation and Impact Effects of Laser Shock Processing. Ph.D. Thesis, Shanghai Jiao Tong University, Shanghai, China, 2008.
31. Wu, J.J.; Zhao, J.B.; Qiao, H.C.; Hu, X.L.; Yang, Y.Q. The New Technologies Developed from Laser Shock Processing. *Materials* **2020**, *13*, 1453. [CrossRef]
32. Fabbro, R.; Peyre, P.; Berthe, L.; Scherpereel, X. Physics and applications of laser-shock processing. *J. Laser Appl.* **1998**, *10*, 265–279. [CrossRef]
33. Clauer, A.H.; Fairand, B.P.; Wilcox, B.A. Pulsed laser induced deformation in an Fe-3 Wt Pct Si alloy. *Metall. Trans. A* **1977**, *8*, 119–125. [CrossRef]

34. Montross, C.; Wei, T.; Lin, Y.; Clark, G.; Mai, Y. Laser shock processing and its effects on microstructure and properties of metal alloys: A review. *Int. J. Fatigue* **2002**, *24*, 1021–1036. [CrossRef]
35. Anderholm, N.C. Laser-Generated Stress Waves. *Appl. Phys. Lett.* **1970**, *16*, 113–115. [CrossRef]
36. Dai, F.; Pei, Z.; Ren, X.; Huang, S.; Hua, X.; Chen, X. Effects of different contact film thicknesses on the surface roughness evolution of LY2 aluminum alloy milled surface subjected to laser shock wave planishing. *Surf. Coat. Technol.* **2020**, *403*, 126391. [CrossRef]
37. Johnon, J.N.; Rhode, R.W. Dynamic Deformation Twinning in Shock-Loaded Iron. *J. Appl. Phys.* **1971**, *42*, 4171. [CrossRef]
38. Dhakal, B.; Swaroop, S. Effect of laser shock peening on mechanical and microstructural aspects of 6061-T6 aluminum alloy. *J. Mater. Process. Technol.* **2020**, *282*, 116640. [CrossRef]
39. Hu, Y.X.; Yao, Z.Q. Overlapping rate effect on laser shock processing of 1045 steel by small spots with Nd: YAG pulsed laser. *Surf. Coat. Technol.* **2008**, *202*, 1517–1525. [CrossRef]
40. Wang, J.F. Study on Fretting Wear Behavior of Titanium Alloy in Aero-Engine. Ph.D Thesis, University of Science and Technology of China, Hefei, China, 2022.
41. Yin, M.G.; Cai, Z.B.; Li, Z.Y.; Zhou, Z.R.; Wang, W.J.; He, W.F. Improving impact wear resistance of Ti-6Al-4V alloy treated by laser shock peening. *Trans. Nonferrous Metals Soc.* **2019**, *29*, 1439–1448. [CrossRef]
42. Anandavel, K.; Prakash, R.V. Effect of three-dimensional loading on macroscopic fretting aspects of an aero-engine blade–disc dovetail interface. *Tribol. Int.* **2011**, *44*, 1544–1555. [CrossRef]
43. Li, Y.Q.; Wang, X.D.; Yang, Z.F.; Song, F.L. Wear resistance of copper improved by laser shock peening. *High Power Laser Part. Beams* **2016**, *28*, 202–206.
44. Lu, J.Z.; Xue, K.N.; Lu, H.F.; Xing, F. Laser shock wave-induced wear property improvement and deformation mechanism of laser cladding Ni25 coating on H13 too steel. *J. Mater. Process. Technol.* **2021**, *296*, 117202. [CrossRef]
45. Luo, W.; Selvadurai, U.; Tillmann, W. Effect of residual stress on the wear resistance of thermal spray coating. *J. Therm. Spray Technol.* **2016**, *25*, 321–330. [CrossRef]
46. Geng, Y.X.; Mo, Y.; Zheng, H.Z.; Li, G.F.; Wang, K.D. Effect of laser shock peening on the hot corrosion behavior of Ni-based single-crystal superalloy at 750 °C. *Corros. Sci.* **2021**, *185*, 109419. [CrossRef]
47. Wang, J.; Zhou, L.; Sheng, L.; Guo, J. The microstructure evolution and its effect on the mechanical properties of a hot-corrosion resistant Ni-based superalloy during long-term thermal exposure. *Mater. Des.* **2012**, *39*, 55–62. [CrossRef]
48. Huang, G.H.; Dai, J.W.; Wang, Z.K.; Mou, R.D.; He, L.M. High temperature corrosion behavior of Co-Al coating on the single crystal superalloy. *Vacuum* **2016**, *53*, 13–17.
49. He, D.G.; Lin, Y.C.; Tang, Y.; Li, L.; Chen, J.; Chen, M.S.; Chen, X.M. Influences of solution cooling on microstructures, mechanical properties and hot corrosion resistance of a nickel-based superalloy. *Mater. Sci. Eng. A* **2019**, *746*, 372–383. [CrossRef]
50. Cao, J.; Zhang, J.; Hua, Y.; Chen, R.; Li, Z.; Ye, Y. Microstructure and hot corrosion behavior of the Ni-based superalloy GH202 treated by laser shock processing. *Mater. Charact.* **2017**, *125*, 67–75. [CrossRef]
51. Wei, B.X.; Xu, J.; Gao, L.Q.; Feng, H.; Wu, J.J.; Sun, C.; Wang, Z.Y.; Ke, W. Nanosecond pulsed laser-assisted modified copper surface structure: Enhanced surface microhardness and microbial corrosion resistance. *J. Mater. Sci. Technol.* **2022**, *107*, 111–123. [CrossRef]
52. Chen, L.; Zhang, X.; Gan, S. Microstructure and Hot Corrosion of GH2036 Alloy Treated by Laser Shock Peening. *JOM* **2020**, *72*, 754–763. [CrossRef]
53. Karthik, D.; Swaroop, S. Laser shock peening enhanced corrosion properties in a nickel based Inconel 600 superalloy. *J. Alloys Compd.* **2016**, *694*, 1309–1319. [CrossRef]
54. Ralston, K.D.; Birbilis, N.; Davies, C.H.J. Revealing the relationship between grain size and corrosion rate of metals. *Scr. Mater.* **2010**, *63*, 1201–1204. [CrossRef]
55. Orlowska, M.; Ura-Bińczyk, E.; Olejnik, L.; Lewandowska, M. The effect of grain size and grain boundary misorientation on the corrosion resistance of commercially pure aluminium. *Corros. Sci.* **2019**, *148*, 57–70. [CrossRef]
56. Shadangi, Y.; Chattopadhyay, K.; Rai, S.B.; Singh, V. Effect of LASER shock peening on microstructure, mechanical properties and corrosion behavior of interstitial free steel. *Surf. Coat. Technol.* **2015**, *280*, 216–224. [CrossRef]
57. Jia, Y.J.; Chen, H.N.; Zhang, J.Q.; Lei, J.B. Research on the Wear Resistance and Electrochemical Corrosion Properties of CoCrBiNbW High Entropy Alloy by Laser Melting Deposited. *Surf. Technol.* **2022**, *51*, 350–357, 370.

Article

Molecular Understanding of the Interfacial Interaction and Corrosion Resistance between Epoxy Adhesive and Metallic Oxides on Galvanized Steel

Shuangshuang Li [1], Yanliang Zhao [2], Hailang Wan [1,*], Jianping Lin [1,*] and Junying Min [1]

[1] School of Mechanical Engineering, Tongji University, Shanghai 201804, China
[2] Baoshan Iron & Steel Co., Ltd., Shanghai 201900, China
* Correspondence: hailang_wan@tongji.edu.cn (H.W.); jplin58@tongji.edu.cn (J.L.)

Abstract: The epoxy adhesive-galvanized steel adhesive structure has been widely used in various industrial fields, but achieving high bonding strength and corrosion resistance is a challenge. This study examined the impact of surface oxides on the interfacial bonding performance of two types of galvanized steel with Zn–Al or Zn–Al–Mg coatings. Scanning electron microscopy and X-ray photoelectron spectroscopy analysis showed that the Zn–Al coating was covered by ZnO and Al_2O_3, while MgO was additionally found on the Zn–Al–Mg coating. Both coatings exhibited excellent adhesion in dry environments, but after 21 days of water soaking, the Zn–Al–Mg joint demonstrated better corrosion resistance than the Zn–Al joint. Numerical simulations revealed that metallic oxides of ZnO, Al_2O_3, and MgO had different adsorption preferences for the main components of the adhesive. The adhesion stress at the coating–adhesive interface was mainly due to hydrogen bonds and ionic interactions, and the theoretical adhesion stress of MgO adhesive system was higher than that of ZnO and Al_2O_3. The corrosion resistance of the Zn–Al–Mg adhesive interface was mainly due to the stronger corrosion resistance of the coating itself, and the lower water-related hydrogen bond content at the MgO adhesive interface. Understanding these bonding mechanisms can lead to the development of improved adhesive-galvanized steel structures with enhanced corrosion resistance.

Keywords: galvanized steel; adhesive bonding; interfacial interaction; corrosion resistance; molecular dynamics (MD); density functional theory (DFT)

Citation: Li, S.; Zhao, Y.; Wan, H.; Lin, J.; Min, J. Molecular Understanding of the Interfacial Interaction and Corrosion Resistance between Epoxy Adhesive and Metallic Oxides on Galvanized Steel. *Materials* **2023**, *16*, 3061. https://doi.org/10.3390/ma16083061

Academic Editors: Costica Bejinariu and Raul D. S. G. Campilho

Received: 18 February 2023
Revised: 21 March 2023
Accepted: 10 April 2023
Published: 13 April 2023

1. Introduction

Galvanized steel is utilized in various intricate shapes and structures, such as automobile bodies, due to its outstanding corrosion resistance [1–3]. In industrial manufacturing, adhesive bonding of metal materials is increasingly replacing conventional joining methods, including welding, diffusion bonding, and riveting, owing to its merits of balanced load, high joint stiffness, and reduced galvanic corrosion between jointed components [4–6]. However, the adhesive interface may fail in a humid environment due to moisture-induced desorption [7]. To design high-performance adhesive bonding structures for galvanized steel, it is crucial to understand interfacial interaction and corrosion resistance mechanisms at the atomic and molecular levels. This understanding can also provide theoretical guidance for coating preparation to achieve superior interfacial bonding strength and joint durability.

In recent years, molecular dynamics (MD) simulation and density functional theory (DFT) research has become essential to understand the complex interfacial phenomena between metal and polymer [8–11]. Bahlakeh et al. [12] investigated the effects of a new cerium–lanthanum (Ce–La) nanofilm-treated steel surfaces on the interfacial bonding mechanism of an epoxy adhesive. They found that the electrostatic interactions between epoxy adhesive and the nanofilm (consisting of CeO_2 and LaO_3) were stronger than those on an untreated steel surface. In addition to investigations on flat metal surfaces, some research focuses on the interfacial interactions of metal–adhesive interfaces in nanostructures. Liu et al. [13] used MD simulations to calculate interfacial interactions and bonding processes between polymers and metals (aluminum and copper). They proposed that the viscoelasticity and polarity of the polymers influence the interfacial interactions, which determines the final performance of the bonded structure by influencing the wall-slip behavior. Li et al. [14] investigated the influence of nanopit structures on the interactive behavior and bonding performance between metal and polymer. Compared with rectangular, cylindrical, and pyramidal nanopits, the conical nanopits were found to be beneficial for the bonding performance of the Cu-PPS interface due to their enhanced interfacial energy and wettability. According to the DFT calculation results of Lee et al. [15], the horizontal orientation of epoxy adhesive on Fe (100) metal surface is stronger than the vertical orientation, and the hydroxyl group and benzene ring of the epoxy adhesive are main functional groups that generate adhesion forces in metal–adhesive interface. Moreover, adhesive components adsorbed on the oxide surface significantly affect the minimum energy path and reaction energy. Knaup et al. [16] found that the adhesion promoter, 3-aminopropylmethoxysilane, exhibits a preference for adsorption over bisphenol A diglycidyl ether on the Al_2O_3 surface, whereas the adsorption of the curing agent (diethyltriamine) is poor. To better understand the dynamic behavior of polymers on metal surfaces, Semoto et al. [17] showed that hydrogen bonds were generated between the hydroxyl groups of the epoxy polymer and aluminum oxide, which are the main forces contributing to the adhesion force. Additionally, Tsurumi [18] found that the epoxy cresol novolac and phenol novolac fragments form physical bonds to the Cu surface through dispersion forces, while chemical bonds to the surface of Cu_2O through σ-bonds and hydrogen bonds. The maximum adhesion stress was 1.6 and 2.2 GPa for the Cu and Cu_2O surfaces, respectively. The hot-dip Zn–Al or Zn–Al–Mg coatings on steel sheets can provide excellent corrosion resistance and alter the adhesion of the steel sheet to adhesive [1,19]. However, existing research is insufficient to explain the interfacial phenomena between the epoxy adhesive and the galvanizing coating, and there is a lack of understanding of the adhesion mechanism of the adhesive interface on the galvanized steel sheet at the molecular level.

Water at metal–adhesive interface is a common cause of adhesion loss in moist environments, but its effects on metal–adhesive adhesion are not fully understood [7,20–23]. Semoto et al. [24] proposed that the water molecules in the interface generate a hydrogen bond network and interact with the epoxy resin and the substrate surface, providing a weak adhesion interaction. However, the study of Higuchi et al. [24] on the interface between silica and epoxy resin revealed that adsorbed water molecules reduced the interfacial adhesion energy and force. This reduction may be due to the deformation and flexibility of the H_2O molecules and the hydrogen bond network. In addition, some scholars have studied the influence of the thickness of water layer on interfacial adhesion energy and force [25]. DFT calculations of the effect of water molecules on the interface between aluminum oxide and epoxy resin revealed that when an H_2O molecule resides in close proximity to the Al–O bond, it enhances the dissociation of the O atom from the epoxy group, causing the water layer in the interfacial area to become alkaline. This alkaline environment damages the interfacial bonding and breaks the bisA ether groups, leading to a significant reduction in adhesion strength [26,27]. Galvanized steel sheets have complex elements, making it a significant research topic to use the density functional theory (DFT) method to explore the effect of surface oxides on the adhesion loss of the coating–adhesive interface caused by H_2O molecules.

In this study, it is intended to assess the effect of surface oxides on the interfacial interaction and corrosion resistance between epoxy adhesive and galvanized steel. The physical and chemical properties of Zn–Al and Zn–Al–Mg galvanized steel surfaces were characterized, and the adhesion strength in dry and wet condition between the coatings and the epoxy adhesive was investigated. Molecular models of the adhesive interface were established based on the surface characterization. Through MD simulations, nonbonded interactions between epoxy adhesive molecules and oxide surfaces (consisting of ZnO, Al_2O_3, and MgO) were elucidated. Periodic DFT calculations were then carried out on slab models consisting of epoxy fragments and the three oxide surfaces, with and without water molecules at the interface. The calculation results provided electronic characteristics, chemical bonding, and adhesion force information for the interfaces between the adhesives and the oxides. Finally, by combining DFT calculation with EIS test, the corrosion resistance of ZnO, Al_2O_3, and MgO adhesive interfaces was compared and analyzed under water conditions.

2. Experimental Procedures

2.1. Materials

Two types of zinc galvanized steel sheets, with Zn–Al and Zn–Al–Mg coatings and a thickness of 0.8 mm, were selected as the bonding materials in this study. The zinc galvanized steel sheets were cut into $100 \times 25 \times 0.8$ mm^3 substrates by laser cutting. All specimens were cleaned with anhydrous ethanol before adhesive bonding. Henkel TEROSON EP 5089 (Düsseldorf, Germany), a hot-cured epoxy adhesive (mixing bisphenol A diglycidyl ether, DGEBA, and modified polyurethane resin, MPUR) was selected to prepare the joints.

2.2. Characterization

The chemical compositions of the zinc galvanized steel surfaces were analyzed by using an X-ray photo electron spectroscopy (XPS, ESCALAB 250, Thermo Fisher Scientific (Waltham, MA, USA)). An Al-Kα source operating at 15 kV and 30 mA was used to obtain the XPS spectrum. The basic pressure of the analysis chamber was 1×10^{-9} Torr.

A scanning electron microscope (SEM, Nova NanoSEM 450, FEI (Hillsboro, OR, USA)) was used to observe the surface physical morphology of the zinc galvanized steel sheets. Element concentration detection was conducted using an energy dispersive X-ray spectroscope (EDS).

The open circuit potential and electrochemical impedance spectroscopy (EIS) tests were performed on the Zn–Al and Zn–Al–Mg coating samples using the CorrTest CS310 electrochemical measuring system. A three-electrode system was used, with the galvanized steel sample serving as the working electrode, a platinum sheet electrode (20 mm × 20 mm) as the counter electrode, and Ag/AgCl (saturated KCl) as the reference electrode. The frequency range for the EIS was 10^5–10^{-2} Hz, during which the open circuit potential (OCP) remained stable. The EIS data were further fitted using the ZView 3.0a software.

2.3. Water Soak

To simulate the effect of interfacial water on the interfacial bonding performance, the fully cured galvanized steel/adhesive joints were soaked in deionized water at 55 °C for 21 days, following the standard GMW15200. After the 21-day immersion period, the joints were immediately removed from the water and subjected to lap-shear strength testing.

2.4. Lap-Shear Strength Testing

Lap-shear strength testing was performed to determine the interfacial bonding performance of the joints. The joints were prepared in accordance with ISO4587 with a lapped area of 12.5×25 mm^2. Glass balls with a diameter of 0.25 mm were used to control the thickness of the adhesive layer. Due to the low thickness of the galvanized steel sheet, to avoid its plastic deformation during the joint strength testing, a 1.0 mm thick TC4 titanium alloy sheet was bonded to the back of the lapped side to strengthen the joint. To prevent adverse effects on the adhesive interface caused by corrosion mediums of the galvanized coating in non-lapped area during the water soak, the non-lapped area on the lapped side was painted to protect the galvanized coating. The joints were cured at 170 °C for 20 min to ensure full curing of the adhesive. The schematic diagram and photograph of the joints are shown in Figure 1. Five repeat joints were pulled off on a universal testing machine (MTS E45.105) at a speed of 10 mm/min.

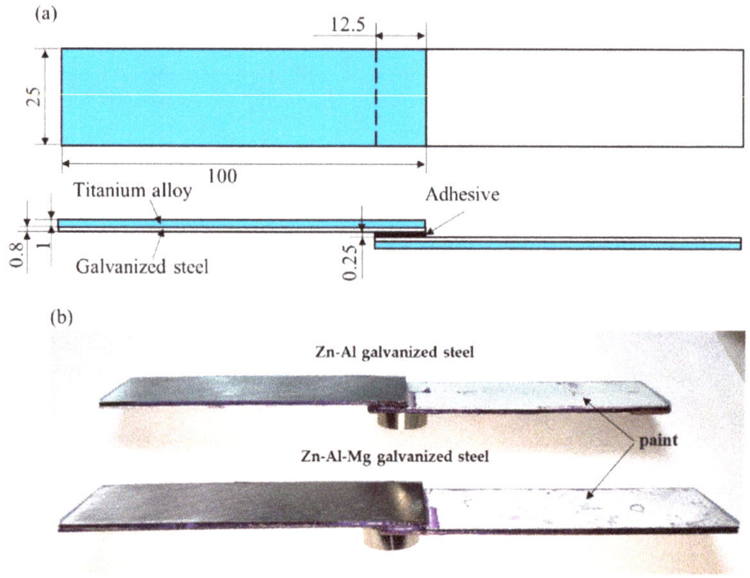

Figure 1. (**a**) Schematic diagram and (**b**) photograph of the galvanized steel joints.

3. Modeling Details

The simulation of this work was completed with Materials Studio 2018 software. The Henkel adhesive is a complex epoxy system consisting of epoxy monomers/oligomers, curing agents, and additives. According to the references [9,28,29], the molecular structure of the Henkel adhesive was simplified into two main components: the epoxy monomer/oligomer (DGEBA) and the curing agent (MPUR). The optimized geometries of the monomers of DGEBA, MPUR, and the reaction product of MPUR and DGEBA (abbreviated as MUPR-modified DGEBA) are shown in Figure 2a–c. Surface models of ZnO (10-10), Al_2O_3 (100), and MgO (100) were constructed with similar number of atoms of 1296, 1200, and 1296, respectively. After adding a vacuum layer with a thickness of 35 Å on top of the oxide surfaces, the dimensions of the three models were 29.2 Å × 31.2 Å × 50.0 Å, 27.9 Å × 25.2 Å × 49.5 Å, and 26.8 Å × 26.8 Å × 49.7 Å for ZnO (10-10), Al_2O_3 (100), and MgO (100), respectively. All adhesive molecular chains were placed 10 Å above the surfaces to ensure a similar initial distance between the adhesive molecules and the oxide surfaces. The main chains were oriented parallel to the surface to maintain the same distance. All oxide–adhesive geometries were optimized and equilibrated for 500 ps with fixed oxide surface atoms at room temperature to ensure that the oxide–adhesive system reached equilibrium. The simulations were carried out under the NVT ensemble and the COMPASS force field [30]. The Andersen thermostat was used for temperature control. By using atom-based cutoff and Ewald methods, nonbonded van der Waals (vdW) and electrostatic interactions were considered, respectively.

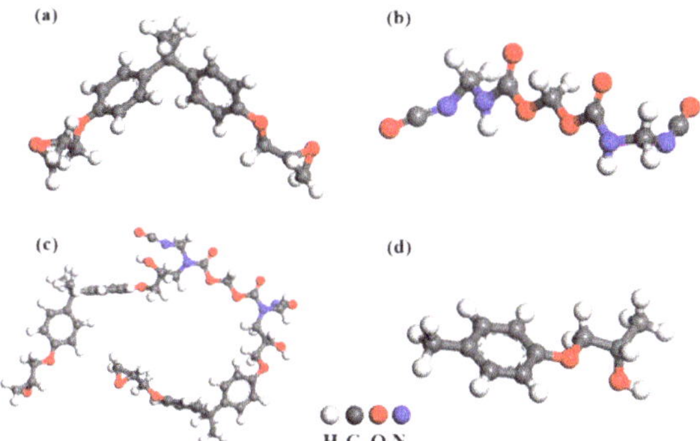

Figure 2. Optimized geometries of the monomer of (**a**) DGEBA, (**b**) MUPR, and (**c**) MUPR-modified DGEBA in MD simulation, and (**d**) DGEBA segment in DFT calculation [16,28,29].

A DGEBA molecule segment was used as the adhesive model in DFT calculations, as shown in Figure 2d. Supercells of ZnO (10-10), Al_2O_3 (100), and MgO (100) with 128, 120, and 120 atoms, respectively, and a vacuum layer thickness of 15 Å were placed above the oxide surfaces. The final dimensions of the ZnO (10-10), Al_2O_3 (100), and MgO (100) supercells were 13.0 Å × 10.4 Å × 24.4 Å, 12.6 Å × 8.4 Å × 23.4 Å, and 11.2 Å × 8.4 Å × 26.3 Å, respectively. Two adsorption models were established to investigate the effect of water on the interfacial interaction between the adhesive and the galvanized coating: a dry model, in which the DGEBA segment was placed parallel to the oxide surfaces and then geometrically optimized; and a water model, in which the stable structure of five water molecules adsorbed on the surface was calculated, and then the DGEBA segment was placed above the structure of water molecules for geometric optimization. In all adsorption models, the oxide surfaces were fixed, while the adsorbates (adhesive/water molecules) were allowed to relax.

The interaction energy, E_{int}, in MD simulation and adsorption energy, E_{ad}, in DFT calculation can be calculated as follows [17]:

$$E_{\text{int or ad}} = E_{\text{substrate/adsorbate}} - (E_{\text{substrate}} + E_{\text{adsorbate}}) \tag{1}$$

where $E_{\text{substrate/adsorbate}}$ represents the total energy of the substrate–adsorbate system, and $E_{\text{substrate}}$ and $E_{\text{adsorbate}}$ represent the energies of the substrate and the adsorbate molecule, respectively.

The adhesion force can be calculated as follows:

$$F = \frac{dE_{ad}}{d\Delta r} \tag{2}$$

where Δr is the distance from the stable equilibrium position of the adhesive molecule, and E_{ad} is the total energy of the surface–adhesive system. The value of Δr changes in the range of −0.8~1.2 Å. The system energy containing adhesive molecule and surface was calculated every 0.1 Å in dry/water models. Morse potential approximation fitting of the energy–distance plots was performed using the least square method in the range of −0.8~1.2 Å, where the Morse potential is written as follows:

$$E = E_0\left(e^{-2\frac{\Delta r}{\lambda}} - 2e^{\frac{-\Delta r}{\lambda}}\right) \tag{3}$$

where E_0 is the minimum value of the potential, and λ is a constant that determines the range of the interaction force.

All DFT calculations were performed in $Dmol^3$ module. The double-numerical plus polarization (DNP) functions were used [31]. A 3 × 3 × 1 k-point grid was used in the calculations. The Perdew–Burke–Ernzerhof (PBE) generalized gradient approximation (GGA) was employed to treat exchange-correlation interactions of electrons and optimize each adsorption mode. The energy and displacement convergence criteria of geometric optimization were set to 2×10^{-5} and 5×10^{-3}, respectively.

4. Results and Discussion

4.1. Physical and Chemical Properties of Galvanized Steel Surface

The chemical compositions of the Zn–Al and Zn–Al–Mg coatings were investigated using XPS. Table 1 shows that the main metallic elements of the Zn–Al coating were Zn and Al, while the Zn–Al–Mg coating contained an additional 2.16 at. % Mg. The low Fe content on the surface of both coatings indicates complete coverage of the coatings on the steel substrate. The XPS results show that the coating surface contains significant amounts of C and O. C is possibly produced by carbon-containing contaminants during the production, storage, and transportation of galvanized steel. Part of the O element comes from the contaminants and part from metal oxides on the coating. High-resolution XPS spectra of Zn 2p, Al 2p, and Mg 2p measured for Zn–Al and Zn–Al–Mg coatings are presented in Figure 3. The Zn $2p^{3/2}$ core level of both coatings were deconvoluted into two peaks at 1022.1 eV (oxidized zinc species, Zn^{2+}) and 1020.9 eV (metallic zinc, Zn^0), respectively. The Al 2p peak reflects the oxidation states (Al^{3+}) of the Al on the Zn–Al coating, while metallic aluminum (Al^0) is found in the Zn–Al–Mg coating. The Mg1s has one peak at a binding energy of 1304.1 eV, corresponding to the presence of oxidized magnesium species (Mg^{2+}). The composition of these elements is similar to that reported in the literature [1]. Based on the analysis of metal species on the Zn–Al and Zn–Al–Mg coatings, the coatings have a high concentration of metal oxides [32]. Therefore, the surfaces characterized by metal oxides, namely ZnO, Al_2O_3, and MgO, will be the focus of the research on the adhesion characteristics between the coatings and epoxy adhesives.

Table 1. Chemical composition (at. %) on the surface of Zn–Al and Zn–Al–Mg coatings.

Element	Atomic %	
	Zn–Al	**Zn–Al–Mg**
Zn	13.10	8.37
Al	9.00	7.41
Mg	0.05	2.16
Fe	0.69	0.36
O	37.47	42.27
C	39.70	39.43

Figure 4 presents SEM and EDS maps of the Zn–Al and Zn–Al–Mg coatings. The Zn–Al coating is mainly composed of a massive microstructure (Figure 4a). As shown in Figure 4b–d, Zn and Al elements are uniformly distributed on the surface, while the higher concentration of O at the edge of the massive microstructure reveals the local enrichment of zinc oxide and aluminum oxide. The Zn–Al–Mg coating presents a mixed surface morphology of convex microstructure and dendritic microstructure (Figure 4e). EDS results show that Zn is uniformly distributed, while Al, Mg, and O are segregated in the dendritic microstructure area [33], as shown in Figure 4f–i, which reflects the uneven distribution of aluminum oxide and magnesium oxide on the surface of the coating.

4.2. Interfacial Bonding Performance between Epoxy Adhesive and Galvanized Steel

To evaluate the interfacial bonding performance between Zn–Al or Zn–Al–Mg galvanized steel and adhesive and the effect of water soak on the bonding performance, the lap-shear strengths were measured under both dry and water-soaked conditions. As shown in Figure 5, under dry conditions, the Zn–Al joint exhibits a higher adhesion strength (32.3 MPa) than the Zn–Al–Mg joint (29.2 MPa). Despite the difference in adhesion strengths, both joints show excellent interfacial bonding performance with cohesive fracture surfaces [34,35]. SEM observation of the fracture surface of the Zn–Al–Mg joint reveals a large number of holes, which are few in the fracture surface of the Zn–Al joint. The cross section of the adhesive joint (Figure 5b) reveals the presence of bubbles in the adhesive layer of Zn–Al–Mg joint. In contrast, fewer and smaller bubbles are observed in the adhesive layer of the Zn–Al joint. These bubbles weakened the mechanical properties of the adhesive and contributed to the lower bonding strength of the Zn–Al–Mg joint. The formation of bubbles suggests that the Zn–Al–Mg coating exhibited poor infiltration to the adhesive compared to the Zn–Al coating, and air could not be expelled outside the adhesive structure when the adhesive was in contact with the Zn–Al–Mg coating.

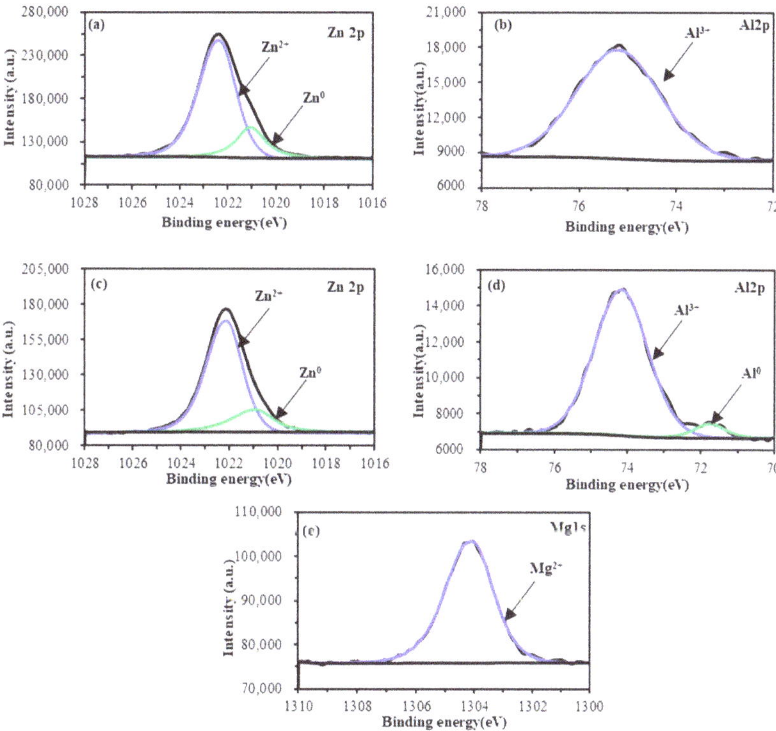

Figure 3. High resolution elemental spectra of (**a**) Zn 2p, (**b**) Al 2p of the Zn–Al coating, (**c**) Zn 2p, (**d**) Al 2p, and (**e**) Mg 1s of the Zn–Al–Mg coating.

The experimental results indicate that after 21 days of water soaking, the adhesion strength of galvanized steel sheets with Zn–Al and Zn–Al–Mg coatings decreased by 14.7% and 7.6%, respectively. The fracture surface of the Zn–Al joint shows large-area interfacial failure, which is related to the diffusion of corrosive medium from the edge of the interface [19,36,37]. In contrast, less and sporadic interfacial failure areas are observed on the fracture surface of the Zn–Al–Mg joint. These findings suggest that the Zn–Al–Mg coating has better corrosion resistance in a water environment than the Zn–Al coating. To understand the difference in their corrosion resistance, it is crucial to investigate the interfacial interaction between the two coatings and the adhesive, as well as the behavior of water molecules in the interfaces [38].

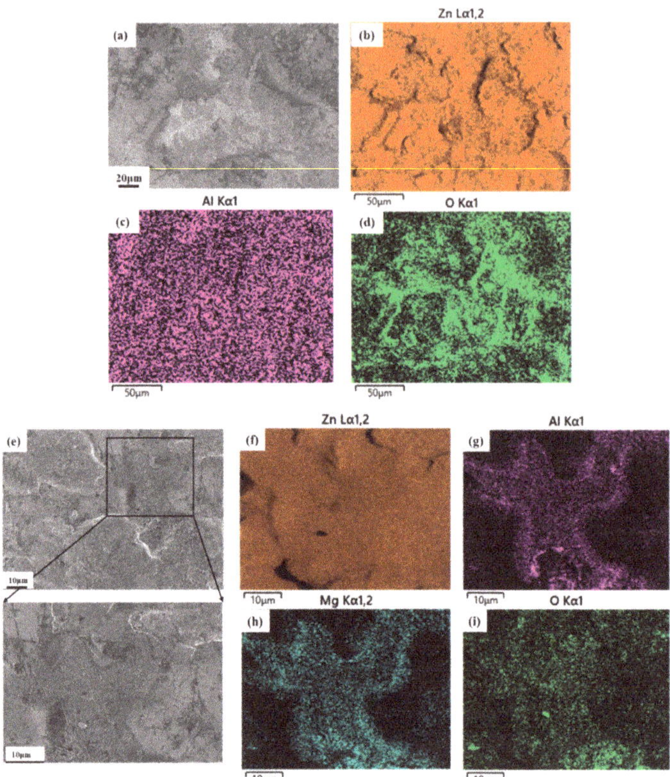

Figure 4. SEM micrographs and EDS maps of elements acquired on (**a**–**d**) Zn–Al and (**e**–**i**) Zn–Al–Mg coating surfaces.

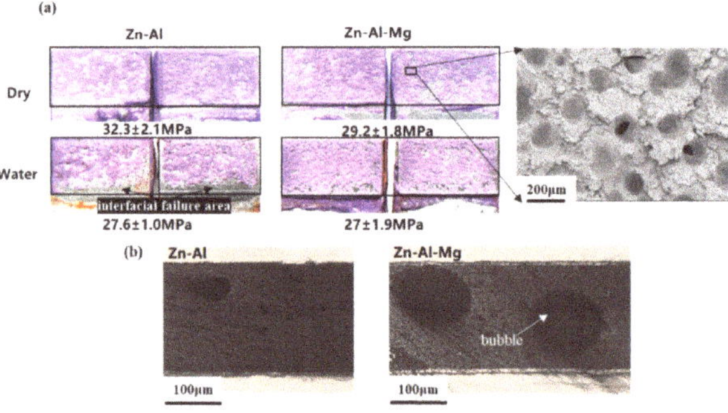

Figure 5. (**a**) Adhesion strength and fracture surfaces of dry and water joint and (**b**) cross section of dry joints of the Zn–Al and Zn–Al–Mg galvanized steel sheets.

4.3. Molecular Behavior and Adhesion Force at Epoxy Adhesive/Galvanizing Coating Interface

This section focuses on the role of nonbonded and chemical interactions in the interaction between epoxy adhesive and galvanized steel. Nonbonded interactions mainly refer to van der Waals forces and electrostatic interactions, while chemical interactions include ionic bonding, covalent bonding, and coordination bonding, etc. [39]. The physical adsorption resulting from nonbonded interactions can provide some initial adhesion strength, but it is usually not sufficient to form a durable bonding [40]. In contrast, chemical interactions can form a relatively strong and durable bonding between the adhesive and metal surface during the curing process [41].

Upon contact with a metal surface, the adhesive can form a uniform thin film on the surface, driven by nonbonded interactions [41,42]. These forces reduce the distance between the adhesive and the metal surface, promoting infiltration and interfacial contact. Investigating the interfacial nonbonded interactions between metal oxides on the coating surface and epoxy adhesive molecules can help to understand the infiltration behavior of epoxy adhesive on the coating surface at the molecular/atomic scale [43–45]. Figure 6 shows the molecular structure and adsorption energy of DGEBA, MPUR, and MPUR-modified DGEBA before and after adsorption on ZnO (10-10) surfaces. After adsorption, all three adsorbents moved to the ZnO substrate, with the distance between the adsorbents and substrate remaining constant as the equilibrium time increased. This process reflects the phenomenon of adhesive infiltration on the ZnO (10-10) surface under the action of nonbonded interaction forces. According to the adsorption results, the ZnO (10-10) surface has adsorption effects on DGEBA, MPUR, and MPUR-modified DGEBA. The simulation of MPUR-modified DGEBA show that the hydroxyl group points to the surface, and the methyl group deviates from the surface, indicating that the hydroxyl group is a functional group that is easily adsorbed, while the methyl group is a functional group that is difficult to be adsorbed. In addition, the adsorption energy of the three adsorbed substances, which is composed of van der Waals and electrostatic interaction energies, is quantitatively analyzed. The nonbonded interaction energies of DGEBA, MPUR, and MPUR-modified DGEBA with ZnO (10-10) surfaces are −49.1 kcal/mol, −42.8 kcal/mol, and −112.6 kcal/mol, respectively. It can be concluded from the comparison of interaction energy that ZnO (10-10) prefers to adsorb DGEBA rather than MUPR. Negative interaction energies further indicate the spontaneous occurrence of the adsorption and infiltration of the adhesive on ZnO (10-10) surfaces, and the van der Waals force dominates the process as demonstrated by the relatively high interaction.

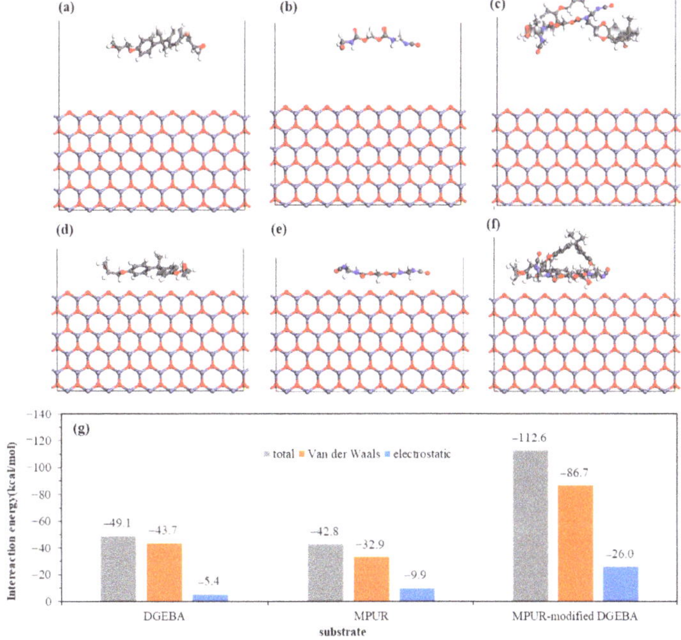

Figure 6. Molecular structure of ZnO(10-10) surface before and after interaction with (**a**,**b**) DGEBA, (**c**,**d**) MPUR, and (**e**,**f**) MPUR-modified DGEBA, and (**g**) their adsorption energy.

Figure 7a–f shows the molecular structures and adsorption energies of the adsorbates before and after adsorption on the Al_2O_3 (100) surface. After adsorption, the adsorbates are found to be close to the Al_2O_3 (100) surface, and the distance of the adsorbates in the longitudinal direction decreased, indicating their spreading on the substrate. The nonbonded interaction energies of DGEBA, MPUR, and MPUR-modified DGEBA with the Al_2O_3 (100) surface are −67.7 kcal/mol, −61.7 kcal/mol, and −147.1 kcal/mol, respectively. Similar to the ZnO (10-10) surface, the Al_2O_3 (100) surface also prefers to adsorb DGEBA. The negative nonbonded interaction energies again demonstrate the interfacial infiltration of the adsorbates on the Al_2O_3 (100) surface. Compared with the ZnO (10-10) surface, the Al_2O_3 (100) surface has stronger interfacial interactions with the three adsorbates, which is confirmed by the higher interaction energy. However, considering that chemical interactions also play a crucial role in determining the final adhesion effect, this result does not necessarily indicate that the adhesive force on the Al_2O_3 (100) surface is stronger than that on the ZnO (10-10) surface.

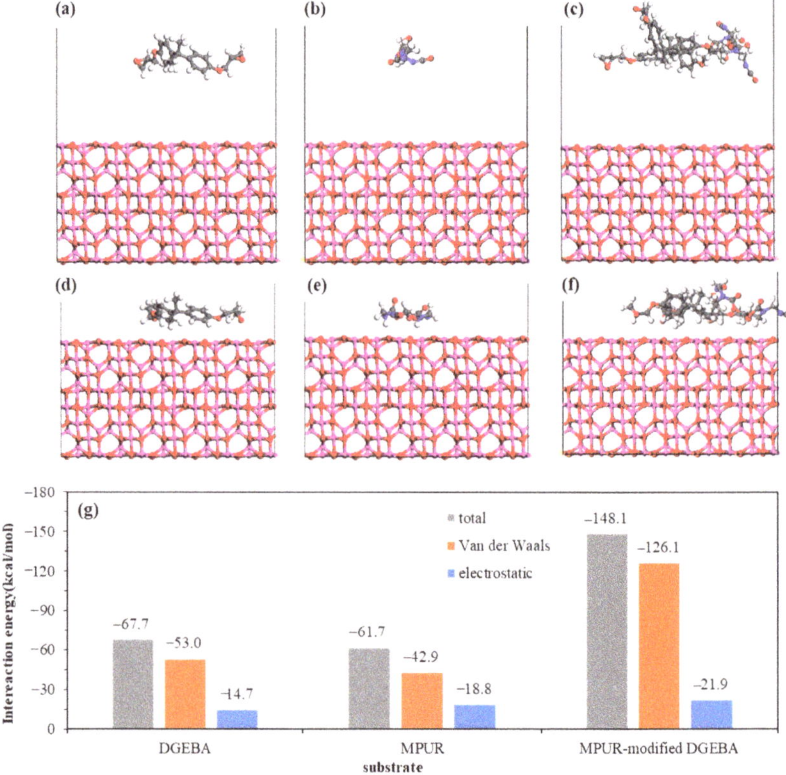

Figure 7. Molecular structure of Al_2O_3 (100) surface before and after interaction with (**a**,**b**) DGEBA, (**c**,**d**) MPUR, and (**e**,**f**) MPUR-modified DGEBA, and (**g**) their adsorption energy.

The molecular structures and adsorption energies of DGEBA, MPUR, and MPUR-modified DGEBA on the MgO (100) surface before and after adsorption are shown in Figure 8. After adsorption, all three adsorbents are found to be in close proximity to the MgO substrate. However, unlike on the ZnO (10-10) and Al_2O_3 (100) surfaces, the longitudinal size of MPUR-modified DGEBA did not decrease significantly after adsorption on the MgO (100) surface, indicating that its infiltration into the MgO surface is lower. The nonbonded interaction energies of DGEBA, MPUR, and MPUR-modified DGEBA with the

ZnO (10-10) surface are −40.7 kcal/mol, −61.6 kcal/mol, and −89.2 kcal/mol, respectively. Compared with DGEBA, the MgO (100) surface shows a stronger preference for the adsorption of MPUR due to its electrostatic force. Furthermore, the difference in interaction energy between DGEBA and MPUR on the MgO (100) surface (51.4%) is significantly higher than that on the ZnO (10-10) (12.8%) and Al_2O_3 (100) (8.9%) surfaces, indicating a more obvious preference for MPUR adsorption on the MgO (100) surface. However, as previously analyzed, the adhesion force of the adhesive on the coating surface is not only directly determined by the nonbonded interaction but also by chemical interaction.

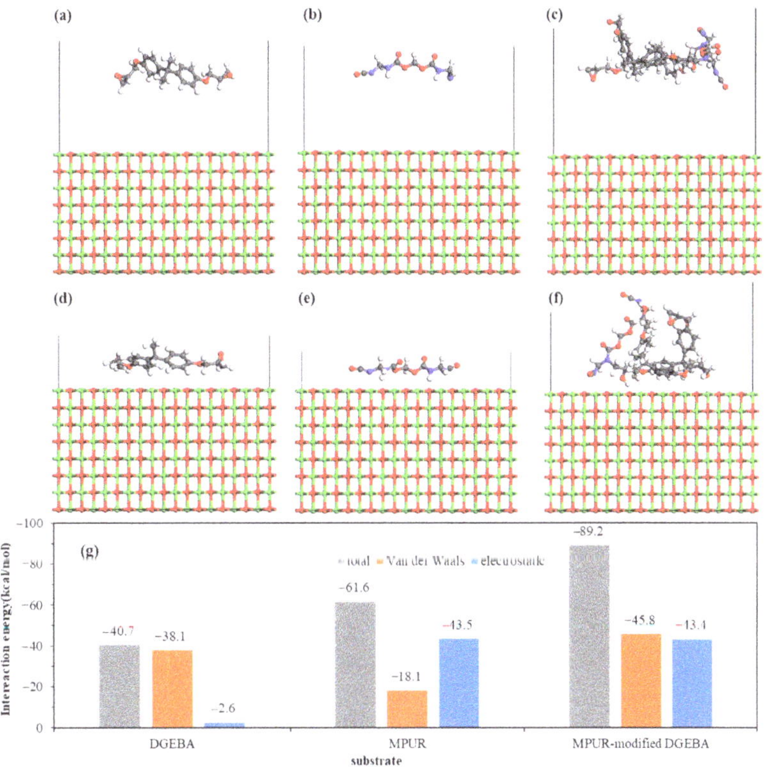

Figure 8. Molecular structure of MgO(100) surface before and after interaction with (**a**,**b**) DGEBA, (**c**,**d**) MPUR, and (**e**,**f**) MPUR-modified DGEBA, and (**g**) their adsorption energy.

To quantitatively investigate the influence of ZnO, Al_2O_3, and MgO on the adhesion performance of the galvanized steel sheet, DFT calculations were employed to evaluate the adhesion force and chemical bonding of DGEBA fragment on three oxide surfaces. In a dry environment, the calculated energy–displacement curve is shown in Figure 9a. The curve conforms to the Morse potential in the range of −0.8 eV to 1.2 eV. The distance between the DGEBA fragment and the surface was defined as 0 Å, where the system reached a stable adsorption state. As the distance exceeds 0, the adsorption energy tends to 0, and as the distance becomes less than 0, the adsorption energy tends to infinity. The force–distance curves are shown in Figure 9b. To compare $F_{\max}$ with the macroscopic adhesion strength, $F_{\max}$ was converted into adhesion stress, which can be calculated as follows:

$$S_{\max} = F_{\max}/A \qquad (4)$$

where A is mean value of surface areas of oxide models, 1.12×10^{-18} m^{-2}.

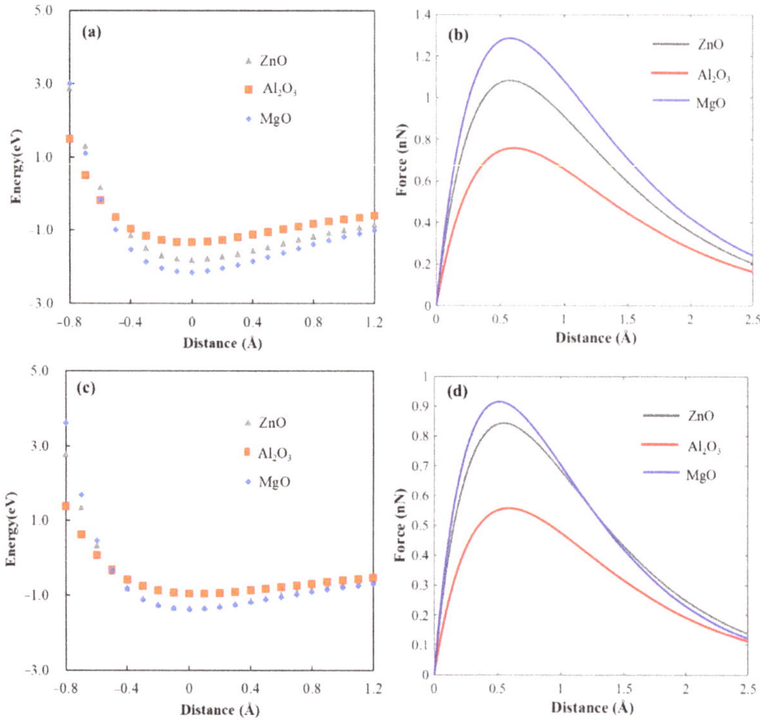

Figure 9. Energy−displacement plots and force−displacement plots for (**a**,**b**) dry environment and (**c**,**d**) water environment.

The fitting parameters obtained are shown in Table 2. As can be seen, the adhesion stresses of the DGEBA fragment on ZnO(10-10), Al_2O_3(100), and MgO(100) surfaces are 0.96 GPa, 0.68 GPa, and 1.14 GPa, respectively, which are in good agreement with the order of magnitude reported in the literature [17,18,25]. The calculated adhesion stresses are two orders of magnitude larger than the actual joint strengths (Zn–Al: 32.3 MPa, Zn–Al–Mg: 29.2 MPa). This difference is attributed to the fact that the actual joint strength is affected by more complex factors, such as surface micro-nano structure, surface contaminants, internal stress, adhesive infiltration on the bonded surface, etc. [46–48]. Moreover, it was found that cohesion failure occurred in both the experimental Zn–Al and Zn–Al–Mg joints, but the true value of the interfacial strength was not obtained. Therefore, the calculated adhesion stress is a theoretical value of the interfacial strength, which is greater than the actual joint strength. As previously analyzed, the calculated adhesion stress between the adhesive and MgO is significantly higher than that of the ZnO and Al_2O_3 models, which is a significant factor contributing to the excellent interfacial bonding performance of the Zn–Al–Mg joint.

To elucidate the chemical interaction between the adhesive and the oxide substrates, the molecular structure and difference charge density plots of the ZnO, Al_2O_3, and MgO systems after chemisorption equilibrium were analyzed [49], as shown in Figure 10. The adhesion behavior of the adhesive molecules was found to be different on the three oxide surfaces. The straight adhesive molecules are adsorbed on the ZnO (10-10) and MgO (100) surfaces at angles of 7.4° and 4.6°, respectively, while the bent adhesive molecule is adsorbed on the Al_2O_3 (100) surface. It is well known that the more parallel the adhesive is to the surface, the more atoms participate in the interfacial chemical interaction, which is the reason why the MgO (100) surface can obtain a relatively higher adhesion force. In addition, the adhesive molecules generated different chemical bonding with the three surfaces.

On the ZnO (10-10) surface, the O atom on the surface is surrounded by the electron accumulation region, while the H atom of hydroxyl group in the adhesive is surrounded by the electron depletion region, which indicates the formation of hydrogen bond. On the Al_2O_3 (100) surface, the 5-coordinated Al atom loses electrons and is surrounded by electron depletion regions, while the O atom of ether group is surrounded by electron accumulation regions. Based on these observations, it can be concluded that there is ionic interaction between the 5-coordinated Al atom and the O atom of the adhesive. In addition, a stronger ionic interaction was generated between the MgO (100) surface and the hydroxyl group of the adhesive than in the ZnO model, which is demonstrated by the darker color of the electron accumulation and depletion regions between the Mg atoms and hydroxyl O. The above analysis shows that hydrogen and ionic bonds formed between the adhesive and oxides are the origin of interfacial adhesion force. Compared with the ZnO (10-10) and Al_2O_3 (100) surfaces, the MgO (100) surface has a chemical adsorption advantage on adhesive molecules in terms of molecular configuration and electronic interaction, which is the reason for its high adhesion.

Table 2. Theoretical adhesion properties and fitting parameters for the dry and water surface models.

Model	Oxides in Coating	E_0/eV	λ/Å^{-1}	F_{max}/nN	S_{max}/GPa in This Work	S_{max}/GPa in Reference
dry	ZnO	1.80	0.8331	1.08	0.96	0.62 [17], 0.61 [17], 1.55 [18], 2.18 [18], 1.546 [25]
	Al_2O_3	1.33	0.8748	0.76	0.68	
	MgO	2.14	0.8324	1.28	1.14	
water	ZnO	1.34	0.7929	0.84	0.75	0.51 [24], 0.50 [24], 1.38 [46]
	Al_2O_3	0.95	0.8512	0.56	0.50	
	MgO	1.36	0.7413	0.91	0.81	

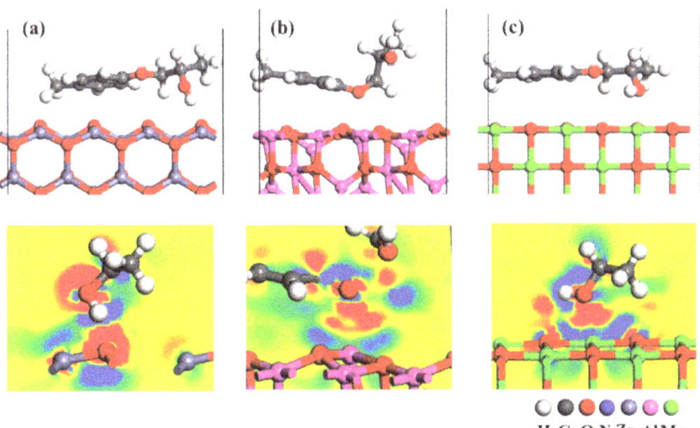

Figure 10. Molecular structures and difference charge density plots calculated for the adhesive interface of (**a**) ZnO, (**b**) Al_2O_3, and (**c**) MgO models (electron failure depletion = blue; electron accumulation = red).

4.4. Corrosion Resistance Mechanism at Epoxy Adhesive/Galvanizing Coating Interface

The immersion and diffusion of water molecules at the adhesive interface, as well as their corrosion on the coating surface, can accelerate the degradation of interfacial bonding performance [50]. To assess the corrosion resistance of Zn–Al and Zn–Al–Mg coatings, the electrochemical impedance spectroscopy (EIS) data were analyzed. Figure 11a shows the Nyquist plots of the two coatings, which take the form of a semicircle, indicating the occurrence of corrosion on the coating surfaces. The radius of the Nyquist curve is indicative of the corrosion resistance of the coating, with a larger radius corresponding to a lower corrosion rate and stronger corrosion resistance. The Nyquist curve radius of the Zn–Al–Mg coating is greater than that of the Zn–Al coating, demonstrating that it possesses stronger corrosion resistance. The equivalent circuit diagram of the EIS fitting is shown in Figure 11b, where R_s represents the resistance of the corrosive solution, R_{ct} represents the resistance of charge transfer, and a constant phase element (CPE) replaced the capacitive element of coating to obtain the best fit [51]. The fitting data for the EIS equivalent circuit are provided in Table 3, where the R_s values for both samples are similar (with a relative error of no more than 5%). CPE-T and CPE-P represent the characteristic and index parameters of the CPE, respectively [52]. In comparison to the Zn–Al coating, the Zn–Al–Mg coating exhibits a lower CPE-T value and a higher R_{ct} value, which further indicates its stronger corrosion resistance. As a result, it is relatively difficult for water molecules to cause corrosion at the adhesive interface of the Zn–Al–Mg coating compared to the Zn–Al coating, resulting in better interfacial bonding performance between the Zn–Al–Mg coating and the adhesive.

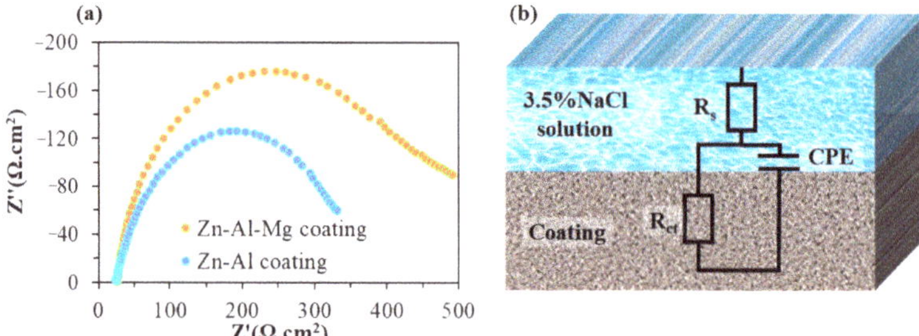

Figure 11. (**a**) Nyquist plot and (**b**) equivalent circuit used to fit the EIS data of the galvanized coating surface.

Table 3. EIS equivalent fitting data.

Sample	R_s ($\Omega \cdot cm^2$)	R_{ct} ($\Omega \cdot cm^2$)	CPE-T ($s^{-n}/\Omega \cdot cm^2$)	CPE-P	χ^2
Zn–Al coating	23.65	342.10	4.66×10^{-5}	0.79	0.0007
Zn–Al–Mg coating	24.83	461.10	3.24×10^{-5}	0.85	0.0048

To further investigate the corrosion resistance mechanism of the adhesive interface on galvanized steel sheets, the interfacial adhesion forces of oxide–adhesive systems were calculated in a water environment. The energy–displacement and force–displacement curves of the ZnO, Al_2O_3, and MgO adhesion systems in water are shown in Figure 9c,d, and the fitting parameters, maximum adhesion force, and adhesion stress of the energy–displacement curves are summarized in Table 2. As shown in Figure 9c,d, the force–displacement curves of the three oxide systems are significantly reduced in the presence of water molecules, indicating a decrease in interfacial bonding performance. The adhesion stress of the ZnO, Al_2O_3, and MgO systems decreased by 21.9%, 26.5%, and 28.9%, respectively. This reduction reflects the sensitivity of adhesion strength to water molecules, with the ZnO system showing the lowest sensitivity, followed by the Al_2O_3 system, and the MgO system showing the highest sensitivity. Even with a decrease in adhesion stress, the MgO system still exhibited the highest adhesion stress among the three systems, which is mainly due to the molecular structure of the interface.

Figure 12 displays the stable adsorption structure of the adhesive on the oxide surfaces in the water model. As illustrated in Figure 12, the presence of water molecules in the interface increased the spatial distance between adhesive molecules and the oxide surfaces, which negatively impacted the generation of adhesion force at the interface. Conversely, hydrogen bonds formed between water molecules and O atoms of adhesives and oxides, creating an interfacial hydrogen-bonding network. This network broke the chemical interaction between the adhesives and the oxide surfaces, thus reducing the adhesion performance of the coatings. The hydrogen bond network contents at the three oxide–adhesive interfaces were quantitatively analyzed as the number of hydrogen bonds divided by the number of hydrogen atoms in water molecules. It was found that the value was lower in the MgO–adhesive interface (0.4) than in the ZnO–adhesive (0.8) and Al_2O_3–adhesive (0.7) interfaces, resulting in higher adhesion stress in the MgO system than in the ZnO and Al_2O_3 systems.

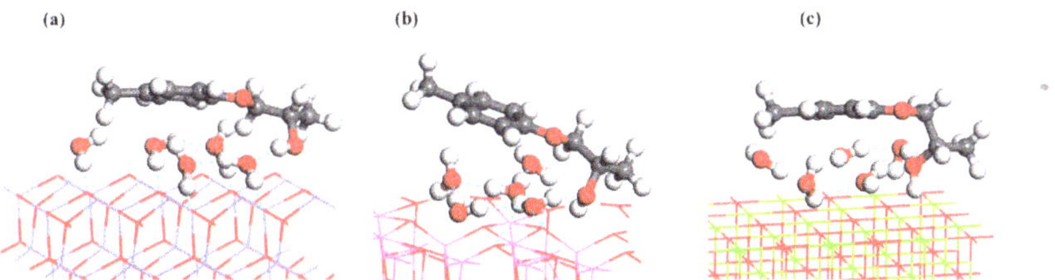

Figure 12. Stable adsorption structure of adhesive on (**a**) ZnO(10-10), (**b**) Al_2O_3(100), and (**c**) MgO(100) surfaces in the water model.

Based on the above analysis, the addition of Mg element in the coating has a positive effect on the corrosion resistance of the adhesive interface, which is in agreement with the experimental results that indicate a stronger adhesive interface corrosion resistance of the Zn–Al–Mg coating compared to the Zn–Al coating. As shown in Figure 3, the dendritic structure containing magnesium is extensively distributed on the surface of the Zn–Al–Mg coating, providing numerous sites for the interaction between the coating and the adhesive. Water molecules are relatively difficult to diffuse at MgO–adhesive interface. Moreover, the corrosion resistance of the Zn–Al–Mg coating is better than that of the Zn–Al coating, thus further improving the corrosion resistance of the Zn–Al–Mg adhesive interface when compared to the Zn–Al coating adhesive interface.

5. Conclusions

This study investigated the interfacial interaction of epoxy adhesive on galvanized steel (Zn–Al and Zn–Al–Mg) and its interfacial corrosion resistance through a combination of experiments and simulations. It was found that the bubbles in the adhesive layer of the Zn–Al–Mg joint reduced its mechanical properties, resulting in the joint strength being approximately 10% lower than that of the Zn–Al joint in the case of cohesive failure. However, the Zn–Al–Mg joint has better interfacial corrosion resistance than the Zn–Al joint after 21 days water soak. Since the coating surfaces were shown to be covered by metal oxides (ZnO, Al_2O_3 and MgO), the influence of these oxides on the interfacial interaction and corrosion resistance between galvanized steel and adhesive was analyzed. The nonbonded interaction analysis revealed that the ZnO and Al_2O_3 surfaces preferentially adsorbed epoxy resin, while the surface of MgO showed a preference for adsorbing curing agent. The hydrogen bond and ionic interaction between oxide surfaces and the adhesive contributes to the interfacial adhesion stress. The theoretical adhesion stress of MgO is the highest and 1.19 times and 1.68 times of ZnO and Al_2O_3, respectively. The corrosion resistance of the Zn–Al–Mg coating was stronger than that of the Zn–Al coating, as demonstrated by the better EIS results. MgO in the Zn–Al–Mg coating is conducive to the reduction of the hydrogen bond network content related to water molecule at the coating–adhesive interface, resulting in better interfacial corrosion resistance.

Author Contributions: Conceptualization, J.L.; Methodology, S.L. and H.W.; Validation, J.M.; Investigation, S.L. and Y.Z.; Writing—original draft, S.L. All authors have read and agreed to the published version of the manuscript.

Funding: This work was funded by the National Natural Science Foundation of China [No. 51575397].

Institutional Review Board Statement: Not applicable.

Informed Consent Statement: Not applicable.

Data Availability Statement: The data can be provided by authors on request.

Acknowledgments: The authors would like to thank the School of Materials Science and Engineering, Tongji University for providing Materials Studio software for the computational work presented in this article. H.W. thanks the financial support from Shanghai Post-doctoral Excellence Program (2021346).

Conflicts of Interest: The authors declare no conflict of interest.

References

1. Yasakau, K.A.; Giner, I.; Vree, C.; Ozcan, O.; Grothe, R.; Oliveira, A.; Grundmeier, G.; Ferreira, M.G.S.; Zheludkevich, M.L. Influence of stripping and cooling atmospheres on surface properties and corrosion of zinc galvanizing coatings. *Appl. Surf. Sci.* **2016**, *389*, 144–156. [CrossRef]
2. Xie, Y.; Du, A.; Zhao, X.; Ma, R.; Fan, Y.; Cao, X. Effect of Mg on Fe–Al interface structure of hot–dip galvanized Zn–Al–Mg alloy coatings. *Surf. Coat. Technol.* **2018**, *337*, 313–320. [CrossRef]
3. Jiang, L.; Volovitch, P.; Wolpers, M.; Ogle, K. Activation and inhibition of Zn–Al and Zn–Al–Mg coatings on steel by nitrate in phosphoric acid solution. *Corros. Sci.* **2012**, *60*, 256–264. [CrossRef]
4. Wan, H.; Min, J.; Zhang, J.; Lin, J.; Sun, C. Effect of adherend deflection on lap-shear tensile strength of laser-treated adhesive-bonded joints. *Int. J. Adhes. Adhes.* **2020**, *97*, 102481. [CrossRef]
5. Wan, H.; Min, J.; Lin, J.; Carlson, B.E.; Maddela, S.; Sun, C. Effect of laser spot overlap ratio on surface characteristics and adhesive bonding strength of an Al alloy processed by nanosecond pulsed laser. *J. Manuf. Process.* **2021**, *62*, 555–565. [CrossRef]
6. Min, J.; Wan, H.; Carlson, B.E.; Lin, J.; Sun, C. Application of laser ablation in adhesive bonding of metallic materials: A review. *Opt. Laser Technol.* **2020**, *128*, 106188. [CrossRef]
7. Ni, J.; Min, J.; Wan, H.; Lin, J.; Wang, S.; Wan, Q. Effect of adhesive type on mechanical properties of galvanized steel/SMC adhesive-bonded joints. *Int. J. Adhes. Adhes.* **2020**, *97*, 102482. [CrossRef]
8. Bahlakeh, G.; Ghaffari, M.; Saeb, M.R.; Ramezanzadeh, B.; De Proft, F.; Terryn, H. A close-up of the effect of iron oxide type on the interfacial interaction between epoxy and carbon steel: Combined molecular dynamics simulations and quantum mechanics. *J. Phys. Chem. C* **2016**, *120*, 11014–11026. [CrossRef]
9. Krüger, T.; Amkreutz, M.; Schiffels, P.; Schneider, B.; Hennemann, O.-D.; Frauenheim, T. Theoretical Study of the Interaction between Selected Adhesives and Oxide Surfaces. *J. Phys. Chem. B* **2005**, *109*, 5060–5066. [CrossRef]
10. Anand, K.; Duguet, T.; Esvan, J.; Lacaze-Dufaure, C. Chemical Interactions at the Al/Poly-Epoxy Interface Rationalized by DFT Calculations and a Comparative XPS Analysis. *ACS Appl. Mater. Interfaces* **2020**, *12*, 57649–57665. [CrossRef]
11. Semoto, T.; Tsuji, Y.; Tanaka, H.; Yoshizawa, K. Role of Edge Oxygen Atoms on the Adhesive Interaction between Carbon Fiber and Epoxy Resin. *J. Phys. Chem. C* **2013**, *117*, 24830–24835. [CrossRef]
12. Bahlakeh, G.; Ramezanzadeh, B. A detailed molecular dynamics simulation and experimental investigation on the interfacial bonding mechanism of an epoxy adhesive on carbon steel sheets decorated with a novel cerium–lanthanum nanofilm. *ACS Appl. Mater. Interfaces* **2017**, *9*, 17536–17551. [CrossRef] [PubMed]
13. Liu, D.; Zhou, F.; Li, H.; Xin, Y.; Yi, Z. Study on the interfacial interactions and adhesion behaviors of various polymer-metal interfaces in nano molding. *Polym. Eng. Sci.* **2021**, *61*, 95–106. [CrossRef]
14. Li, H.; Cai, Z.; Zhou, F.; Liu, D. MD simulation analysis of the anchoring behavior of injection-molded nanopits on polyphenylene sulfide/Cu interface. *Compos. Interfaces* **2022**, *29*, 431–446. [CrossRef]
15. Lee, J.H.; Kang, S.G.; Choe, Y.; Lee, S.G. Mechanism of adhesion of the diglycidyl ether of bisphenol A (DGEBA) to the Fe(100) surface. *Compos. Sci. Technol.* **2016**, *126*, 9–16. [CrossRef]
16. Knaup, J.M.; Köhler, C.; Frauenheim, T.; Blumenau, A.T.; Amkreutz, M.; Schiffels, P.; Schneider, B.; Hennemann, O.-D. Computational Studies on Polymer Adhesion at the Surface of γ-Al2O3. I. The Adsorption of Adhesive Component Molecules from the Gas Phase. *J. Phys. Chem. B* **2006**, *110*, 20460–20468. [CrossRef]
17. Semoto, T.; Tsuji, Y.; Yoshizawa, K. Molecular understanding of the adhesive force between a metal oxide surface and an epoxy resin. *J. Phys. Chem. C* **2011**, *115*, 11701–11708. [CrossRef]
18. Tsurumi, N.; Tsuji, Y.; Masago, N.; Yoshizawa, K. Elucidation of Adhesive Interaction between the Epoxy Molding Compound and Cu Lead Frames. *ACS Omega* **2021**, *6*, 34173–34184. [CrossRef]

19. Duchoslav, J.; Arndt, M.; Steinberger, R.; Keppert, T.; Luckeneder, G.; Stellnberger, K.H.; Hagler, J.; Riener, C.K.; Angeli, G.; Stifter, D. Nanoscopic view on the initial stages of corrosion of hot dip galvanized Zn–Mg–Al coatings. *Corros. Sci.* **2014**, *83*, 327–334. [CrossRef]
20. Zheng, R.; Lin, J.-p.; Wang, P.-C.; Wu, Y.-R. Effect of hot-humid exposure on static strength of adhesive-bonded aluminum alloys. *Def. Technol.* **2015**, *11*, 220–228. [CrossRef]
21. Wu, Y.; Lin, J.; Wang, P.-C.; Zheng, R.; Wu, Q. Effect of long-term neutral salt spray exposure on durability of adhesive-bonded Zr–Ti coated aluminum joint. *Int. J. Adhes. Adhes.* **2016**, *64*, 97–108. [CrossRef]
22. Wu, Y.; Lin, J.; Wang, P.-C.; Zheng, R. Effect of Thermal Exposure on Static and Fatigue Characteristics of Adhesive-bonded Aluminum Alloys. *J. Adhesion* **2016**, *92*, 722–738. [CrossRef]
23. Wang, S.; Min, J.; Lin, J.; Wu, Y. Effect of neutral salt spray (NSS) exposure on the lap-shear strength of adhesive-bonded 5052 aluminum alloy (AA5052) joints. *J. Adhes. Sci. Technol.* **2019**, *33*, 549–560. [CrossRef]
24. Semoto, T.; Tsuji, Y.; Yoshizawa, K. Molecular Understanding of the Adhesive Force between a Metal Oxide Surface and an Epoxy Resin: Effects of Surface Water. *Bull. Chem. Soc. Jpn.* **2012**, *85*, 672–678. [CrossRef]
25. Yoshizawa, K.; Murata, H.; Tanaka, H. Density-Functional Tight-Binding Study on the Effects of Interfacial Water in the Adhesion Force between Epoxy Resin and Alumina Surface. *Langmuir* **2018**, *34*, 14428–14438. [CrossRef]
26. Ogata, S.; Takahashi, Y. Moisture-Induced Reduction of Adhesion Strength between Surface Oxidized Al and Epoxy Resin: Dynamics Simulation with Electronic Structure Calculation. *J. Phys. Chem. C* **2016**, *120*, 13630–13637. [CrossRef]
27. Ogata, S.; Uranagase, M. Unveiling the Chemical Reactions Involved in Moisture-Induced Weakening of Adhesion between Aluminum and Epoxy Resin. *J. Phys. Chem. C* **2018**, *122*, 17748–17755. [CrossRef]
28. Wu, C.; Xu, W. Atomistic molecular modelling of crosslinked epoxy resin. *Polymer* **2006**, *47*, 6004–6009. [CrossRef]
29. Yao, L. Multi-scale studies of glass transition and uniaxial tensile properties of a commercially available epoxy adhesive using experimental measurements and molecular dynamics simulation. *Chin. J. Polym. Sci.* **2015**, *33*, 465–474. [CrossRef]
30. Amirkhani, F.; Harami, H.R.; Asghari, M. CO_2/CH_4 mixed gas separation using poly(ether-b-amide)-ZnO nanocomposite membranes: Experimental and molecular dynamics study. *Polym. Test.* **2020**, *86*, 106464. [CrossRef]
31. Liu, W.; Tkatchenko, A.; Scheffler, M. Modeling Adsorption and Reactions of Organic Molecules at Metal Surfaces. *Acc. Chem. Res.* **2014**, *47*, 3369–3377. [CrossRef] [PubMed]
32. Duchoslav, J.; Steinberger, R.; Arndt, M.; Keppert, T.; Luckeneder, G.; Stellnberger, K.H.; Hagler, J.; Angeli, G.; Riener, C.K.; Stifter, D. Evolution of the surface chemistry of hot dip galvanized Zn–Mg–Al and Zn coatings on steel during short term exposure to sodium chloride containing environments. *Corros. Sci.* **2015**, *91*, 311–320. [CrossRef]
33. Saarimaa, V.; Markkula, A.; Juhanoja, J.; Skrifvars, B.-J. Novel insight to aluminum compounds in the outermost layers of hot dip galvanized steel and how they affect the reactivity of the zinc surface. *Surf. Coat. Technol.* **2016**, *306*, 506–511. [CrossRef]
34. Foster, R.T.; Schroeder, K.J. Adhesive Bonding to Galvanized Steel: I. Lap Shear Strengths and Environmental Durability. *J. Adhes.* **1987**, *24*, 259–278. [CrossRef]
35. Korta, J.; Mlyniec, A.; Uhl, T. Experimental and numerical study on the effect of humidity-temperature cycling on structural multi-material adhesive joints. *Compos. Pt. B-Eng.* **2015**, *79*, 621–630. [CrossRef]
36. Yasakau, K.A.; Kallip, S.; Lisenkov, A.; Ferreira, M.G.S.; Zheludkevich, M.L. Initial stages of localized corrosion at cut-edges of adhesively bonded Zn and Zn-Al-Mg galvanized steel. *Electrochim. Acta* **2016**, *211*, 126–141. [CrossRef]
37. Legghe, E.; Aragon, E.; Bélec, L.; Margaillan, A.; Melot, D. Correlation between water diffusion and adhesion loss: Study of an epoxy primer on steel. *Prog. Org. Coat.* **2009**, *66*, 276–280. [CrossRef]
38. Cavodeau, F.; Brogly, M.; Bistac, S.; Devanne, T.; Pedrollo, T.; Glasser, F. Hygrothermal aging of an epoxy/dicyandiamide structural adhesive—Influence of water diffusion on the durability of the adhesive/galvanized steel interface. *J. Adhes.* **2020**, *96*, 1027–1051. [CrossRef]
39. Baldan, A. Adhesion phenomena in bonded joints. *Int. J. Adhes. Adhes.* **2012**, *38*, 95–116. [CrossRef]
40. de la Roza, A.O.; DiLabio, G.A. *Non-Covalent Interactions in Quantum Chemistry and Physics: Theory and Applications*; Elsevier: Amsterdam, The Netherlands, 2017.
41. Tardio, S.; Abel, M.-L.; Carr, R.H.; Watts, J.F. The interfacial interaction between isocyanate and stainless steel. *Int. J. Adhes. Adhes.* **2019**, *88*, 1–10. [CrossRef]
42. Doonan, C.; Riccò, R.; Liang, K.; Bradshaw, D.; Falcaro, P. Metal–Organic Frameworks at the Biointerface: Synthetic Strategies and Applications. *Acc. Chem. Res.* **2017**, *50*, 1423–1432. [CrossRef] [PubMed]
43. Pandey, P.R.; Roy, S. Is it Possible to Change Wettability of Hydrophilic Surface by Changing Its Roughness? *J. Phys. Chem. Lett.* **2013**, *4*, 3692–3697. [CrossRef]
44. Ashraf, A.; Wu, Y.; Wang, M.C.; Yong, K.; Sun, T.; Jing, Y.; Haasch, R.T.; Aluru, N.R.; Nam, S. Doping-Induced Tunable Wettability and Adhesion of Graphene. *Nano Lett.* **2016**, *16*, 4708–4712. [CrossRef] [PubMed]
45. Park, J.W.; Jo, W.H. Infiltration of Water Molecules into the Oseltamivir-Binding Site of H274Y Neuraminidase Mutant Causes Resistance to Oseltamivir. *J. Chem. Inf. Model.* **2009**, *49*, 2735–2741. [CrossRef]
46. Higuchi, C.; Tanaka, H.; Yoshizawa, K. Molecular understanding of the adhesive interactions between silica surface and epoxy resin: Effects of interfacial water. *J. Comput. Chem.* **2019**, *40*, 164–171. [CrossRef] [PubMed]
47. Bajat, J.B.; Mišković-Stanković, V.B.; Popić, J.P.; Dražić, D.M. Adhesion characteristics and corrosion stability of epoxy coatings electrodeposited on phosphated hot-dip galvanized steel. *Prog. Org. Coat.* **2008**, *63*, 201–208. [CrossRef]

48. Bajat, J.B.; Mišković-Stanković, V.B.; Bibić, N.; Dražić, D.M. The influence of zinc surface pretreatment on the adhesion of epoxy coating electrodeposited on hot-dip galvanized steel. *Prog. Org. Coat.* **2007**, *58*, 323–330. [CrossRef]
49. Koyanagi, J.; Itano, N.; Yamamoto, M.; Mori, K.; Ishida, Y.; Bazhirov, T. Evaluation of the mechanical properties of carbon fiber/polymer resin interfaces by molecular simulation. *Adv. Compos. Mater* **2019**, *28*, 639–652. [CrossRef]
50. Calvez, P.; Bistac, S.; Brogly, M.; Richard, J.; Verchère, D. Mechanisms of Interfacial Degradation of Epoxy Adhesive/Galvanized Steel Assemblies: Relevance to Durability. *J. Adhes.* **2012**, *88*, 145–170. [CrossRef]
51. Nazeer, A.A.; Al Sagheer, F.; Bumajdad, A. Aramid-Zirconia Nanocomposite Coating With Excellent Corrosion Protection of Stainless Steel in Saline Media. *Front. Chem.* **2020**, *8*, 391. [CrossRef]
52. Olarte, J.; Martinez de Ilarduya, J.; Zulueta, E.; Ferret, R.; Garcia-Ortega, J.; Lopez-Guede, J.M. Online Identification of VLRA Battery Model Parameters Using Electrochemical Impedance Spectroscopy. *Batteries* **2022**, *8*, 238. [CrossRef]

Review

Recent Progress of Polymeric Corrosion Inhibitor: Structure and Application

Xuanyi Wang [1,2], Shuang Liu [1,2], Jing Yan [1,2,*], Junping Zhang [2], Qiuyu Zhang [2] and Yi Yan [1,2,*]

1 Chongqing Technology Innovation Center, Northwestern Polytechnical University, Chongqing 401135, China
2 Department of Chemistry, School of Chemistry and Chemical Engineering, Key Laboratory of Special Functional and Smart Polymer Materials of Ministry of Industry and Information Technology, Northwestern Polytechnical University, Xi'an 710129, China
* Correspondence: yanjing@nwpu.edu.cn (J.Y.); yanyi@nwpu.edu.cn (Y.Y.)

Abstract: An anti-corrosion inhibitor is one of the most useful methods to prevent metal corrosion toward different media. In comparison with small molecular inhibitors, a polymeric inhibitor can integrate more adsorption groups and generate a synergetic effect, which has been widely used in industry and become a hot topic in academic research. Generally, both natural polymer-based inhibitors and synthetic polymeric inhibitors have been developed. Herein, we summarize the recent progress of polymeric inhibitors during the last decade, especially the structure design and application of synthetic polymeric inhibitor and related hybrid/composite.

Keywords: metal corrosion; anti-corrosion inhibitor; polymeric inhibitor; adsorption

Citation: Wang, X.; Liu, S.; Yan, J.; Zhang, J.; Zhang, Q.; Yan, Y. Recent Progress of Polymeric Corrosion Inhibitor: Structure and Application. *Materials* **2023**, *16*, 2954. https://doi.org/10.3390/ma16082954

Academic Editors: Yong Sun and Juan José Santana-Rodríguez

Received: 18 February 2023
Revised: 3 April 2023
Accepted: 4 April 2023
Published: 7 April 2023

1. Introduction

Metal corrosion has become a global problem, which not only induce accident due to the decrease in the mechanical strength, but also causes huge economic losses. Corrosion inhibitor is one of the most useful and economic strategies to protect metal materials toward different media. Generally, there are several kinds of inhibitors, including inorganic inhibitors, organic inhibitors, and polymeric inhibitors [1–3]. Compared with inorganic inhibitors, organic inhibitors and polymeric inhibitors are much cheaper and more powerful. More importantly, both organic inhibitors and polymeric inhibitors can be rationally designed and easily synthesized. It is well-known that the adsorption of inhibitors on the surface of metal and corresponding adhesion property performs important role in the application of inhibitors [4]. Therefore, adsorption groups are broadly used in the structure design of inhibitors. Some pioneered review papers have already summarized the progress of organic inhibitors [5,6].

In comparison with small molecular organic inhibitor, polymeric inhibitors show the following advantages (as illustrated in Scheme 1): (i) more adsorption groups can be introduced to one molecule by tuning the number of the repeat unit; (ii) different adsorption groups can be integrated into one polymer via copolymerization (for example copolymerization of monomer A and monomer B), which may generate a synergetic adsorption effect; (iii) the configuration of supramolecular self-assembly of polymeric inhibitor allows the structure optimization of polymeric inhibitor to achieve best adsorption performance; and (iv) the flexibility and mobility of the polymer chain provides processability, which also allow the formation of hybrid/composite with inorganic inhibitor to achieve improved anti-corrosion performance.

Heterocyclic compounds (as shown in Figure 1) have been considered to be excellent corrosion inhibitors, owing to the dense electron centers of heteroatom; however, their synthetic process is usually very harmful to the environment. The adsorption sites of polymeric inhibitor can be increased by increasing their molecular weight (in other words, the number of repeat unit) and can be a potential candidate for the use of heterocyclic compounds in corrosion protection [7]. The mechanism by which polymers can perform corrosion inhibition is the metal-polymer bonding mechanism. Polymeric corrosion inhibitors dissolved in corrosive electrolyte can be adsorbed on the metal surface to form polymeric film through the metal-polymer bonding mechanism, as shown in Figure 2. The adsorption mechanism of polymers containing heteroatoms discussed in this mini-review mainly follows such mechanisms and forms polymer film on the metal surface through chemisorption or physical-chemical adsorption (mixed adsorption) mechanisms [8,9]. The hydrophilicity and hydrophobicity of the polymer determine whether the polymer film formed on the metal surface can effectively protect against corrosion and whether it can exist stably. A suitable ratio between hydrophilic and hydrophobic segments is essential to the corrosion inhibitive performance of polymeric inhibitor [8,10,11].

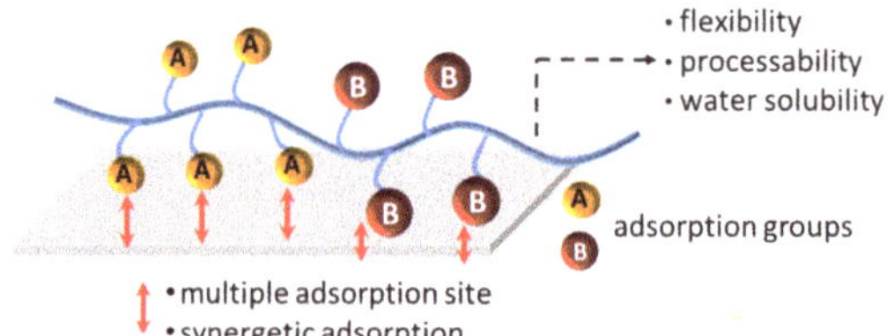

Scheme 1. Schematic illustration of the advantages of polymeric inhibitor.

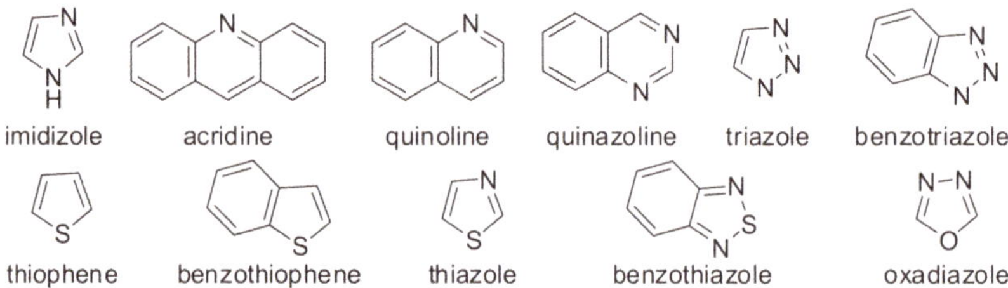

Figure 1. Chemical structures of typical heterocyclic compounds used in the design of inhibitors.

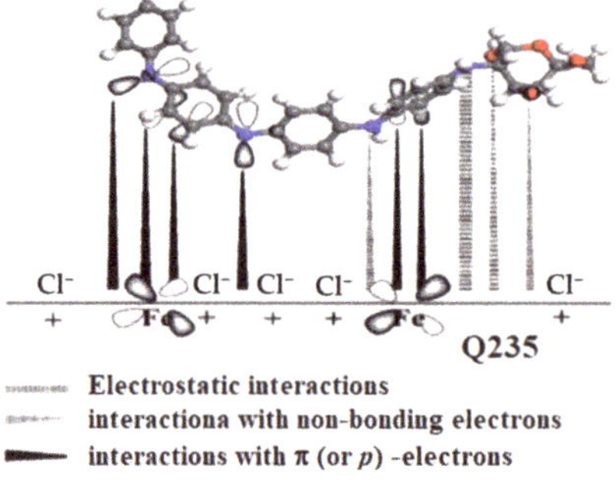

Figure 2. Schematic illustration for the protective effect of polymeric inhibitor on steel sheet in corrosive medium [7]. Reproduced with permission from ref. [7]. Copyright 2022 Elservier.

According to the source and synthetic method, there are generally two kinds of polymeric inhibitors, natural polymer-based inhibitors [12,13] and synthetic polymeric inhibitors [14,15]. With regarding to the design of polymeric inhibitor, on one hand, the presence of heteroatom is crucial for polymeric corrosion inhibitors and greatly affects the corrosion inhibitive properties, which are currently known to be O < N < S < P according to their capability of coordination [16]. One the other hand, most of the polymeric inhibitors belong to polyelectrolyte, both the charged group and its counterion are important.

Herein, we are trying to summarize the progress on the design and application of polymeric inhibitors during the last decade. Generally, both inhibitors from natural polymer and synthetic polymer and their related hybrid/composite will be covered, which may be helpful to researchers, entrepreneurs, and engineers in the fields of functional polymer and metal protection.

2. Corrosion Inhibitor from Natural Polymer

Natural polymers (Figure 3), such as starch, cellulose, chitosan, and other natural polysaccharide substances extracted from natural products, have been historically used as corrosion inhibitor [17] due to the presence of multiple coordination groups in their structures. Generally, the inhibition property of natural polymers depends on the number of their functional groups, which usually varies according to their source. Therefore, different strategies have been developed to extract natural products from high content raw plants, such as solvent extraction, mechanical extrusion, pre-gelatinization treatment, and so on.

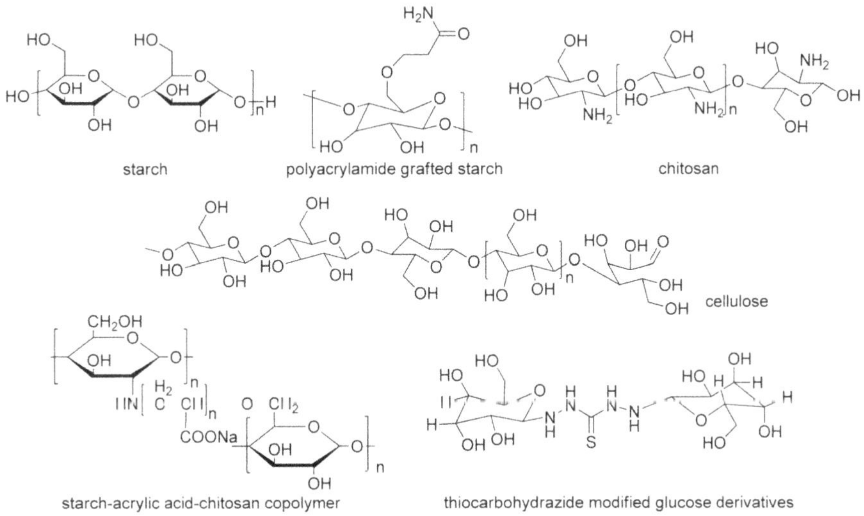

starch

polyacrylamide grafted starch

chitosan

cellulose

starch-acrylic acid-chitosan copolymer

thiocarbohydrazide modified glucose derivatives

Figure 3. Typical chemical structures of natural polymers are used in the design of inhibitors.

Starch-base inhibitors were broadly used due to their abundance in nature. For example, Kaseem et al. found that the addition of starch to the electrolyte can protect Al-Mg-Si alloy via the formation of α-Al_2O_3 on the surface, while the oxide layer of the alloy without the starch addition consists of γ-Al_2O_3 only [18]. Thamer et al. used acid hydrolysis to synthesize starch nanocrystal and investigated its corrosion inhibition for mild steel in 1 M HCl. The maximum inhibition efficiency was 67% when the concentration of the nanocrystal was 0.5 g/L [19]. Physical blend of starch and pectin was developed by Sushmitha and coworkers [20]. They found that such a blend can be used as corrosion inhibitor for mild steel in 1 M HCl. The corrosion inhibition efficiency was up to 74% when the temperature was controlled at 30 °C. Furthermore, the addition of this blend to the epoxy resin can effectively reduce the porosity of the coating and therefore enhance the corrosion inhibition efficiency.

Zhang et al. investigated the corrosion inhibition of steel by maize gluten meal extract in simulated concrete pore solution containing 3% NaCl using electrochemical techniques and showed that the corrosion resistance of steel was improved after the introduction of the extract [21]. The corn protein powder extract is a mixed inhibitor causing an inhibitory effect on both anodic and cathodic reactions, where polar groups, such as N and O in the amide group, perform the main adsorptive effect.

However, the crystalline regions decreased their water solubility and limited corresponding application in metal corrosion protection [22]. Different strategies have been developed to increase the water solubility of starch. For example, *gadong* tuber starch (GTS)-based inhibitor was extracted from *Dioscorea hispida* and dispersed in 90% dimethyl sulfoxide [23]. The corrosion inhibition property was evaluated by using SAE 1045 carbon steel in 0.6 M NaCl, giving inhibition efficiency as high as 86.3% at GTS concentration of 1500 ppm. Extrusion can also be used to physically modify starch by increasing the amorphous region. Anyiam et al. [24] used an alkaline treatment with NaOH to modify starch extracted from sweet potato and investigated its corrosion inhibitor performance on mild steel at 0.25 M H_2SO_4 by gravimetric and potentiodynamic polarization (PDP) techniques. The alkaline treated starch can achieve inhibitive efficiency as high as 84.2%, which was found to be more effective than the unmodified starch in corrosion inhibition. Meanwhile the addition of KI could perform a synergistic role to enhance the inhibition effect. They also modified starch by extrusion and found that the modified starch can give inhibition efficiency up to 64% at concentration as low as 0.7 g/L in 1 M HCl solution [25].

To improve the water solubility of natural starch, Deng et al. prepared cassava starch-acrylamide graft copolymer (CS-AAGC) by grafting cassava starch with acrylamide using $(NH_4)_2S_2O_8$ and $NaHSO_3$ as initiators [26]. Weight loss and electrochemical analysis revealed that CS-AAGC is a hybrid inhibitor for aluminum in 1 M H_3PO_4, which performs a major role in inhibiting the anodic reaction with a maximum inhibition efficiency of 90.6%.

Grafting starch with other kinds of water-soluble monomers can also be used to prepare starch-based inhibitors. For example, Hou et al. [27] modified starch by grafting with acrylic acid (AA) and investigated the corrosion inhibition performance on Q235 carbon steel under HCl environment by weight loss and electrochemistry. It was found that 200 ppm of such terpolymer can give inhibitive efficiency up to 90.1% at 30 °C. Acrylamide is generally used as a co-monomer in the modification of starch. For example, Li et al. [28] grafted acrylamide onto cassava starch and studied the corrosion inhibition effect on aluminum in 1 M HNO_3. According to the results of the weight loss and electrochemical study, the acrylamide-grafted starch showed better anti-corrosion performance than the un-grafted one. While in a similar study, Wang et al. [29] found that the acrylamide (AM)-grafted starch can also be used as inhibitor for the protection of zinc in 1 M HCl with inhibition efficiency as high as 92.2%. Furthermore, they grafted sodium allyl sulfonate and acrylamide onto tapioca starch [30]. It was found that such terpolymer excellent anti-corrosion property on cold-rolled steel, the inhibition efficiency can be achieved as high as 97.2% for 1 M HCl and 90% for 1 M H_2SO_4, respectively.

Comparing with starch, cellulose derivatives are also widely used as inhibitors due to their easy modification [31]. Methyl hydroxyethyl cellulose (MHEC) shows good water solubility and facilitates its adsorption on metal surface [32]. Eid et al. [33] investigated the corrosion inhibition effect of MHEC on copper in 1 M HCl solution by cyclic voltammetry, potentiodynamic polarization, and weight loss techniques. It was found that such inhibitors could prevent the mass and charge transfer on the copper surface by calculating the activation energy (E_a) and heat of adsorption (Q_{ads}) of the adsorption process. The corrosion efficiency can achieve as high as 90% at concentration as low as 4 g/L. In addition to copper, cellulose derivatives can also be used to protect aluminum alloys. Nwanonenyi et al. [34] found that hydroxypropyl cellulose (HPC) was a good inhibitor for aluminum in acidic media (0.5 M HCl and 2 M H_2SO_4) at 30–65 °C.

Wang et al. [35] prepared thiocarbonylhydrazide modified glucose derivatives (Figure 3) through *N*-glycosyl linkage at the C-1 of the saccharide moiety and applied them to the corrosion protection of N80 carbon steel pipelines in the oil and gas industry. It was found that these compounds showed excellent corrosion protection with corrosion inhibition efficiencies of 99.1% and 99.4%, respectively.

Different from starch and cellulose, the presence of amino groups in chitosan (CS) provide additional adsorption groups and enables improved inhibition property. For example, Wen and coworkers investigated the effect of water-soluble chitosan corrosion inhibitor on the corrosion behavior of 2205 duplex stainless steel in NaCl and $FeCl_3$ solutions [36]. When immersed in 0.2 g/L chitosan solution for ca. 4 h, the surface of stainless steel specimen will be covered by a dense and uniform adsorption film, and the corrosion inhibition efficiency can achieve as high as 62%. In order to improve the corrosion inhibition performance of CS, chemical modification, such as esterification and grafting on chitosan, has been developed [37]. For example, Zhang et al. [38]. introduced extra adsorption groups to CS by using facile Schiff base reaction and etherification. As shown in Figure 4, the synthesized CS-1 and CS-2 exhibited excellent anti-corrosion property on Q235 mild steel in 1 M HCl solution. It was found that the introduction of pyridine groups increased the absorption of inhibitors on the metal surface comparing with crude CS. Moreover, the presence of hydrophobic phenyl ring in CS-2 enhanced the electron density at active sites and gave better surface coverage than CS-1, achieving inhibition efficiency as high as 98.0% at a low concentration of 150 mg/L at 25 °C.

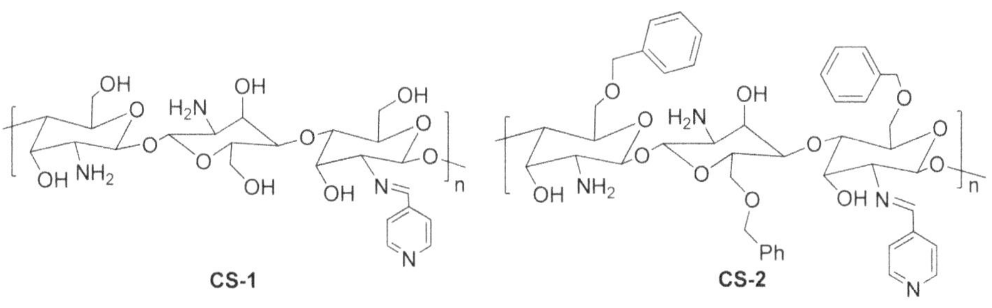

Figure 4. Typical chemical structures of chitosan-based inhibitor: CS-1 and CS-2.

In addition to the above three types of natural polymer-typed inhibitors, other natural polysaccharides can also be used as metal corrosion inhibitors, such as gelatin, pectin, carrageenan, xanthan gum, sodium alginate, and so on. Fares et al. [39] investigated the anti-corrosion property of ι-carrageenan for aluminium in the presence of pefloxacin mesylate and 1.5 M and 2.0 M HCl. It was revealed that pefloxacin acted as zwitterion mediator for the adsorption of carrageenan on the metal surface, and therefore, gave an increase in the inhibition efficiency from 67% in the absence of the medium to 92%. Similarly, pectin is also used as corrosion inhibitor [40]. Grassino et al. [41] extracted pectin from tomato peel waste (TPP) and used it as tin corrosion inhibitor. Compared with the corrosion inhibitor of commercial apple pectin, it was found that TPP mainly performed a cathodic protection role and was an effective corrosion inhibitor for tin in NaCl, acetic acid and citric acid solutions.

To compare the corrosion inhibition property of different natural polymers, Chen et al. investigated the scale and corrosion inhibition properties of polyaspartic acid (PASP), polyepoxysuccinic acid (PESA), oxidized starch (OS), and carboxymethyl cellulose (CMC) by molecular dynamics simulations and density flooding theory calculations, and the binding energies of these substances to Fe surfaces were: PASP > OS > CMC > PESA [42].

Owing to their abundant resources, the conversion of natural polymers to carbon dots (CDs) became an emerging research topic in the design of novel inhibitors. For example, He et al. [43] developed a green and high-yielding CDs-based pickling solution using a range of low-cost and easily available sugars, such as glucose, soluble starch, fructose, or sucrose, and using concentrated H_2SO_4 as the oxidant and an acid source for the preparation of CDs. The corrosion inhibition performance of CDs on Q235 carbon steel in 0.5 M H_2SO_4 solution was subsequently determined by weight loss test, electrochemical impedance spectroscopy (EIS), and PDP measurement. It was found that the corrosion inhibition of carbon steel could reach about 95% at a low precursor concentration of 0.26%. Based on the electrochemical and corrosion surface analysis, it can be reasonably assumed that the corrosion inhibition and protection effect of CDs is exerted by forming a protective film on the metal surface.

Overall, although natural polymer-based corrosion inhibitors (Table 1) show the advantages of wide source, environmentally friendly, easy to obtain and degrade, their chemical modification and modulation of adsorption groups, and corrosion inhibition efficiency is still more challenging As an alternative strategy, synthetic polymeric corrosion inhibitors provide precision structure and architecture design; therefore, the concept of synthetic polymer-based corrosion inhibitor was also developed.

Table 1. Inhibition property of typical polymeric inhibitor from natural polymer.

Inhibitor	Metal	Corrosion Medium	TEST METHOD	Highest IE (%)	Reference
starch	Al-Mg-Si alloy	3.5% NaCl	PDP, EIS	/	[18]
starch nanocrystals	mild steel (ST37-2)	1 M HCl	weight loss, PDP, EIS	67.0	[19]
starch-pectin blend	mild steel	1 M HCl	PDP, EIS	88.9	[20]
maize gluten meal extract	steel	simulated concrete pore solution with 3.0% NaCl	PDP, EIS	88.1	[21]
alkaline modified starch	mild steel	0.25 M H_2SO_4	PDP, weight loss	84.2	[24]
AM-grafted cassava starch	aluminum	1 M H_3PO_4	weight loss, PDP, EIS	91.9	[26]
starch-AA-CS copolymer	Q235 carbon steel	1 M HCl	weight loss, EIS, PDP	90.1	[27]
AM grafted cassava starch	aluminum	1 M HNO_3	weight loss, EIS, PDP	97.8	[28]
AM grafted starch	Zn	1 M HCl	weight loss, EIS, PDP	92.2	[29]
AM grafted starch	Zn	1 M HCl	weight loss, EIS, PDP	97.2	[30]

3. Synthetic Polymeric Corrosion Inhibitor

In principle, both coordination group and charged group can be potential adsorption sites for metal surface. To design effective inhibitors, both the adsorption site and the polymeric structure perform important roles. The development of polymer chemistry, especially controlled radical polymerization techniques, provides accurate architecture and structure control over the number of repeat units, functional groups, and the topology. In this section, different kinds of synthetic polymeric inhibitors will be discussed according to their polymer skeletons and adsorption sites, such as phosphorus, sulfur, nitrogen, and so on.

3.1. Phosphorus-Containing Synthetic Polymeric Inhibitor

Phosphorus can coordinate with metal surfaces more strongly than oxygen, which has been introduced to polymeric inhibitor as additional adsorption site in the form of both coating and solution. Generally, there are two kinds of strategies to prepare phosphorus-containing synthetic polymeric inhibitor: (i) post-modification of natural polymer [44]; and (ii) direct polymerization of phosphorus-containing monomer [45].

Grafting phosphorus-containing groups to natural polymers not only can introduce additional adsorption sites, but also improve their water solubility. David et al. functionalized chitosan with phosphonic acids (P-1 in Figure 5) via the Kabachnik-Fields reaction and subsequent hydrolysis [44]. The introduction of phosphonic acid groups greatly decreased the dynamic viscosity of chitosan. Moreover, compared with the native chitosan, P-1 showed improved adsorption on the carbon steel surface and better corrosion inhibition property. Meanwhile, such material can also be used in metallic aluminum corrosion protection. David et al. [44] fabricated inhibitive coating by the layer-by-layer (L-b-L) technique from native chitosan or synthesized phosphorylated chitosan (P-1 in Figure 5) combined with alginate functionalized chitosan. It was found that the coating can create a physical barrier that acts mainly by reducing the active surface area and has the effect of blocking the penetration of the aggressive species into the metal substrate. According to the result of electrochemical impedance spectroscopy (EIS) in 0.1 M Na_2SO_4 solution, the as prepared coating showed improved corrosion resistance for aluminum alloy 3003.

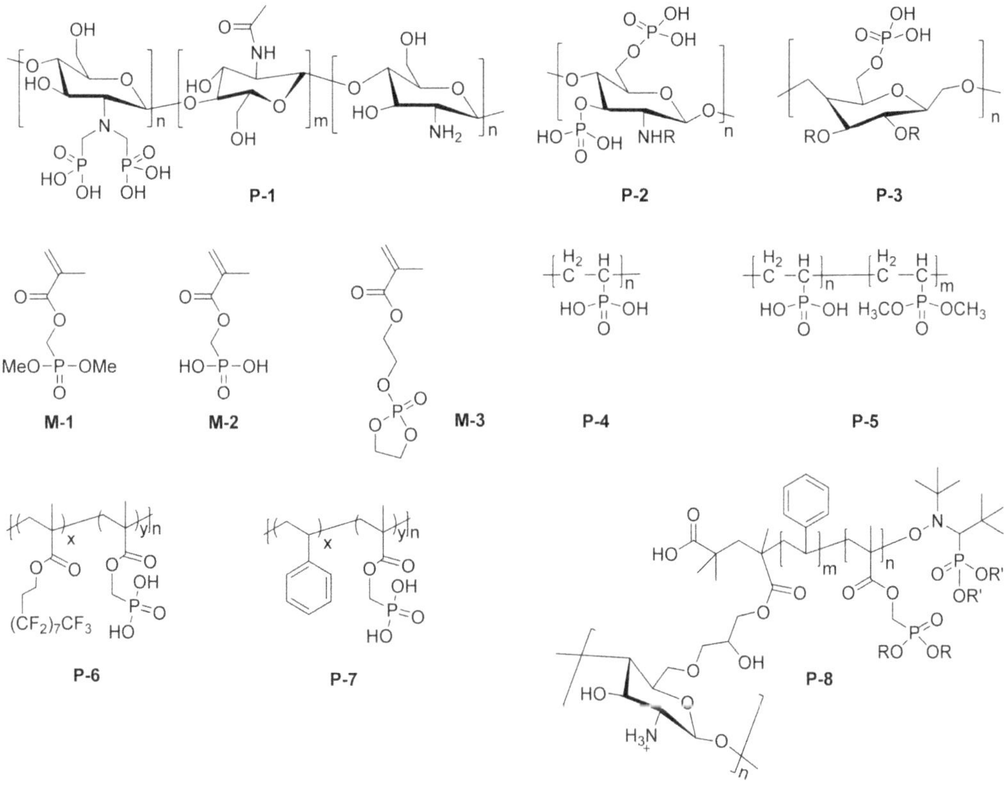

Figure 5. Chemical structures of typical phosphorus-containing polymeric inhibitors (P1–P8) and monomers (M1–M3).

Phosphonic acids can also be introduced to chitin. As demonstrated by Hebalkar and coworkers, they synthesized phosphorylated chitin (P-2 in Figure 5) to solve the water solubility of native chitin and improve their coordination ability with metal surface [46]. According to the gravimetric analysis in NaCl solution, only 200 ppm P-2 can protect copper very well with maximum inhibition efficiency as high as 92%.

By using simple phosphorylation, phosphonic acids can be introduced to ethyl cellulose. Ben Youcef et al. [47] reported the method of microcapsulation to encapsulate almond oil as inhibitor by using phosphorylated ethyl cellulose (P-3 in Figure 5). It was found that the P-O^- functionalized groups strongly interacted with the metal ions in the metal substrate, and therefore, generated synergetic anti-corrosion effect for almond oil.

Comparing with the post-functionalization of natural polymer, the direct polymerization of phosphorus-containing monomer provides accurate structure control and rational design of polymeric inhibitors [48]. The most commonly used phosphorus-containing methacrylate-base monomers are shown in Figure 5 (M-1 and M-2). For example, Keil et al. [49] reported UV-cured polyester acrylate coatings on zinc and iron as anti-corrosion surface. Moreover, Pebere et al. [50] extended the study on the UV polymerization of M-1 and M-2 to prepare effective corrosion protection. It was found that coatings containing phosphonic acid methacrylate (M-2) showed better anti-corrosive performance than those containing methacrylate phosphonic dimethyl ester (M-1), owing to the improved adsorption of phosphonic acid to metal surface than corresponding ester. Similarly, Ilia et al. [51] copolymerized vinylphosphonic acid (VPA) and dimethyl vinylphosphonate (DMVP) to prepare PVPA-*co*-PDMVP copolymer P-5 (Figure 5) via free radical photopolymerization. Interestingly, they found that the presence of phosphonate groups from DMVP in copolymers was beneficial and a molar ratio VPA:DMVP 4:1 and 3:1 enhanced the anticorrosion for iron surface in comparison with homopolymer of vinylphosphonic acid (P-4). In other words, the coordination between P-OH and metal surface competes with the formation rate of uniform protective layer. While copolymers with VPA:DMVP 4:1 may show higher diffusion coefficient, and therefore, faster formation of protective film than other polymers.

In addition to introducing additional adsorption sites to polymeric inhibitors, low surface energy monomers have also been used in the design of efficient inhibitor. Moratti et al. [52] prepared block copolymer P-6 (Figure 5) via free radical polymerization of heptafluorodecyl methacrylate and (dimethoxyphosphoryl) methyl methacrylate. The copolymers were then immobilized as a monolayer film to the surface of 316L stainless steel by treatment of dilute solutions in trifluoroacetic acid for 30 min followed by rinsing. Owing to the presence of fluorinated block and the presence of adsorption site from the phosphorus block, the resulting polymeric inhibitor exhibited excellent anti-corrosion property and long-term stability.

Champagne et al. [45] reported the nitroxide-mediated polymerization (NMP) to prepared polystyrene-*b*-poly(dimethyl(methacryloyloxy)methyl phosphonate) and corresponding polystyrene-*b* poly(dimethyl(methacryloyloxy)methyl phosphonic acid) (P-7 in Figure 5) and its graft onto polysaccharide chitosan (P-8 in Figure 5). The controlled radical polymerization feature of NMP allowed rational design of the repeat unit and the topology of the resulting polymer. It can be anticipated that the resulting grafted polymers are potential candidate for excellent inhibitor due to the coexistence of chitosan and phosphorus-containing polymers (Table 2).

In addition to the radical-based polymerization, sol-gel method has also been developed for the preparation of phosphorus-containing polymeric inhibitors. Mandler et al. [53] incorporated phenylphosphonic acid (PPA) to a sol-gel film to enhance the corrosion protection of metallic aluminum; however, it was found that such method resulted in aggregation and phase separation. To overcome such problem, Choudhury et al. [54] proposed the design of networked methacrylate- hybrid by copolymerizing 2-(methacryloyloxy)ethyl phosphate (M-3 in Figure 5), containing a polymerizable methacrylate group and functional phosphate group with 3-[(methacryloyloxy)propyl] trimethoxysilane [55]. It was found that the proposed network not only enhanced the binding of the coating to the metal substrate via the acid-base interaction of the P-O− group of the phosphate with the Mn+ of the metal substrate, but also resulted excellent anti-corrosion property.

Table 2. Inhibition property of typical phosphorus-containing synthetic polymeric inhibitor.

Inhibitor	Metal	Corrosion Medium	Test Method	Highest IE (%)	Reference
P-1	aluminum alloy 3003	pH = 5 acetic acid	EIS	/	[44]
P-2	copper	200×10^{-3} g/L NaCl	weight loss, EIS	92	[46]
P-3	mild steel	3.5% NaCl	salt spray test	/	[47]
M-1/M-2	low-carbon steel	0.1 M NaCl	EIS	85	[49]
M-3	mild steel	3.5% NaCl	PDP, EIS		[54]
P-4/P-5	iron coin	3% NaCl	PDP, EIS	83.5	[51]
P-6	316L stainless steel	trifluoroacetic acid	long term stability tests	/	[52]

3.2. Sulfur-Containing Synthetic Polymeric Inhibitor

As a classic coordination atom, sulfur can form a stable complex with different metal ions. Generally, there are two kinds of sulfur-containing polymeric inhibitors: (i) polythiophene and (ii) polysulfone.

Normally, intrinsically conducting polymers or conjugated polymers, such as polythiophene, polypyrrole, and polyaniline, have been widely used as protective coating for corrosion protection of steel [56]. Polythiophene was first reported as corrosion protection in 1989 [57]. The facile electropolymerization of thiophene and its derivatives allows the preparation of homogeneous polymer films on the surface of different metals with good electrical properties and chemical stability [58]. Gonzalez-Rodriguez and coworkers [59] compared the anti-corrosion property of poly(3-octyl thiophene) (P3OT) and poly(3-hexylthiophene) (P3HT) (P-9 in Figure 6). The plate they used was a commercially available 1018 carbon steel sheets, and copper wires were welded to the plate, which was used as a reaction platform for electrode deposition. The polymer solution was then deposited on the electrode, evaporated solvent, dried, and annealed to afford the P3OT/P3HT coating. Both polymer films were found to be effective in protecting the substrate from corrosion by decreasing the critical current necessary to passive the substrate, increasing the pitting potential, and broadening the passive interval, and P3HT was found to be more effective due to a much lower number of defects than P3OT films. Interestingly, P3HT gave better protection than P3OT, because of the lower defects in the film of P3HT than that for the P3OT films. Thiophene can also be copolymerized with other monomers to improve the anti-corrosion property. For example, Branzoi et al. [60] investigated the anti-corrosive properties of poly(*N*-methylpyrrole-Tween20/3-methylthiophene) coatings on carbon steel type OLC 45 in 0.5 M H_2SO_4 medium. The surfactant Tween 20 was a dopant used in the electropolymerization process, which could improve the anti-corrosive properties by hindering the corrosive sulfate ion penetration. The corrosion rate of PNMPY-TW20/P3MT-coated OLC 45 has been indicated to be ~10 times reduced in comparison with uncoated OL 45, and the corrosion protection efficiency of the coating is above 90%. More importantly, the anti-corrosion property of such coating can be tuned by the condition of electropolymerization, such as electrodeposition current and time, highlighting the potential application of such technique.

Furthermore, the anti-corrosion performance of polythiophene can be improved by blending with other polymers [61]. Meanwhile, blending can also improve the processability and mechanical strength of the material and reduce the cost of expensive conductive polymers [62]. For example, Nicho et al. [63] blended P3OT with polystyrene (PS) and deposited it onto stainless steel sheets using the drop-casting technique, where a solution of the blend is added dropwise to the steel sheet, the solvent is evaporated, then dried and annealed. Subsequently, the room temperature corrosion behavior of the prepared P3OT/PS coated 304 stainless steel was studied under 0.5 M NaCl. According to their study on the temperature effect, it was found that high temperature (e.g., 100 °C) can increase the adhesion degree between coating and substrate, making the coating less porous and defective to give a denser surface, therefore giving it better inhibition performance. Furthermore, they systematically investigated the anti-corrosion property of P3HT, P3HT/PS and P3HT/PMMA (polymethyl methacrylate) blends coatings on A36 steel corrosion protection in 0.5 M H_2SO_4 solution. It was found that blends of P3HT with PMMA and PS improved the protection of steel in comparison with native P3HT. P3HT/PMMA blend gave the best protraction to the steel. To enhance the interface interaction between different polymers in the blend, Huang et al. [64] reported the preparation of P3HT/poly(styrene-*co*-hydroxystyrene) blend (P-10 in Figure 6) via the intermolecular hydrogen bonding between thiophene and phenol group. The hydrogen bonding not only improved the miscibility between two polymers, but also enhanced the adhesion force between iron and coating layer. Compared with native P3HT, the inhibition performance of the blend improved and the decreased upon thermal treatment.

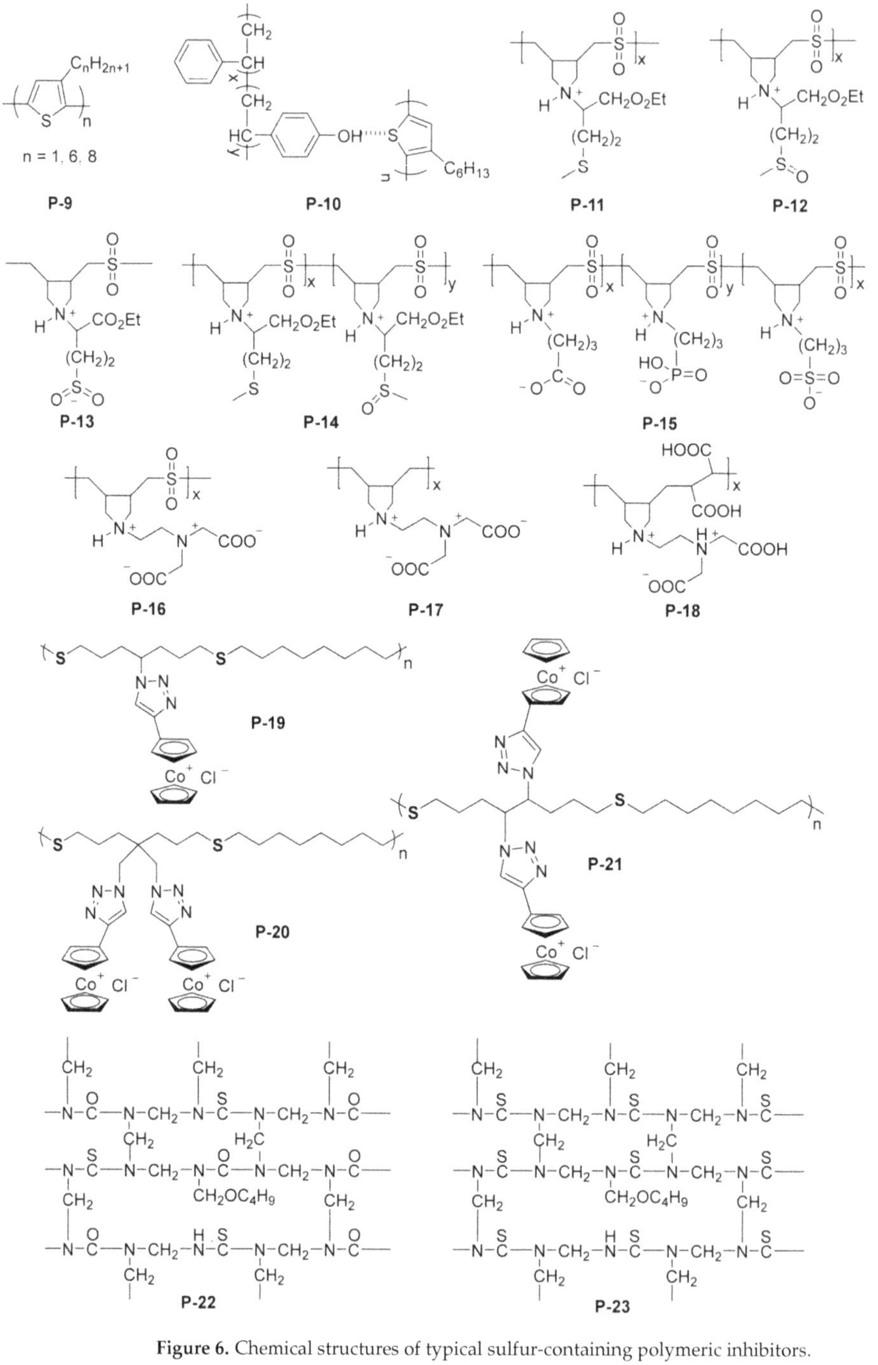

Figure 6. Chemical structures of typical sulfur-containing polymeric inhibitors.

In addition to polythiophene-based inhibitor, polysulfone represents another kind of sulfur-containing polymeric inhibitor. For example, the non-toxic amino acid methionine was used as a sulfur-containing corrosion inhibitor for mild steel because of the coexistence of N, O, and S atoms in one molecule. Moreover, the corrosion inhibition performance of methionine, methionine sulfoxide, and methionine sulfone in HCl for mild steel has been studied previously and moderate inhibition efficiency has been achieved. To enhance the adsorption of inhibitor and metal substrate, polysulfone may be potential strategy [65]. Ali and coworker carried out systematical study on the anti-corrosion performance of series of polysulfones. Butler's cyclopolymerization of diallyl ammonium salts and their copolymerization with SO_2 was used to synthesize a series of polysulfones with residues of essential amino acid methionine (P-11 and P-12 in Figure 6) [66]. Especially, in the copolymer P-14 (Figure 6), half of the sulfide was oxidized to corresponding sulfone, which greatly improved the water solubility. More importantly, the copolymer P-14 demonstrated superior inhibition of mild steel in 1 M HCl at 60 °C with inhibition efficiency of 99% at concentration of 25 ppm, while corresponding monomer can only achieve inhibition efficiency of 31% at the same concentration, highlighting the contribution of polymer configuration. In another study, they treated polymer P-11 with H_2O_2 to generate corresponding sulfone P-12 and sulfoxide P-13 [67]. All of these polymers can achieve inhibition efficiency (IE) up to 87%, even at a very low concentration of 6 ppm in 1 M HCl. It revealed that the sulfoxide (S=O) base sequence was more effective in mitigating mild steel corrosion in comparison with sulfide (S) and sulfone (O=S=O). The copolymerization methodology also allowed the introduction of multiple adsorption groups to one inhibitor. For example, they [67] also synthesized a new tripolymer P-15 (Figure 6) consisting of carboxylate, sulfonate, and phosphonate using the Butler cyclopoymerization technique and copolymerization with sulfur dioxide. They evaluated the performance of P-15 as a corrosion inhibitor for St37 carbon steel. It was found that the as prepared inhibitor demonstrated protection efficiency of 79.5% and 61.1% in HCl and H_2SO_4 media at a concentration of 1000 mg/L, respectively. Interestingly, it was found that the addition of KI can greatly enhance the performance to give IE as high as 93.5%, which may due to the synergistic effect of the cooperative co-adsorption of I^- on the metal surface. They also synthesized series of poly(bis-zwitterion) (P-16, P-17, and P-18 in Figure 6) with chelating motifs of $[NH^+(CH_2)_2NH^+(CH_2CO_2^-)_2]$ via the same strategy. These polymers were found to be very good inhibitors of mild steel corrosion in 1 M HCl. Similarly, the addition of KI (400 ppm) can generate synergistic effect to achieve 98% inhibition of mild steel corrosion for a duration 24 h at 60 °C [68].

Compared with polysulfone, polythioether can bind with metal surface more strongly due to the S-Fe bond. In order to improve the anti-corrosion property of native polythioether, our group designed a series of cobaltocenium-containing polythioether type metallo-polyelectrolytes (P-19, P-20, and P-21 in Figure 6). The synthesized diolefin monomers were first sulfonated to introduce sulfonate groups, followed by photo-induced thiol-ene polymerization to synthesize the polymer, then azidation and copper-catalyzed post-click modification to graft the cobalt dichloride groups to the polysulfide backbone to obtain the target polymer (The synthesis process is shown in Figure 7). It can be found that there are multiple interactions between these polymers and metal surface, such as coordination between S and metal, triazole and metal, electrostatic interactions, and the potential ion-π interaction. According to the weight loss experiments and electrochemical study, all these polymers were found to be effective inhibitors, which can achieve inhibitive efficiency as high as 95% at concentration as low as 10 mg/L. Moreover, our study also revealed the structure–property relationship for the design of new polymeric inhibitor, highlighting the important role of flexible linkage between the polymer main-chain and the charged group and the number of charged groups [69].

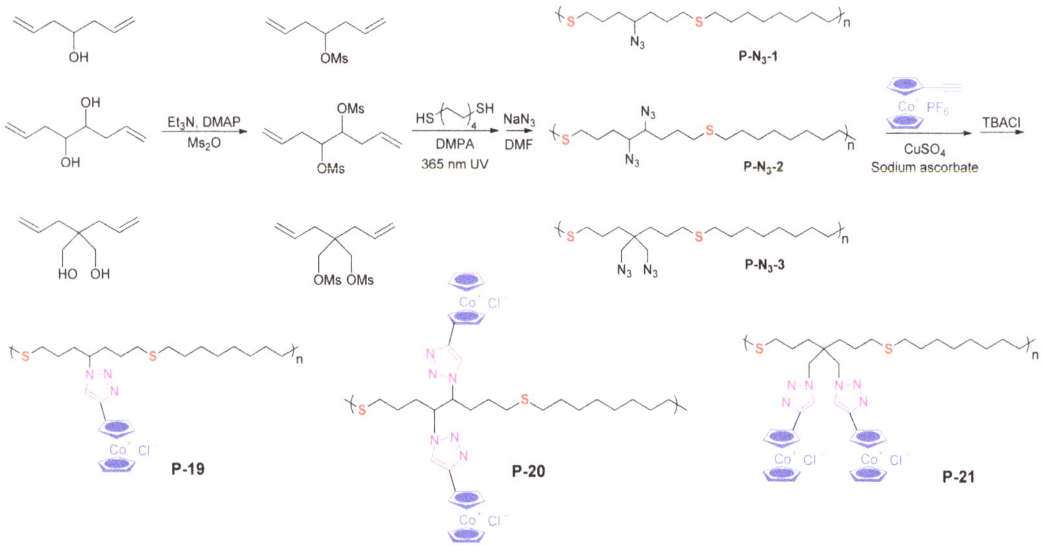

Figure 7. Synthetic procedure for cobaltocenium-containing polythioether type inhibitors: P-19/P-20/P-21 [69].

In addition to the above sulfur-containing polymeric inhibitors (Table 3), sulfur has also been introduced to resin coating. For example, Mohammad El-Sawy and coworkers compared the inhibition performance of modified urea (P-22) with thiourea formaldehyde (P-23) resins for steel surfaces. As expected, owing to the presence of sulfur atoms in thiourea resin, P-23 demonstrated the best protection performance and adhesion [61].

Table 3. Inhibition property of typical sulfur-containing synthetic polymeric inhibitor.

Inhibitor	Metal	Corrosion Medium	Test Method	Highest IE (%)	Reference
P-9	1018 carbon steel	0.5 M H_2SO_4	EIS	/	[59]
P-10	iron	3.5% NaCl	EIS, PDP	96	[64]
P-11/P-14	mild steel	1 M HCl	EIS	99	[66]
P-12/P-13	mild steel	1 M HCl	Weight loss	P-12: 94 P-13: 87	[67]
P-15	St37 carbon steel	15% HCl/15% H_2SO_4	EIS, PDP, linear polarization resistance, electrochemical frequency modulation	79.5/61.1	[62]
P-16/P-17/P-18	mild steel	1 M HCl	Weight loss	P-16: 92.3 P-18: 95.7	[68]
P-19/P-20/P-21	mild steel	5% HCl	weight loss, EIS, PDP	95	[69]
P-22/P-23	cold-rolled mild steel	3.5% NaCl	weight loss,	/	[61]

3.3. *Nitrogen-Containing Synthetic Polymeric Inhibitor*

Nitrogen-containing synthetic polymeric inhibitor represents a major class of polymeric inhibitors. Most of the nitrogen-containing polymeric inhibitors belong to polyelectrolyte, such as poly(quaternary ammonium), polyethyleneimine, polyaniline, and so on.

3.3.1. Poly (Quaternary Ammonium)

Generally, polymeric corrosion inhibitors tend to outperform their monomer counterparts due to advantages, such as increased adsorption sites and entropy processes that displace water molecules from the metal surface. Since quaternary ammonium salts-based small molecular corrosion inhibitors have been widely used in metal corrosion inhibition, their corresponding poly(quaternary ammonium) salts have also been investigated as potential inhibitor owing to their advantages of high water solubility [70].

As discussed in the section of sulfur-containing polymeric inhibitors, polyquaternary ammonium salts are usually synthesized via the Butler's cyclopolymerization of diallyl ammonium salts. For example, as shown in Figure 8, Ali et al. [66] prepared a variety of unsaturated *N*,*N*-diallyl compounds from 1,6-hexanediamine and performed cycliopolymerization to synthesize a series of water-soluble polyquaternium oligomers (P-24, P-25, and P-26). All these inhibitors demonstrated good corrosion inhibition property with IE as high as 93%.

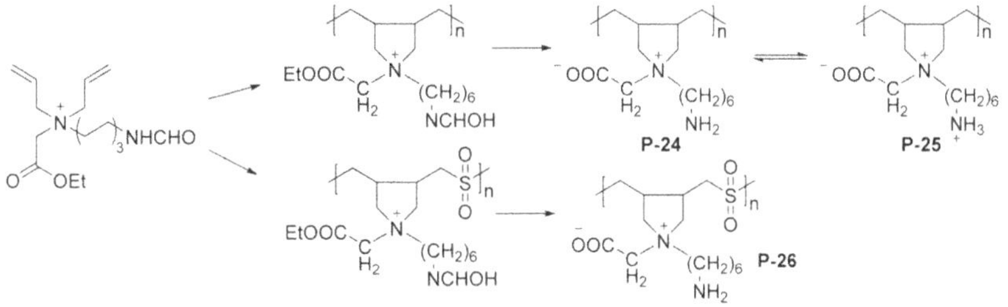

Figure 8. Typical synthetic route of polyquaternary ammonium type inhibitors via Butler's cyclopolymerization.

3.3.2. Polyethyleneimine

In addition to the polyquaternary ammonium salts, the commercially available polyethyleneimine (PEI) and its derivatives have also been broadly used as inhibitor. In principle, the corrosion inhibition efficiency of PEI is related to the number of sub-methyl groups and the number of vinyl and amino groups, which may influence the adhesion and adsorption properties with metal substances. Moreover, the different types of nitrogen atoms in PEI may also be protonated under the condition of acid corrosion, further enhancing the adsorption affinity [66].

Milošev et al. investigated the corrosion protection of PEI (P-27 in Figure 9) for ASTM 420 stainless steel [71] and the relationship between anti-corrosion property and the molecular sizes of PEI [72]. It was found that PEI is a corrosion inhibitor against pitting corrosion, and PEI with an average molar mass of 2000 g/mol shows the best effect against localized corrosion. Moreover, the amine groups in PEI can be protonated by CO_2 in water, PEI can also be used as green inhibitor under the condition of saturated CO_2 solution. Umer and coworkers studied the effect of temperature on the corrosion behavior of API X120 steel in a saline solution saturated with CO_2 in absence and presence of PEI [72]. It was found that PEI significantly decreases the corrosion rate of API X120 steel with inhibition efficiency of 94% at a concentration of 100 μmol/L. This is attributed to the fact that PEI molecules adsorb onto the metal surface through heteroatoms, forming a dense protective film that limits the transfer of aggressive ions to the metal surface and reduces the corrosion rate. While the addition of PEI with saturated CO_2 works best because PEI adsorbs on the steel surface and has the ability to trap CO_2 in the brine solution, with the end result being a reduction in fouling and a smoother steel surface.

Figure 9. Typical chemical structures of PEI-based nitrogen-containing polymeric inhibitors.

PEI can also be used to form hybrids with inorganic material, such as grapheme oxide (GO), to further improve the inhibition performance. As shown in Figure 10, Quraishi et al. [73] modified GO with PEI via facile amidation and investigated its corrosion inhibition of carbon steel in a solution of 15% HCl. According to the weight loss experiment, the PEI-GO alone provided a high corrosion inhibition efficiency of 88.2% at a temperature of 65 °C. Moreover, KI can further increase IE to 95.8% due to their synergistic effect.

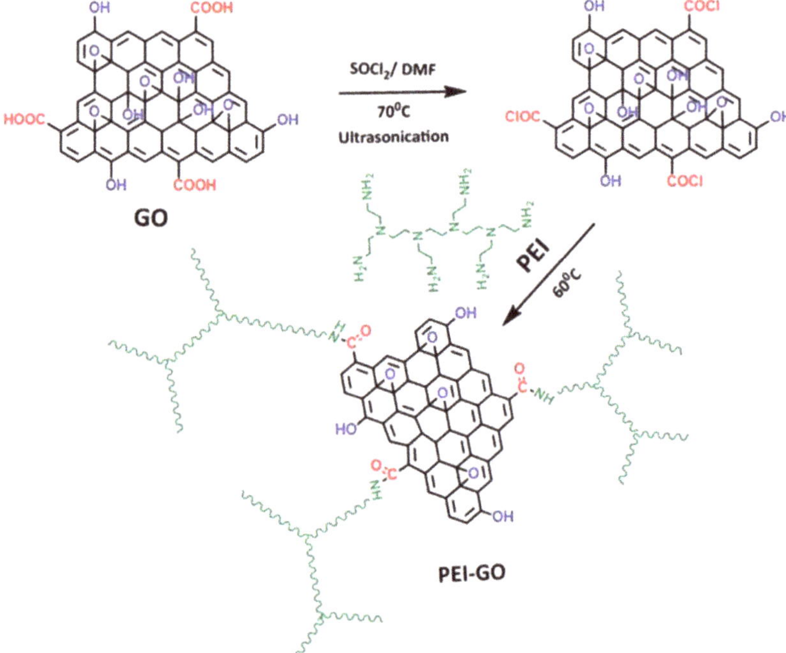

Figure 10. Schematic illustration of the synthesis of PEI-GO inhibitor [73], Reproduced with permission from ref. [73]. Copyright 2020 Royal Society of Chemistry.

As mentioned, PEI can be potentially quaternized to generate poly(quaternary ammonium) salt. Gao et al. [74] prepared quaternary polyethyleneimine (Q-PEI, P-28 in Figure 9) via tertiary amination reaction and subsequent quaternization reaction, and investigated the corrosion inhibition performance of Q-PEI on mild steel (A_3 steel) in H_2SO_4 solution. It was found that Q-PEI is a very effective corrosion inhibitor for A_3 steel, and the corrosion inhibition rate was up to 92% when immersed in 0.5 M H_2SO_4 solution for 72 h at a concentration of only 5 mg/L. Another study [75] further investigated the corrosion protection mechanism of PEI and Q-PEI on Q235 carbon steel under different acid mediums. It was found that Q-PEI is more cationic compared with PEI. The chemisorption that occurs between Q-PEI and carbon steel surface is mainly influenced by three aspects: (i) the electrostatic interaction between the positive charge of the amine nitrogen atom and the electronegative carbon steel surface; (ii) the covalent bond formed by the electrons on the phenylbenzene that can enter the three-dimensional orbitals of the iron atom; and (iii) the nitrogen atom of Q-PEI and iron atoms on the surface of carbon steel to form chelate complexes.

From the previous results of Ingrid et al. [71], it can be seen that there is still a possibility to improve the corrosion protection performance of PEI as a corrosion inhibitor for mild steel in neutral media. To further enhance the performance of PEI, Fu et al. [76] modified PEI with phosphorus groups to give polyethyleneimine phosphorous acid (PEPA) (P-29 in Figure 9). Owing to the synergetic effect from nitrogen atoms and the phosphorus groups, P-29 can be used as corrosion inhibitor in neutral medium for mild steel. Subsequently, they investigated the influence of molecular weights of PEPA on the anti-corrosion property [77]. Generally, three kinds of PEI with molecular weight of 600, 1800, and 10,000 were used. The results showed that PEPA is an excellent corrosion inhibitor, and the best inhibition performance was achieved at a dosage of 200 mg/L for PEPA synthesized from PEI with a molecular weight of 1800, with an inhibition rate of 94.6%.

3.3.3. Polyaniline

As a typical conductive polymer, polyaniline (PANI, P-30 in Figure 11) and its derivatives have been widely used as corrosion inhibitors in the form of either solution or coatings. Generally, the mechanism of PANI corrosion protection can be attributed to three effects: (i) the physical corrosion protection due to PANI coating separating the metal from the corrosive medium [78]; (ii) the passivation effect of PANI catalyzing the formation of oxide film on the metal surface [79]; and (iii) the heteroatom in PANI can form some kind of bond with metal atoms and perform a role in corrosion inhibition. As early as 1985, Deberry [80] reported the application of electrochemically synthesized polyaniline coating for corrosion protection of ferritic stainless steel in H_2SO_4. It was found that polyaniline could provide anodic protection for stainless steel and significantly reduce the corrosion rate of stainless steel. In a later study, Wessling found that polyaniline coating synthesized from dispersion may lead to significant shift of the corrosion potential and generated so-called passivation of metal. Similarly, Mirmohseni et al. [81] synthesized polyaniline polymers chemically and casted from 1-methyl-2-pyrrolidone (NMP) solution to iron samples. Such coating was found to be an effective inhibitor for iron in various corrosive environments, such as NaCl (3.5 wt%), tap water, and acids. Meanwhile, polyaniline was also found to be a more powerful inhibitor than conventional polymer, such as polyvinyl chloride.

The effect of substituents, such as $-OCH_3$, $-COOH$, and $-CH_3$, in the structure of polyaniline on corrosion inhibition has been investigated [82]. For example, Bereket and coworkers prepared poly(*N*-ethylaniline) (P-31 in Figure 11) coatings via cyclic voltammetry method on copper [83].

In addition to the polyaniline and substituted polyaniline (Table 4), Venkatachari et al. [84] synthesized water-soluble poly(*p*-phenylenediamine) (P-32 in Figure 11) and investigated its corrosion inhibition of iron in different concentrations of 1 M HCl using polarization technique and electrochemical impedance spectroscopy. The corrosion inhibition efficiency of poly(*p*-phenylenediamine) was found to be 85% at a concentration of 50 ppm, and even when the concentration of its monomer was scaled up to 5000 ppm, the corrosion inhibition efficiency of the monomer was only 73%. Compared with polyaniline, the corrosion inhibition performance of poly(*p*-phenylenediamine) is also found to be superior. Venkatachari et al. [85] found that poly(diphenylamine) also has good corrosion inhibition properties and the corrosion inhibition efficiency can be as high as 96% even at a very low concentration of 10 ppm. The analysis of FT-IR revealed that poly(diphenylamine) has a strong adsorption effect on the iron surface, which can improve the passivation of iron in sulfuric acid.

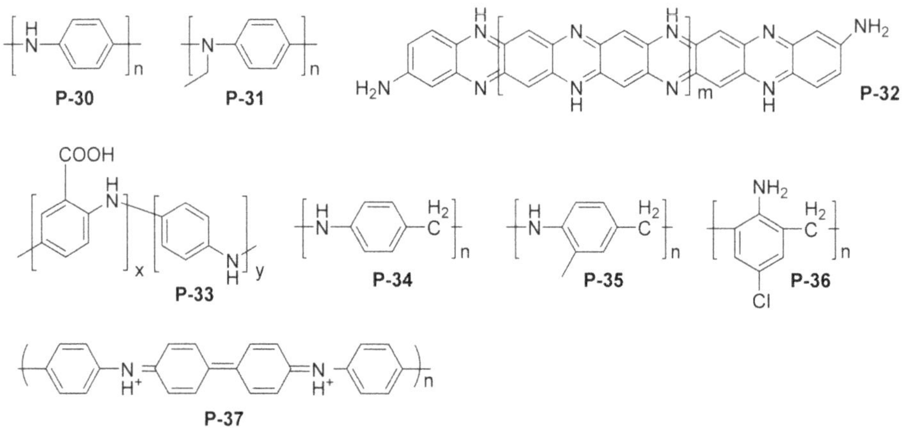

Figure 11. Typical chemical structure of polyaniline-based inhibitors.

Table 4. Inhibition property of typical nitrogen-containing synthetic polymeric inhibitor.

Inhibitor	Metal	Corrosion Medium	Test Method	Highest IE (%)	Reference
P-24/P-25/P-26	mild steel	1 M HCl	weight loss	P-25: 82 P-26: 82	[66]
P-27	ASTM 420 stainless steel	3% NaCl	linear polarization, cyclic polarization	81.9	[71]
P-28	mild steel (A3 steel)	0.5 M H_2SO_4	weight loss, PDP	92	[74]
P-29	mild steel	simulated neutral medium	weight loss tests, PDP, EIS	88	[76]
P-30	iron	3.5% NaCl/0.1 M HCl	PDP,	/	[78]
P-31	copper	0.1 M H_2SO_4	PDP, EIS	/	[83]
P-32	iron	1 M HCl	PDP, EIS	92.7	[84]
P-33	carbon steel	1 M HCl	weight loss, PDP, EIS	90.5	[86]
P-34/P-35/P-36	mild steel	1 M HCl	weight loss, PDP, EIS	P-34: 97.23 P-35: 98.46 P-36: 98.96	[87]

Polyaniline can also be copolymerized with other monomers to exert corrosion inhibition. Shi et al. [86] obtained poly(aniline-co-o-anthranilic acid) (P-33 in Figure 11) by copolymerizing polyaniline with anthranilic acid and investigated the corrosion inhibition efficiency of polyaniline copolymer in HCl solution on carbon steel. The results showed that the polyaniline copolymer solution in HCl solution had an effective corrosion inhibitor effect on carbon steel, and the adsorption on the metal surface followed the Langmuir adsorption isotherm with an IE value of 90.5% when the concentration was 20 ppm. Shukla et al. [87] investigated the slow-release properties of a series of anilines-formaldehyde polymers in 1 M HCl solution on mild steel. The three copolymers synthesized, poly(aniline-formaldehyde) (P-34), poly(o-tolylene-formaldehyde) (P-35), and poly(p-chloroaniline-formaldehyde) (P-36), showed excellent corrosion inhibition, and poly(p-chloroaniline-formaldehyde) had the best corrosion inhibition efficiency due to the presence of Cl in the structure and the interaction of the lone electron pair of chlorine atom with the metal surface, with an IE value of 98% at a concentration of 10 ppm.

Seegmiller et al. [88] prepared coatings by blending the conductive polymer polyaniline with the vinyl polymer poly(methyl methacrylate) and investigated the anticorrosive properties of the blended films on different metals in 1 mol/L H2SO4. The as-prepared conductive blends were effective in protecting Cu, Ni, Fe, and other metals [89].

3.3.4. Inorganic Mineral-Doped Polyaniline-Based Inhibitor

The doping of inorganic materials, such as montmorillonite, zeolite, and silica, in polyaniline can enhance the gas barrier properties of the composite [90], which can make the path that corrosive substances, such as oxygen and moisture, pass through the coating more tortuous, making the corrosive medium less corrosive and thus achieving anti-corrosion effects. Wei et al. [91] investigated the effect of alkali and acid doping forms of polyaniline on cold rolled steel using electrochemical corrosion measurements and compared it with undoped polyaniline coatings. They found that the polyaniline-based and epoxy topcoats treated with zinc nitrate had better overall protection compared to other coating systems, which is consistent with the observations of Wessling et al. [79] Yeh et al. [92] prepared polyaniline-clay nanocomposites (PCN) by doping montmorillonite (MMT) clay into polyaniline. It was found that PCN has 400% lower permeability and better corrosion resistance than conventional polyaniline. In a similar study, Olad et al. [93] prepared two kinds of nanocomposites, PANI/Na-MMT and PANI/O-MMT, by doping polyaniline with organophilic montmorillonite (O-MMT) and hydrophilic montmorillonite (Na-MMT). Although the conductivity of the nanocomposite was lower than that of polyaniline films, both hydrophilic and organophilic PANI/MMT nanocomposite coatings outperformed the pure polyaniline coating for corrosion protection of iron samples.

Alloys, such as stainless steel and magnesium alloys, can also be protected from corrosion with PANI/MMT nanocomposite coatings. The release of metal ions, such as iron, chromium, and nickel from the biological environment surrounding the alloy leads to reduced biocompatibility and susceptibility to local attack, which limits its use in biomedical applications [94]. Stainless steel is protected by a passive film formed on the surface, and the most serious problem in practical application is localized corrosion [95]. Mehdi et al. [96] used direct electrochemical galvanostatic method to electrochemically synthesize polyaniline and PANI/MMT nanocomposite coatings on the surface of 316L stainless steel, and the studied the corrosion protection performance in 0.5 M HCl medium. It was found that the inhibition efficiency was up to 99.8%, which indicated that the coatings have outstanding potential in protecting 316 L SS against corrosion in acidic media. Shao et al. [97] investigated the corrosion protection of AZ91D magnesium alloy by epoxy coating containing PANI/OMMT powder in 3.5% NaCl solution. It was found that the lamellar structure of PANI/OMMT could improve the barrier ability of the coating to electrolyte solution and PANI could form an oxide layer to enhance the corrosion resistance of magnesium alloy. Even after 6000 h of immersion, PANI/OMMT coating could still maintain the good corrosion resistance of AZ91D magnesium alloy.

In addition to montmorillonite, carbon-based inorganic materials, such as graphene, carbon nanotubes, and glass fibers (GB), can also be doped with polyaniline to afford composites for corrosion protection. Conductive graphene has also attracted interest in recent years due to its high aspect ratio of about 500 [98], and the combination of graphene with polymers has also been found to have good mechanical [99] and galvanic properties [100], and potential barrier properties [101]. Yeh et al. [102] doped graphene with polyaniline to obtain polyaniline/graphene composites (PAGCs). In comparison with pure polyaniline, the hybrid PAGCs were found to have excellent barrier properties to O_2 and H_2O.

The good electronic conductivity of carbon nanotube endows the resulting hybrid excellent corrosion protection performance. Zhu et al. [103] prepared dendrite-like polyaniline/carbon nanotube (PANI/CNT) nanocomposites via the interfacial π-π interactions between carbon nanotubes and polyaniline. The PANI/CNT nanocomposites have good redox ability in acidic and neutral media with excellent physical barrier effect of water and oxygen and metal passivation catalytic effect. Hou et al. [104] prepared polyaniline/glass fiber (PANI/GB, PGB) composites by in situ oxidative polymerization and used these materials to modify epoxy coatings and investigated the effect on their corrosion resistance. The results showed that the addition of PGB composites led to a significant increase in the corrosion resistance of the coatings, and the coatings had the best corrosion resistance when the PANI to GB mass ratio was 1:1.

3.3.5. Polyaniline-Based Inhibitors from Doping with Protonic Acids

The doping of polyaniline with different mineral and organic acids affects the degree of protonation of polyaniline, which affects the anions in the structure of polyaniline, and both the type and concentration of the doped acid affect the conductivity of polyaniline conductive form emeraldine salt (P-37 in Figure 11) [105]. Kohl et al. [106] prepared five polyaniline salts (PANI-HA) using five acids, namely, phosphoric acid (H_3PO_4), sulfuric acid (H_2SO_4), hydrochloric acid (HCl), *p*-toluenesulfonic acid, and 5-sulfosalicylic acid. The results showed that the type of PANI dopants and pigment volume concentration (PVC) perform an important role in the corrosion resistance of the coating, and the PVC corresponding to the best corrosion inhibition effect is different for different PANI-HA. The general rule of PVC effect on the corrosion resistance of the coating is at low PVC values (PVC = 0.1–5%), the corrosion resistance is independent of the pigment used; when the PVC increases up to 10%, the corrosion resistance becomes significantly worse with the increase in PVC. Generally, the incorporation of PANI pigments into epoxy resins often resulted in severe agglomeration and required sonication [107]. Sadegh et al. [108] prepared glycol-modified epoxy coatings containing polyaniline-camphorsulfonic acid particles on mild steel surfaces using a novel one-pot method without further dispersion of the particles. Zhang et al. [91] doped two acids, hydrofluoric acid and camphorsulfonic acid, as doping acids into PANI subsequently added to epoxy coatings. The EIS results showed that both acids of doped PANI had an improved effect on the retardation performance of epoxy coating, especially the anticorrosion effect of camphorsulfonic acid doped PANI coatings was greatly enhanced. The anticorrosion mechanism of different forms of PANI coatings can be summarized as following: (i) through barrier properties; (ii) through the formation of passivation layer by redox reaction; and (iii) through the formation of insoluble counter anionic salts. Ali et al. [109] prepared self-assembled polyaniline nanotubes using sodium dodecylbenzene sulfonate as the dopant acid and compared their corrosion performance with that of conventional polyaniline. It was found that the corrosion currents of iron sheets coated with polyaniline nanotubes were much lower than those of iron sheets coated with conventional polyaniline. Grgur et al. [110] conducted a more practical study on benzoate doped polyaniline coatings, which were tested for cathodic protection under 3% NaCl, atmospheric and Sahara desert conditions. The results show that the coating can effectively protect mild steel even in a short period of time, and they propose a "switching zone mechanism" (Figure 12).

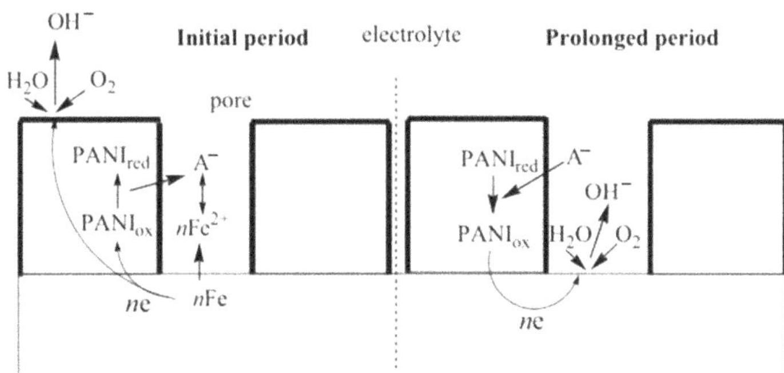

Figure 12. Proposed mechanism of the mild steel corrosion protection with PANI coatings [110].

3.3.6. Polyaniline-Based Inhibitor from Doping with Metals Oxides

In addition to the inorganic substances and protonic acids, metals and their oxides can also be doped with polyaniline to improve the anodic protection of polyaniline, owing to their sacrificial oxidation. The most commonly used metal materials are iron, zinc, titanium, and their corresponding oxides. Venkatachari et al. [111] prepared polyaniline-Fe_2O_3 composites with aniline to Fe_2O_3 ratios of 2:1, 1:1, and 1:2, respectively, using chemical oxidative method in phosphoric acid medium with ammonium persulfate as the oxidizing agent, combined the composites with an acrylic binder to prepare a primer coated on steel specimens. They found that the best corrosion resistance to 3% NaCl solution of the composite coating was achieved at a ratio of 1:1 of aniline to Fe_2O_3. The good anti-corrosion effect of polyaniline-Fe_2O_3 composites can be mainly attributed to the barrier effect due to the formation of a passive film and iron-phosphate salt film on the iron surface. The size of the filler particle size in polymer composites has a great influence on the performance of the composite [112], and when the size of the filler particles is reduced to the nanoscale, the performance is usually very different from that of the micron size [113]. It was found that the doping of zinc also effectively improved the electrical conductivity and corrosion resistance of PANI [114]. The highest electrical conductivity of PANI/Zn nanocomposites was obtained when the zinc content was 4 wt%, and the performance exhibited by PANI/Zn nanocomposite films was more excellent than the composite films. Radhakrishnan et al. [115] synthesized polyaniline/nano-TiO_2 composites using in situ polymerization. In comparison with conventional polyaniline, the addition of nanosized TiO_2 improved the corrosion resistance of polyaniline coatings, and the corrosion resistance of polyaniline prepared by 4.18% nano-TiO_2 could be increased to more than 100 times. Al-Masoud et al. [116] prepared bimetallic oxide nanocomposite functionalized with polyaniline, by combining ZnO and TiO_2 simultaneously with polyaniline (PANI). It was found that the as prepared ZnTiO@PANI is a strong acidic corrosion inhibitor, and its corrosion inhibition rate can reach 98.9% at a concentration of 100 ppm in acidic chloride solution (1.0 M HCl + 3.5% NaCl).

3.4. Other Type of Polymeric Inhibitors

In addition to the above well-studied systems, other kinds of polymeric inhibitors have also been investigated because modern polymer chemistry allows the introduction of hydrophilic and adsorption groups to the side chains on the basis of the main chain structure and copolymerization with other monomers (Table 5).

Polyacrylic acid (PAA) is the most well-known vinyl polymer corrosion inhibitor used in the early study of anti-corrosion. Recently, polyacrylate or acrylamide copolymer corrosion inhibitors have become more popular. Lin et al. [117] prepared poly(methyl acrylate)-*co*-poly(acrylic acid imidazoline) (MA-ACI, P-38 in Figure 13) from methyl acrylate and acrylic imidazoline with azo diisobutyronitrile as initiator. According to the rotating hanging plate method, P-38 showed inhibition efficiency as high as 82% at a concentration of 0.10 g/L in 1 mol/L H_2SO_4 at 30 °C. Taghi et al. [118] prepared neodymium-poly acrylic acid complex (Nd-PAA, P-39 in Figure 13) by adding neodymium to PAA and applied them to the corrosion protection of ST-12 type in 0.1 M NaCl. Due to the anionic nature of PAA, they deposited densely and crack-free ultrafine Nd-PAA films on the steel surface.

Figure 13. Chemical structure of the poly methyl acrylate-acrylic acid imidazoline, and cobaltocenium-containing waterborne polymeric inhibitors.

Moreover, chemical modification or grafting of inorganic material by synthetic polymer has also been developed to afford efficient inhibitors. For example, as shown in Figure 14, Yu et al. [119] modified graphene oxide with polystyrene by in situ microemulsion polymerization. The resulted hybrid demonstrated significant improvement in the corrosion resistance in comparison with native graphene and polystyrene, with the corrosion resistance efficiency increasing from 37.9% to 99.5%.

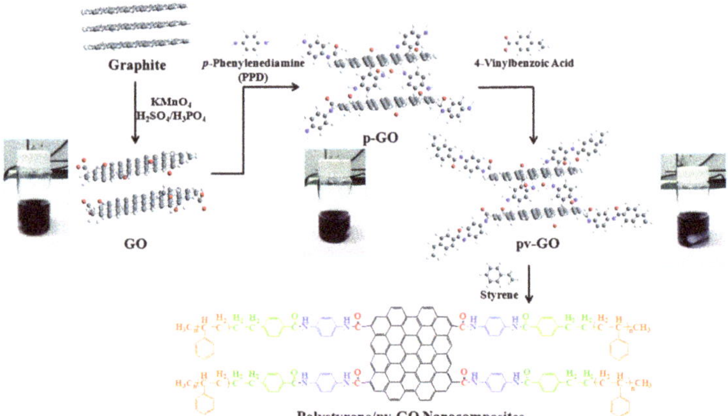

Figure 14. Synthesis of organo-functionalized graphene oxide [119]. Reproduced with permission from ref. [119]. Copyright 2014 Royal Society of Chemistry.

Table 5. Inhibition property of other types of polymeric inhibitor.

Inhibitor	Metal	Corrosion Medium	Test Method	Highest IE (%)	Reference
P-38	N80 steel sheet	1 M H_2SO_4	PDP, EIS	90.2	[117]
P-39	ST-12 type steel sheets	0.1 M NaCl	PDP, EIS	/	[118]
P-40/P-41/P-42/P-43	Mild steel	4 M HCl	weight loss, EIS, PDP	98	[120]

Since water solubility is a crucial problem in the design and application of inhibitors, water soluble polymers, such as waterborne polyurethane, have also been used as inhibitors. Our group developed a series of cobaltocenium-containing polyurethanes (P-40, P-41, P-42, and P-43 in Figure 13) via the reaction between hydroxyl-terminated cobaltocenium monomer and different kinds of commercially available diisocyanates (the synthetic process is shown in Figure 15). The presence of charged cobaltocenium group and its counterion endow the resulting polyurethane good water solubility. Owing to the multiple interactions between these waterborne polyurethanes and metal surface and the inter/intramolecular hydrogen bonding between urethane groups, these polymers can strongly adsorb on the metal surface and therefore demonstrated excellent corrosion protection property. According to the weight loss experiment, the inhibition efficiency can achieve as high as 98.0% at a concentration as low as 20 mg/mL toward mild steel in 4 M HCl [120].

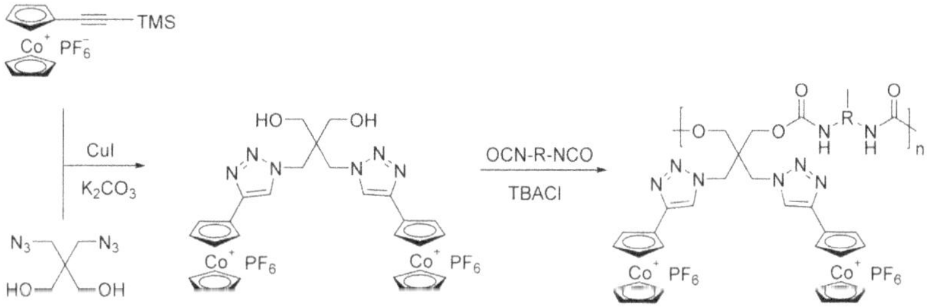

Figure 15. Synthetic procedure for cobaltocenium-containing polyurethane-type inhibitors: P-40/P-41/P-42/P-43 [120].

4. Conclusions

Development and application of inhibitors can not only prolong the lifetime of metal materials, but also provide a possible solution to the energy and economic loss caused by metal corrosion. The development of polymer chemistry provides versatile tools to the rational design and facile syntheses of powerful polymeric inhibitors. As discussed in this review, there are several advantages of polymeric inhibitors:

(i) First, to increase the inhibition efficiency, different adsorption groups with heteroatoms, such as P, S, N, and O, have been introduced to the skeleton of polymer;

(ii) Second, the supramolecular structure of polymeric inhibitors also provides environment adapted configuration of adsorption groups on the metal surface, which enables the synergetic anti-corrosion performance;

(iii) Third, the soft matter nature of polymers enables their integration with different kinds of micro/nano fillers, which can further enhance the anti-corrosion property and therefore improve the actual performance of corrosion inhibitor.

Though a flourishing future of polymeric inhibitor could be envisioned, there are several challenges that limit the further development of polymeric inhibitor and its practical applications. First, a systematic study on the structure-property relationship, especially the relationship between adsorption properties and the chemical structure of inhibitor, should be established as a guide for the design of high efficient inhibitor. Different from small molecular weight inhibitors, the influence of the structure of main-chain and side groups and the molecular weight (number of repeat unit) should be well studied. More importantly, although quantum chemistry, such as density functional theory density, has been used to understand the process of anti-corrosion, artificial intelligence, and computer-aided structural design should be developed to the design of polymeric inhibitors. Second, since most of the polymeric inhibitor has been used as anti-corrosion coatings, the integration of polymeric inhibitors and inorganic fillers and the method to prepare composites should also be focused. Moreover, suitable in-situ characterization of the corrosion process and the structure characterization for the change of inhibitor and the metal surface should be developed. Third, since the broad application of inhibitor has already caused environment pollution, green inhibitors, such as polymeric inhibitors from natural sources, should be investigated. While for the synthetic polymeric inhibitor, facile synthetic methodology and economic issue should be considered. Fourth, with regarding to the practical application of inhibitor in the form of coating, the integration of stimulus-responsive property, such as self-healing, with polymeric inhibitor may be a further direction for the development of smart inhibitors.

Author Contributions: Conceptualization, Y.Y., J.Y., J.Z. and Q.Z.; writing—original draft preparation, X.W., J.Y., S.L. and J.Z.; writing—review and editing, X.W., S.L., Y.Y. and Q.Z.; supervision, Y.Y. All authors have read and agreed to the published version of the manuscript.

Funding: We acknowledge the support by the NSFC (NO. 21975205, 21504068), Natural Science Foundation of Chongqing (cstc2020jcyj-msxmX1078), Shaanxi National Science Foundation (2020JM-138), and Open Project of State Key Laboratory of Supramolecular Structure and Materials (sklssm2022022).

Institutional Review Board Statement: Not applicable.

Informed Consent Statement: Not applicable.

Data Availability Statement: The data presented in this study are available on request from the corresponding author.

Conflicts of Interest: The authors declare no conflict of interest.

References

1. Goyal, M.; Kumar, S.; Bahadur, I.; Verma, C.; Ebenso, E.E. Organic corrosion inhibitors for industrial cleaning of ferrous and nonferrous metals in acidic solutions: A review. *J. Mol. Liq.* **2018**, *256*, 565–573. [CrossRef]
2. Verma, C.; Ebenso, E.E.; Quraishi, M.A. Ionic liquids as green and sustainable corrosion inhibitors for metals and alloys: An overview. *J. Mol. Liq.* **2017**, *233*, 403–414. [CrossRef]
3. Dwivedi, D.; Lepkova, K.; Becker, T. Carbon steel corrosion: A review of key surface properties and characterization methods. *RSC Adv.* **2017**, *7*, 4580–4610. [CrossRef]
4. Finsgar, M.; Jackson, J. Application of corrosion inhibitors for steels in acidic media for the oil and gas industry: A review. *Corros. Sci.* **2014**, *86*, 17–41. [CrossRef]
5. Fateh, A.; Aliofkhazraei, M.; Rezvanian, A.R. Review of corrosive environments for copper and its corrosion inhibitors. *Arab. J. Chem.* **2020**, *13*, 481–544. [CrossRef]
6. El Ibrahimi, B.; Jmiai, A.; Bazzi, L.; El Issami, S. Amino acids and their derivatives as corrosion inhibitors for metals and alloys. *Arab. J. Chem.* **2020**, *13*, 740–771. [CrossRef]
7. Verma, C.; Quraishi, M.A.; Rhee, K.Y. Aqueous phase polymeric corrosion inhibitors: Recent advancements and future opportunities. *J. Mol. Liq.* **2022**, *348*, 118387. [CrossRef]
8. Verma, C.; Olasunkanmi, L.O.; Ebenso, E.E.; Quraishi, M.A. Substituents effect on corrosion inhibition performance of organic compounds in aggressive ionic solutions: A review. *J. Mol. Liq.* **2018**, *251*, 100–118. [CrossRef]
9. Verma, C.; Ebenso, E.E.; Quraishi, M.A. Molecular structural aspects of organic corrosion inhibitors: Influence of -CN and -NO2 substituents on designing of potential corrosion inhibitors for aqueous media. *J. Mol. Liq.* **2020**, *316*, 113874. [CrossRef]
10. Maeztu, J.D.; Rivero, P.J.; Berlanga, C.; Bastidas, D.M.; Palacio, J.F.; Rodriguez, R. Effect of graphene oxide and fluorinated polymeric chains incorporated in a multilayered sol-gel nanocoating for the design of corrosion resistant and hydrophobic surfaces. *Appl. Surf. Sci.* **2017**, *419*, 138–149. [CrossRef]
11. Chang, K.-C.; Hsu, M.-H.; Lu, H.-I.; Lai, M.-C.; Liu, P.-J.; Hsu, C.-H.; Ji, W.-F.; Chuang, T.-L.; Wei, Y.; Yeh, J.-M.; et al. Room-temperature cured hydrophobic epoxy/graphene composites as corrosion inhibitor for cold-rolled steel. *Carbon* **2014**, *66*, 144–153. [CrossRef]
12. Alrefaee, S.H.; Rhee, K.Y.; Verma, C.; Quraishi, M.A.; Ebenso, E.E. Challenges and advantages of using plant extract as inhibitors in modern corrosion inhibition systems: Recent advancements. *J. Mol. Liq.* **2021**, *321*, 114666. [CrossRef]

13. Salleh, S.Z.; Yusoff, A.H.; Zakaria, S.K.; Taib, M.A.A.; Abu Seman, A.; Masri, M.N.; Mohamad, M.; Mamat, S.; Sobri, S.A.; Ali, A.; et al. Plant extracts as green corrosion inhibitor for ferrous metal alloys: A review. *J. Clean. Prod.* **2021**, *304*, 127030. [CrossRef]
14. Umoren, S.A.; Eduok, U.M. Application of carbohydrate polymers as corrosion inhibitors for metal substrates in different media: A review. *Carbohydr. Polym.* **2016**, *140*, 314–341. [CrossRef] [PubMed]
15. Montemor, M.F. Functional and smart coatings for corrosion protection: A review of recent advances. *Surf. Coat. Technol.* **2014**, *258*, 17–37. [CrossRef]
16. Verma, C.; Verma, D.K.; Ebenso, E.E.; Quraishi, M.A. Sulfur and phosphorus heteroatom-containing compounds as corrosion inhibitors: An overview. *Heteroat. Chem.* **2018**, *29*, e21437. [CrossRef]
17. Mobin, M.; Rizvi, M. Polysaccharide from Plantago as a green corrosion inhibitor for carbon steel in 1 M HCl solution. *Carbohydr. Polym.* **2017**, *160*, 172–183. [CrossRef]
18. Kaseem, M.; Ko, Y.G. Effect of starch on the corrosion behavior of Al-Mg-Si alloy processed by micro arc oxidation from an ecofriendly electrolyte system. *Bioelectrochemistry* **2019**, *128*, 133–139. [CrossRef]
19. Thamer, A.N.; Kadham, L.H.; Alwaan, I.M. Preparation and Characteristics of Starch Nanocrystals as Corrosion Inhibitor: Experimental and Theoretical Studies. *Asian J. Chem.* **2022**, *34*, 2854–2864. [CrossRef]
20. Sushmitha, Y.; Rao, P. Material conservation and surface coating enhancement with starch-pectin biopolymer blend: A way towards green. *Surf. Interfaces* **2019**, *16*, 67–75.
21. Zhang, Z.; Ba, H.; Wu, Z. Sustainable corrosion inhibitor for steel in simulated concrete pore solution by maize gluten meal extract: Electrochemical and adsorption behavior studies. *Constr. Build. Mater.* **2019**, *227*, 117080. [CrossRef]
22. Burrell, M.M. Starch: The need for improved quality or quantity—An overview. *J. Exp. Bot.* **2003**, *54*, 451–456. [CrossRef]
23. Othman, N.K.; Salleh, E.M.; Dasuki, Z.; Lazim, A.M. *Dimethyl Sulfoxide-Treated Starch of Dioescorea hispida as a Green Corrosion Inhibitor for Low Carbon Steel in Sodium Chloride Medium*; IntechOpen: London, UK, 2018.
24. Anyiam, C.K.; Ogbobe, O.; Oguzie, E.E.; Madufor, I.C. Synergistic Study of Modified Sweet Potato Starch and KI for Corrosion Protection of Mild Steel in Acidic Media. *J. Bio- Tribo-Corros.* **2020**, *6*, 1–10. [CrossRef]
25. Anyiam, C.K.; Ogbobe, O.; Oguzie, E.E.; Madufor, I.C.; Nwanonenyi, S.C.; Onuegbu, G.C.; Obasi, H.C.; Chidiebere, M.A. Corrosion inhibition of galvanized steel in hydrochloric acid medium by a physically modified starch. *SN Appl. Sci.* **2020**, *2*, 520. [CrossRef]
26. Deng, S.; Li, X.; Du, G. An efficient corrosion inhibitor of cassava starch graft copolymer for aluminum in phosphoric acid. *Chin. J. Chem. Eng.* **2021**, *37*, 222–231. [CrossRef]
27. Hou, D.; Yang, W.; Dong, S. Corrosion inhibition of carbon steel in acid solution by starch-AA-CS terpolymer. *Chem. Ind. Eng.* **2022**, *39*, 100–107.
28. Li, X.; Deng, S.; Lin, T.; Xie, X. Cassava starch graft copolymer as a novel inhibitor for the corrosion of aluminium in HNO3 solution. *J. Mol. Liq.* **2019**, *282*, 499–514. [CrossRef]
29. Wang, Y.; Wang, K.; Gao, P.; Liu, R.; Zhao, D.; Zhai, J.; Qu, G. Inhibition for Zn Corrosion by Starch Grafted Copolymer. *J. Chin. Soc. Corros. Prot.* **2021**, *41*, 131–138.
30. Li, X.; Deng, S.; Lin, T.; Xie, X.; Du, G. Cassava starch ternary graft copolymer as a corrosion inhibitor for steel in HCl solution. *J. Mater. Res. Technol.* **2020**, *9*, 2196–2207. [CrossRef]
31. Arukalam, I.O.; Madufor, I.C.; Ogbobe, O.; Oguzie, E.E. Experimental and Theoretical Studies of Hydroxyethyl Cellulose as Inhibitor for Acid Corrosion Inhibition of Mild Steel and Aluminium. *Open Corros. J.* **2014**, *6*, 1–10. [CrossRef]
32. Eid, S.; Abdallah, M.; Kamar, E.M.; El-Etre, A. Corrosion Inhibition of Aluminum and Aluminum Silicon Alloys in Sodium Hydroxide Solutions by Methyl Cellulose. *J. Mater. Environ. Sci.* **2015**, *6*, 892–901.
33. Sobhi, M.; Eid, S. Chemical, Electrochemical and Morphology Studies on Methyl Hydroxyethyl Cellulose as Green Inhibitor for Corrosion of Copper in Hydrochloric acid Solutions. *Prot. Met. Phys. Chem. Surf.* **2018**, *54*, 893–898. [CrossRef]
34. Nwanonenyi, S.C.; Obasi, H.C.; Eze, I.O. Hydroxypropyl Cellulose as an Efficient Corrosion Inhibitor for Aluminium in Acidic Environments: Experimental and Theoretical Approach. *Chem. Afr. -A J. Tunis. Chem. Soc.* **2019**, *2*, 471–482. [CrossRef]
35. Wang, X.; Lei, Y.; Jiang, Z.N.; Zhang, Q.H.; Li, Y.Y.; Liu, H.F.; Zhang, G.A. Developing two thiocarbohydrazide modified glucose derivatives as high-efficiency green corrosion inhibitors for carbon steel. *Ind. Crops Prod.* **2022**, *188*, 115680. [CrossRef]
36. Yang, S.F.; Wen, Y.; Yi, P.; Xiao, K.; Dong, C.F. Effects of chitosan inhibitor on the electrochemical corrosion behavior of 2205 duplex stainless steel. *Int. J. Miner. Metall. Mater.* **2017**, *24*, 1260–1266. [CrossRef]
37. Liu, Y.; Zou, C.; Yan, X.; Xiao, R.; Wang, T.; Li, M. β-Cyclodextrin Modified Natural Chitosan as a Green Inhibitor for Carbon Steel in Acid Solutions. *Ind. Eng. Chem. Res.* **2015**, *54*, 5664–5672. [CrossRef]
38. Zhang, Q.H.; Hou, B.S.; Li, Y.Y.; Zhu, G.Y.; Liu, H.F.; Zhang, G.A. Two novel chitosan derivatives as high efficient eco-friendly inhibitors for the corrosion of mild steel in acidic solution. *Corros. Sci.* **2020**, *164*, 108346. [CrossRef]
39. Fares, M.M.; Maayta, A.K.; Al-Mustafa, J.A. Corrosion inhibition of iota-carrageenan natural polymer on aluminum in presence of zwitterion mediator in HCl media. *Corros. Sci.* **2012**, *65*, 223–230. [CrossRef]
40. Caffall, K.H.; Mohnen, D. The structure, function, and biosynthesis of plant cell wall pectic polysaccharides. *Carbohydr. Res.* **2009**, *344*, 1879–1900. [CrossRef]
41. Halambek, J.; Cindric, I.; Grassino, A.N. Evaluation of pectin isolated from tomato peel waste as natural tin corrosion inhibitor in sodium chloride/acetic acid solution. *Carbohydr. Polym.* **2020**, *234*, 115940. [CrossRef]

42. Chen, X.; Chen, Y.; Cui, J.; Li, Y.; Liang, Y.; Cao, G. Molecular dynamics simulation and DFT calculation of "green" scale and corrosion inhibitor. *Comput. Mater. Sci.* **2021**, *188*, 110229. [CrossRef]
43. He, C.; Li, X.-Q.; Feng, G.-L.; Long, W.-J. A universal strategy for green and in situ synthesis of carbon dot-based pickling solution. *Green Chem.* **2022**, *24*, 5842–5855. [CrossRef]
44. Coquery, C.; Negrell, C.; Causse, N.; Pebere, N.; David, G. Synthesis of new high molecular weight phosphorylated chitosans for improving corrosion protection. *Pure Appl. Chem.* **2019**, *91*, 509–521. [CrossRef]
45. Solimando, X.; Kennedy, E.; David, G.; Champagne, P.; Cunningham, M.F. Phosphorus-containing polymers synthesised via nitroxide-mediated polymerisation and their grafting on chitosan by grafting to and grafting from approaches. *Polym. Chem.* **2020**, *11*, 4133–4142. [CrossRef]
46. Coquery, C.; Carosio, F.; Negrell, C.; Causse, N.; Pebere, N.; David, G. New bio-based phosphorylated chitosan/alginate protective coatings on aluminum alloy obtained by the LbL technique. *Surf. Interfaces* **2019**, *16*, 59–66. [CrossRef]
47. Ouarga, A.; Noukrati, H.; Iraola-Arregui, I.; Elaissari, A.; Barroug, A.; Ben Youcef, H. Development of anti-corrosion coating based on phosphorylated ethyl cellulose microcapsules. *Prog. Org. Coat.* **2020**, *148*, 70. [CrossRef]
48. Macarie, L.; Ilia, G. Poly(vinylphosphonic acid) and its derivatives. *Prog. Polym. Sci.* **2010**, *35*, 1078–1092. [CrossRef]
49. Posner, R.; Sundell, P.E.; Bergman, T.; Roose, P.; Heylen, M.; Grundmeier, G.; Keil, P. UV-Curable Polyester Acrylate Coatings: Barrier Properties and Ion Transport Kinetics Along Polymer/Metal Interfaces. *J. Electrochem. Soc.* **2011**, *158*, C185–C193. [CrossRef]
50. Millet, F.; Auvergne, R.; Caillol, S.; David, G.; Manseri, A.; Pebere, N. Improvement of corrosion protection of steel by incorporation of a new phosphonated fatty acid in a phosphorus-containing polymer coating obtained by UV curing. *Prog. Org. Coat.* **2014**, *77*, 285–291. [CrossRef]
51. Macarie, L.; Pekar, M.; Simulescu, V.; Plesu, N.; Iliescu, S.; Ilia, G.; Tara-Lunga-Mihali, M. Properties in Aqueous Solution of Homo- and Copolymers of Vinylphosphonic Acid Derivatives Obtained by UV-Curing. *Macromol. Res.* **2017**, *25*, 214–221. [CrossRef]
52. Kousar, F.; Moratti, S.C. Synthesis of fluorinated phosphorus-containing copolymers and their immobilization and properties on stainless steel. *RSC Adv.* **2021**, *11*, 38189–38201. [CrossRef]
53. Sheffer, M.; Groysman, A.; Starosvetsky, D.; Savchenko, N.; Mandler, D. Anion embedded sol-gel films on Al for corrosion protection. *Corros. Sci.* **2004**, *46*, 2975–2985. [CrossRef]
54. Kannan, A.G.; Choudhury, N.R.; Dutta, N.K. Electrochemical performance of sol-gel derived phospho-silicate-methacrylate hybrid coatings. *J. Electroanal. Chem.* **2010**, *641*, 28–34. [CrossRef]
55. Kannan, A.G.; Choudhury, N.R.; Dutta, N.K. Synthesis and characterization of methacrylate phospho-silicate hybrid for thin film applications. *Polymer* **2007**, *48*, 7078–7086. [CrossRef]
56. Tuken, T.; Yazici, B.; Erbil, M. The use of polythiophene for mild steel protection. *Prog. Org. Coat.* **2004**, *51*, 205–212. [CrossRef]
57. Zhi, D.; Smyrl, W.H. Application of Electroactive Films in Corrosion Protection.2.Metal Hwxacyanometalate Films on TiO2/Ti Surfaces. *J. Electrochem. Soc.* **1991**, *138*, 1911–1918.
58. Wang, W.L.; Lai, Y.H. Synthesis and characterization of a novel 1,4-naphthalene-based thiophene copolymers. *Thin Solid Film.* **2002**, *417*, 211–214. [CrossRef]
59. Medrano-Vaca, M.G.; Gonzalez-Rodriguez, J.G.; Nicho, M.E.; Casales, M.; Salinas-Bravo, V.M. Corrosion protection of carbon steel by thin films of poly (3-alkyl thiophenes) in 0.5 M H2SO4. *Electrochim. Acta* **2008**, *53*, 3500–3507. [CrossRef]
60. Branzoi, F.; Mihai, M.A.; Petrescu, S. Corrosion Protection Efficacy of the Electrodeposit of Poly (N-Methyl Pyrrole-Tween20/3-Methylthiophene) Coatings on Carbon Steel in Acid Medium. *Coatings* **2022**, *12*, 1062. [CrossRef]
61. Emira, H.S.A.; Abu-Ayana, Y.M.; El-Sawy, S.M. Modified amino resins for corrosion prevention in organic coatings. *Pigment Resin Technol.* **2013**, *42*, 298–308. [CrossRef]
62. McAndrew, T.P. Corrosion prevention with electrically conductive polymers. *Trends Polym. Sci.* **1997**, *5*, 7–12.
63. Leon-Silva, U.; Nicho, M.E. Poly(3-octylthiophene) and polystyrene blends thermally treated as coatings for corrosion protection of stainless steel 304. *J. Solid State Electrochem.* **2010**, *14*, 1487–1497. [CrossRef]
64. Tsai, M.-C.; Yang, C.-R.; Tsai, J.-H.; Yu, Y.-H.; Huang, P.-T. Enhanced Corrosion Protection of Iron by Poly(3-hexylthiophene)/Poly (styrene-co-hydroxystyrene) Blends. *Coatings* **2018**, *8*, 383. [CrossRef]
65. Al-Muallem, H.A.; Mazumder, M.A.J.; Estaitie, M.K.; Ali, S.A. A novel cyclopolymer containing residues of essential amino acid methionine: Synthesis and application. *Iran. Polym. J.* **2015**, *24*, 541–547. [CrossRef]
66. Ali, S.A.; Saeed, M.T. Synthesis and corrosion inhibition study of some 1,6-hexanediamine-based N,N-diallyl quaternary ammonium salts and their polymers. *Polymer* **2001**, *42*, 2785–2794. [CrossRef]
67. Ali, S.A.; Goni, L.; Mazumder, M.A.J. Butler's cyclopolymerizaton protocol in the synthesis of diallylamine salts/sulfur dioxide alternate polymers containing amino acid residues. *J. Polym. Res.* **2017**, *24*, 1022–9760. [CrossRef]
68. Haladu, S.A.; Umoren, S.A.; Ali, S.A.; Solomon, M.M.; Mohammed, A.R.I. Synthesis, characterization and electrochemical evaluation of anticorrosion property of a tetrapolymer for carbon steel in strong acid media. *Chin. J. Chem. Eng.* **2019**, *27*, 965–978. [CrossRef]
69. Liu, S.; Yan, J.; Shi, J.Q.; Li, X.Z.; Zhang, J.P.; Wang, X.Y.; Cai, N.J.; Fang, Q.H.; Zhang, Q.Y.; Yan, Y. Rational design of cobaltocenium-containing polythioether type metallo-polyelectrolytes as HCl corrosion inhibitors for mild steel. *Polym. Chem.* **2023**, *14*, 330–342. [CrossRef]

70. Braun, R.D.; Lopez, E.E.; Vollmer, D.P. Low-Molecular-Weight Straight-Chain Amines as Corrosion-Inhibitors. *Corros. Sci.* **1993**, *34*, 1251–1257. [CrossRef]
71. Finsgar, M.; Fassbender, S.; Nicolini, F.; Milosev, I. Polyethyleneimine as a corrosion inhibitor for ASTM 420 stainless steel in near-neutral saline media. *Corros. Sci.* **2009**, *51*, 525–533. [CrossRef]
72. Finsgar, M.; Fassbender, S.; Hirth, S.; Milosev, I. Electrochemical and XPS study of polyethyleneimines of different molecular sizes as corrosion inhibitors for AISI 430 stainless steel in near-neutral chloride media. *Mater. Chem. Phys.* **2009**, *116*, 198–206. [CrossRef]
73. Ansari, K.; Chauhan, D.S.; Quraishi, M.A.; Adesina, A.Y.; Saleh, T.A. The synergistic influence of polyethyleneimine-grafted graphene oxide and iodide for the protection of steel in acidizing conditions. *RSC Adv.* **2020**, *10*, 17739–17751. [CrossRef]
74. Gao, B.; Zhang, X.; Sheng, Y. Studies on preparing and corrosion inhibition behaviour of quaternized polyethyleneimine for low carbon steel in sulfuric acid. *Mater. Chem. Phys.* **2008**, *108*, 375–381. [CrossRef]
75. Zhang, X.; Wu, X.; Gao, B. Studies on Cationic Property of Quaternary Polyethyleneimine. *Mater. Sci. Forum* **2011**, *689*, 432–439. [CrossRef]
76. Chen, T.; Chen, M.; Chen, Z.; Fu, C. Comprehensive investigation of modified polyethyleneimine as an efficient polymeric corrosion inhibitor in neutral medium: Synthesis, experimental and theoretical assessments. *J. Mol. Liq.* **2021**, *339*, 116803. [CrossRef]
77. Chen, T.; Chen, M.; Fu, C. Effect of molecular weight on inhibition performance of modified polyethyleneimine as polymer corrosion inhibitor for carbon steel in neutral medium. *J. Appl. Polym. Sci.* **2022**, *139*, e51922. [CrossRef]
78. Mcandrew, T.P.; Miller, S.A.; Gilicinski, A.G.; Robeson, L.M. Polyaniline in Corrision-Resistant Coatings. *ACS Symp. Ser.* **1998**, *689*, 396–408.
79. Wessling, B. Passivation of Metals by Coating with Polyaniline-Corrosion Potential Shift and Morphological-Changes. *Adv. Mater.* **1994**, *6*, 226–228. [CrossRef]
80. Deberry, D.W. Modification Of The Electrochemical And Corrosion Behavior Of Stainless-Steels With An Electroactive Coating. *J. Electrochem. Soc.* **1985**, *132*, 1022–1026. [CrossRef]
81. Mirmohseni, A.; Oladegaragoze, A. Anti-corrosive properties of polyaniline coating on iron. *Synth. Met.* **2000**, *114*, 105–108. [CrossRef]
82. Kalendova, A.; Vesely, D.; Stejskal, J. Organic coatings containing polyaniline and inorganic pigments as corrosion inhibitors. *Prog. Org. Coat.* **2008**, *62*, 105–116. [CrossRef]
83. Duran, B.; Turhan, M.C.; Bereket, G.; Sarac, A.S. Electropolymerization, characterization and corrosion performance of poly(N-ethylaniline) on copper. *Electrochim. Acta* **2009**, *55*, 104–112. [CrossRef]
84. Manivel, P.; Sathiyanarayanan, S.; Venkatachari, G. Synthesis of Poly(p-phenylene diamine) and Its Corrosion Inhibition Effect on Iron in 1M HC1. *J. Appl. Polym. Sci.* **2008**, *110*, 2807–2814. [CrossRef]
85. Jeyaprabha, C.; Sathiyanarayanan, S.; Phani, K.L.N.; Venkatachari, G. Investigation of the inhibitive effect of poly(diphenylamine) on corrosion of iron in 0.5 M H2SO4 solutions. *J. Electroanal. Chem.* **2005**, *585*, 250–255. [CrossRef]
86. Shi, F.; Wang, X.; Yu, J.; Hou, B. Corrosion inhibition by polyaniline copolymer of mild steel in hydrochloric acid solution. *Anti-Corros. Methods Mater.* **2011**, *58*, 111–115.
87. Shukla, S.K.; Quraishi, M.A. Effect of some substituted anilines-formaldehyde polymers on mild steel corrosion in hydrochloric acid medium. *J. Appl. Polym. Sci.* **2012**, *124*, 5130–5137. [CrossRef]
88. Seegmiller, J.C.; da Silva, J.E.P.; Buttry, D.A.; de Torresi, S.I.C.; Torresi, R.M. Mechanism of action of corrosion protection coating for AA2024-T3 based on poly(aniline)-poly(methylmethacrylate) blend. *J. Electrochem. Soc.* **2005**, *152*, B45–B53. [CrossRef]
89. Torresi, R.M.; de Souza, S.; da Silvaa, J.E.P.; de Torresi, S.I.C. Galvanic coupling between metal substrate and polyaniline acrylic blends: Corrosion protection mechanism. *Electrochim. Acta* **2005**, *50*, 2213–2218. [CrossRef]
90. Lan, T.; Kaviratna, P.D.; Pinnavaia, T.J. On The Nature Of Polyimide Clay Hybrid Composites. *Chem. Mater.* **1994**, *6*, 573–575. [CrossRef]
91. Zhang, Y.J.; Shao, Y.W.; Liu, X.L.; Shi, C.; Wang, Y.Q.; Meng, G.Z.; Zeng, X.G.; Yang, Y. A study on corrosion protection of different polyaniline coatings for mild steel. *Prog. Org. Coat.* **2017**, *111*, 240–247. [CrossRef]
92. Yeh, J.M.; Chen, C.L.; Chen, Y.C.; Ma, C.Y.; Lee, K.R.; Wei, Y.; Li, S.X. Enhancement of corrosion protection effect of poly(o-ethoxyaniline) via the formation of poly (o-ethoxyaniline)-clay nanocomposite materials. *Polymer* **2002**, *43*, 2729–2736. [CrossRef]
93. Olad, A.; Rashidzadeh, A. Preparation and anticorrosive properties of PANI/Na-MMT and PANI/O-MMT nanocomposites. *Prog. Org. Coat.* **2008**, *62*, 293–298. [CrossRef]
94. Tang, Y.-C.; Katsuma, S.; Fujimoto, S.; Hiromoto, S. Electrochemical study of Type 304 and 316L stainless steels in simulated body fluids and cell cultures. *Acta Biomater.* **2006**, *2*, 709–715. [CrossRef]
95. Ganash, A.A.; Al-Nowaiser, F.M.; Al-Thabaiti, S.A.; Hermas, A.A. Comparison study for passivation of stainless steel by coating with polyaniline from two different acids. *Prog. Org. Coat.* **2011**, *72*, 480–485. [CrossRef]
96. Shabani-Nooshabadi, M.; Ghoreishi, S.M.; Jafari, Y.; Kashanizadeh, N. Electrodeposition of polyaniline-montmorrilonite nanocomposite coatings on 316L stainless steel for corrosion prevention. *J. Polym. Res.* **2014**, *21*, 416. [CrossRef]
97. Zhang, Y.J.; Shao, Y.W.; Zhang, T.; Meng, G.Z.; Wang, F.H. High corrosion protection of a polyaniline/organophilic montmorillonite coating for magnesium alloys. *Prog. Org. Coat.* **2013**, *76*, 804–811. [CrossRef]

98. Stankovich, S.; Piner, R.D.; Nguyen, S.T.; Ruoff, R.S. Synthesis and exfoliation of isocyanate-treated graphene oxide nanoplatelets. *Carbon* **2006**, *44*, 3342–3347. [CrossRef]
99. Rafiee, M.A.; Rafiee, J.; Wang, Z.; Song, H.H.; Yu, Z.Z.; Koratkar, N. Enhanced Mechanical Properties of Nanocomposites at Low Graphene Content. *ACS Nano* **2009**, *3*, 3884–3890. [CrossRef]
100. Zhang, W.L.; Liu, Y.D.; Choi, H.J. Fabrication of semiconducting graphene oxide/polyaniline composite particles and their electrorheological response under an applied electric field. *Carbon* **2012**, *50*, 290–296. [CrossRef]
101. Tuberquia, J.C.; Harl, R.R.; Jennings, G.K.; Rogers, B.R.; Bolotin, K.I. Graphene: Corrosion-Inhibiting Coating (vol 6, pg 1102, 2012). *ACS Nano* **2012**, *6*, 4540.
102. Chang, C.H.; Huang, T.C.; Peng, C.W.; Yeh, T.C.; Lu, H.I.; Hung, W.I.; Weng, C.J.; Yang, T.I.; Yeh, J.M. Novel anticorrosion coatings prepared from polyaniline/graphene composites. *Carbon* **2012**, *50*, 5044–5051. [CrossRef]
103. Rui, M.; Jiang, Y.L.; Zhu, A.P. Sub-micron calcium carbonate as a template for the preparation of dendrite-like PANI/CNT nanocomposites and its corrosion protection properties. *Chem. Eng. J.* **2020**, *385*, 123396. [CrossRef]
104. Hou, X.M.; Wang, Y.N.; Hou, J.Z.; Sun, G.E.; Zhang, C.L. Effect of polyaniline-modified glass fibers on the anticorrosion performance of epoxy coatings. *J. Coat. Technol. Res.* **2017**, *14*, 407–415. [CrossRef]
105. Chen, X.; Shen, K.; Zhang, J. Preparation and anticorrosion properties of polyaniline-SiO2-containing coating on Mg-Li alloy. *Pigment Resin Technol.* **2010**, *39*, 322–326. [CrossRef]
106. Kohl, M.; Kalendova, A. Effect of polyaniline salts on the mechanical and corrosion properties of organic protective coatings. *Prog. Org. Coat.* **2015**, *86*, 96–107. [CrossRef]
107. Bagherzadeh, M.R.; Mahdavi, F.; Ghasemi, M.; Shariatpanahi, H.; Faridi, H.R. Using nanoemeraldine salt-polyaniline for preparation of a new anticorrosive water-based epoxy coating. *Prog. Org. Coat.* **2010**, *68*, 319–322. [CrossRef]
108. Pour-Ali, S.; Dehghanian, C.; Kosari, A. In situ synthesis of polyaniline-camphorsulfonate particles in an epoxy matrix for corrosion protection of mild steel in NaCl solution. *Corros. Sci.* **2014**, *85*, 204–214. [CrossRef]
109. Olad, A.; Ramazani, Z. Preparation, Characterization, and Anticorrosive Properties of Polyaniline Nanotubes. *Int. J. Polym. Mater.* **2012**, *61*, 949–962. [CrossRef]
110. Elkais, A.R.; Gvozdenovic, M.M.; Jugovic, B.Z.; Grgur, B.N. The influence of thin benzoate-doped polyaniline coatings on corrosion protection of mild steel in different environments. *Prog. Org. Coat.* **2013**, *76*, 670–676. [CrossRef]
111. Sathiyanarayanan, S.; Azim, S.S.; Venkatachari, G. Preparation of polyaniline-Fe2O3 composite and its anticorrosion performance. *Synth. Met.* **2007**, *157*, 751–757. [CrossRef]
112. Lee, S.Y.; Regnault, W.F.; Antonucci, J.M.; Skrtic, D. Effect of particle size of an amorphous calcium phosphate filler on the mechanical strength and ion release of polymeric composites. *J. Biomed. Mater. Res. Part B-Appl. Biomater.* **2007**, *80*, 11–17. [CrossRef]
113. Siviour, C.R.; Gifford, M.J.; Walley, S.M.; Proud, W.G.; Field, J.E. Particle size effects on the mechanical properties of a polymer bonded explosive. *J. Mater. Sci.* **2004**, *39*, 1255–1258. [CrossRef]
114. Olad, A.; Barati, M.; Shirmohammadi, H. Conductivity and anticorrosion performance of polyaniline/zinc composites: Investigation of zinc particle size and distribution effect. *Prog. Org. Coat.* **2011**, *72*, 599–604. [CrossRef]
115. Radhakrishnan, S.; Siju, C.R.; Mahanta, D.; Patil, S.; Madras, G. Conducting polyaniline-nano-TiO2 composites for smart corrosion resistant coatings. *Electrochim. Acta* **2009**, *54*, 1249–1254. [CrossRef]
116. Al-Masoud, M.A.; Khalaf, M.M.; Gouda, M.; Dao, V.D.; Mohamed, I.M.A.; Shalabi, K.; Abd El-Lateef, H.M. Synthesis and Characterization of the Mixed Metal Oxide of ZnO-TiO2 Decorated by Polyaniline as a Protective Film for Acidic Steel Corrosion: Experimental, and Computational Inspections. *Materials* **2022**, *15*, 7589. [CrossRef] [PubMed]
117. Lin, F.; Zhong, Y.; Zeng, T. Preparation and Performance Evaluation of Poly Methyl Acrylate-acrylic Acid Imidazoline with Corrosion Inhibition and Viscosity Reduction Effect. *Oilfield Chem.* **2018**, *35*, 144–149.
118. Majd, M.T.; Shahrabi, T.; Ramezanzadeh, B.; Bahlakeh, G. Development of a high-performance corrosion protective functional nano-film based on poly acrylic acid-neodymium nitrate on mild steel surface. *J. Taiwan Inst. Chem. Eng.* **2019**, *96*, 610–626. [CrossRef]
119. Yu, Y.-H.; Lin, Y.-Y.; Lin, C.-H.; Chan, C.-C.; Huang, Y.-C. High-performance polystyrene/graphene-based nanocomposites with excellent anti-corrosion properties. *Polym. Chem.* **2014**, *5*, 535–550. [CrossRef]
120. Yan, J.; Li, X.; Zhang, X.; Liu, S.; Zhong, F.; Zhang, J.; Zhang, Q.; Yan, Y. Metallo-Polyelectrolyte-Based Waterborne Polyurethanes as Robust HCl Corrosion Inhibitor Mediated by Inter/intramolecular Hydrogen Bond. *ACS Appl. Polym. Mater.* **2022**, *4*, 3844–3854. [CrossRef]

Article

Corrosion Behavior of Nitrided Layer of Ti6Al4V Titanium Alloy by Hollow Cathodic Plasma Source Nitriding

Lei Zhang [1], Minghao Shao [1], Zhehao Zhang [2], Xuening Yi [3], Jiwen Yan [3], Zelong Zhou [3], Dazhen Fang [2], Yongyong He [2,*] and Yang Li [3,*]

1 School of Electromechanical Automobile Engineering, Yantai University, Yantai 264005, China
2 State Key Laboratory of Tribology, Tsinghua University, Beijing 100084, China
3 Department of Nuclear Equipment, Yantai University, Yantai 264005, China
* Correspondence: heyy@mail.tsinghua.edu.cn (Y.H.); metalytu@163.com (Y.L.)

Abstract: Ti6Al4V titanium alloys, with high specific strength and good biological compatibility with the human body, are ideal materials for medical surgical implants. However, Ti6Al4V titanium alloys are prone to corrosion in the human environment, which affects the service life of implants and harms human health. In this work, hollow cathode plasm source nitriding (HCPSN) was used to generate nitrided layers on the surfaces of Ti6Al4V titanium alloys to improve their corrosion resistance. Ti6Al4V titanium alloys were nitrided in NH_3 at 510 °C for 0, 1, 2, and 4 h. The microstructure and phase composition of the Ti-N nitriding layer was characterized by high-resolution transmission electron microscopy, atomic force microscopy, scanning electron microscopy, X-ray diffraction, and X-ray photoelectron spectroscopy. This modified layer was identified to be composed of TiN, Ti_2N, and α-Ti (N) phase. To study the corrosion properties of different phases, the nitriding 4 h samples were mechanically ground and polished to obtain the various surfaces of Ti_2N and α-Ti (N) phases. The potentiodynamic polarization and electrochemical impedance measurements were conducted in Hank's solution to characterize the corrosion resistance of Ti-N nitriding layers in the human environment. The relationship between corrosion resistance and the microstructure of the Ti-N nitriding layer was discussed. The new Ti-N nitriding layer that can improve corrosion resistance provides a broader prospect for applying Ti6Al4V titanium alloy in the medical field.

Keywords: titanium alloys; plasma nitriding; TEM; HRTEM; electrochemical impedance spectroscopy

Citation: Zhang, L.; Shao, M.; Zhang, Z.; Yi, X.; Yan, J.; Zhou, Z.; Fang, D.; He, Y.; Li, Y. Corrosion Behavior of Nitrided Layer of Ti6Al4V Titanium Alloy by Hollow Cathodic Plasma Source Nitriding. *Materials* **2023**, *16*, 2961. https://doi.org/10.3390/ma16082961

Academic Editor: Javier Gil

Received: 19 February 2023
Revised: 5 April 2023
Accepted: 5 April 2023
Published: 7 April 2023

1. Introduction

For various medical applications, the ideal biomaterial is required to have properties such as excellent corrosion resistance in the body fluid medium, good biocompatibility, high strength, high ductility, low modulus, and no adverse tissue reactions [1–3]. Titanium alloys are used for a wide range of applications in medical implant teeth, artificial hip joints, shoulder joints, knee joints, bone fixing nuts, and screws [4,5]. Although titanium alloys have good biocompatibility, the complexity of the human environment makes titanium alloy human implants vulnerable to corrosion, which seriously affects the service life of implants. Several studies have shown that titanium alloy implants, such as artificial hip joints and knee joints, will age after 10–15 years of use [6]. At the same time, the release of Al, V, and other metal ions in titanium alloy implants is harmful to human health under the combined effect of wear and corrosion [7,8].

Different surface treatment techniques of titanium alloys can effectively provide these surfaces with more desired properties and functionalities for exceptional applications [9–16]. Electrochemical anodic oxidation technology has the advantages of rapid preparation, diversified surface morphology, and low cost. Martinez et al. [17] show that dense oxide films were developed on the surface of Ti6Al4V alloy by constant current anodizing. This film contains amorphous oxides which greatly improve the corrosion resistance of the alloy. However, the low energy utilization and high energy consumption of anodic oxidation result in high treatment costs. Laser cladding technology can form a high hard ceramic phase on the surface of the alloy and significantly improve the wear resistance of titanium alloys [18]. Zhang et al. [19] used laser remelting technology to generate hydroxyapatite (HA) coating on the Ti6Al4V surface and studied its friction, corrosion, and biocompatibility. It was found that the coating had excellent corrosion resistance and wear resistance and could promote the proliferation of bone cells. However, its equipment conditions are demanding, and the treatable area is limited, making mass production difficult. The protection of the substrate is insufficient, and cracks are easily generated. Large stress concentration exists between the modified layer and the substrate. Gordon et al. [20] modified the super elastic nickel-free titanium-based biomedical alloy by nitrogen ion implantation, reducing the material surface's friction coefficient, and improving the material's corrosion resistance, hardness, and biocompatibility. The ion implantation technique has no coating film base bonding problems. The injection process does not require an elevated substrate temperature, and the workpiece is not deformed. The disadvantage is that it is easy to cause lattice damage, and the ion injection layer is shallow—only 0.1~1.0 μm. During Plasma Vapor Deposition (PVD) treatment, single or multiple source materials (including oxides, carbides and nitrides) are evaporated or sputtered in a high vacuum [21,22]. Li et al. [23] prepared TiN and Ti/TiN multilayer coatings on the Ti6Al4V alloy using the PVD technique. These coatings exhibited high hardness in simulated body fluids while showing excellent corrosion resistance and relatively low friction coefficient. However, there are some disadvantages in PVD coating, such as poor adhesion with the coating and spontaneous falling under high stress.

Plasma nitriding technology has high activity, a fast penetration rate, and a deeper penetration layer that is easier to obtain [24–26]. In the process of plasma nitriding, ionized nitrogen bombards the surfaces of components with high cathode potential under the action of the electric field. Nitrogen ions sputter and heat the surfaces of the parts to be treated and diffuse into the parts. From the surface of the components to the substrate, the concentration of nitrogen ions gradually decreases, forming a compound layer from TiN to Ti_2N in turn [12,27,28]. The existence of a compound layer enables titanium nitride alloys to obtain high hardness, wear resistance, and good corrosion resistance. Y.V. Borisyuk et al. [29] performed plasma nitriding treatment of the Ti5Al4V2Mo alloy at different temperatures, and the hardness of the treated samples increased by 1.7 times compared to the untreated surface. A. Sowińska et al. [24] concluded that these TiN + Ti_2N + a-Ti(N)diffusion surface layers on the Ti-6Al-4V alloy by using plasma nitriding are a promising bio-material. Such layers of homogenous structure and thickness, and low surface free energy can be produced even on elements with complicated shapes, which is characteristic for cardiovascular implants. N. Rajendran et al. [30] reported that the nitrided b-21S titanium alloy exhibited a high-percentage of cell viability demonstrating their increased biocompatibility.

However, traditional plasma nitriding technology has unavoidable defects, such as edge effect and surface arc [31,32]. One way to avoid this problem is active screen plasma nitriding (APSN) in recent years [33–35]. In this method, a high negative potential is applied to a cage that generates an inner cavity filled with nitrogen plasma. All sides of the parts in the cage are kept at a floating potential, and the coating produced by exposure to hot plasma is very popular. At the same time, the processed parts are isolated from DC high voltage. However, the inflow of high-energy nitrogen ions will be offset, and the modified layer formed by ASPN will become thinner. It can be considered that the ASPN process still belongs to a linear abnormal glow discharge. If the nitriding treatment adopts hollow cathode discharge, the ionization rate and plasma density generated will be greater than the linear abnormal glow discharge [36,37]. Our previous work concluded that the HCPSN nitrided sample had a higher corrosion resistance in comparison with the bare AISI 4140 steel and the CPN samples [38]. HCPSN treatment is widely used for its fast-processing speed, low energy consumption, low part deformation and no pollution. In other studies, HCPSN was used for nitriding the Ti6Al4V alloy. However, the structure of different phases on the surface of the sample after nitriding has not been investigated further [39].

In this study, HCPSN treatment was used to modify the surface of the Ti6Al4V titanium alloy, and the microstructure of the Ti-N nitriding layer was characterized by various methods. Hank's solution was used to simulate the human environment, and the potentiodynamic polarization and electrochemical impedance tests were conducted to characterize its corrosion resistance. The relationship between the microstructure of the Ti-N nitriding layer and the corrosion resistance was discussed. The results show that the corrosion resistance of the Ti6Al4V titanium alloy can be significantly improved by hollow cathode plasma source nitriding.

2. Materials and Methods

Ti6Al4V alloy samples are supplied in the form of 20 mm diameter round bars, wire cut into samples of 8 mm thickness. The chemical composition of the samples is given in Table 1. Their surfaces were ground with sandpapers and mechanically polished with the 50 nm SiO_2 polishing solution. Then they were cleaned separately with acetone and ethanol using an ultrasonic cleaner.

Table 1. Chemical composition of the Ti6Al4V in wt. %, Ti balance.

Element	Fe	O	C	N	H	V	Al
wt. %	≤0.25	≤0.18	≤0.05	≤0.05	≤0.012	3.5–4.5	5.5–6.75

The hollow cathode device consists of a double-layered hollow cathode cylinder with a top cover, as shown in Figure 1. The material of the hollow cathode barrel is Ti6Al4V, with the regular arrangement of small holes of 5 mm diameter. The Ti6Al4V samples were placed in the hollow cathodic device and connected to the cathode in the nitriding furnace (LDMC-20F, WHRCLS, Wuhan, China). The plasma nitriding process was started by pumping the air pressure inside the furnace to less than 10 Pa. After that, NH_3 was gradually introduced into the furnace until a pressure of 300 Pa was reached. The nitriding furnace was maintained at a pulse frequency of 1000 Hz, a duty cycle of 70% and a voltage of 700 V which activated the glow discharge process. The process was conducted at a temperature of 510 °C. Finally, the samples were cooled with the furnace in vacuum. The preparation process of different samples is shown in Table 2.

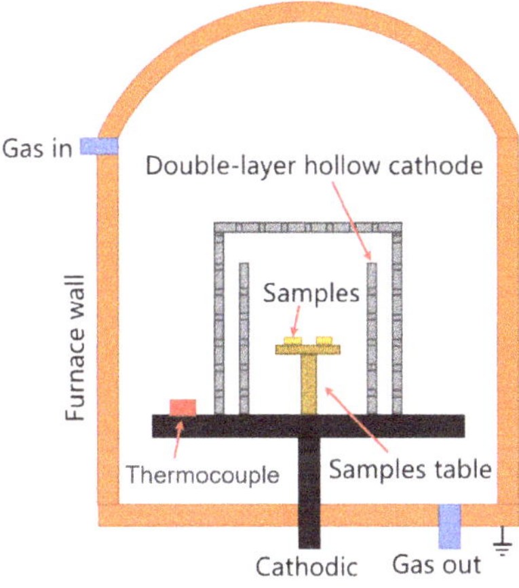

Figure 1. Schematic diagrams of hollow cathodic plasma source nitriding furnace device.

Table 2. The preparation process of different samples.

Sample	Temperature	Heating Up Time	Holding Time	Mechanical Process
Ti6Al4V	-	-	-	grinding + polishing
T0	510 °C	45 min	0 h	-
T1	510 °C	45 min	1 h	-
T2	510 °C	45 min	2 h	-
T4	510 °C	45 min	4 h	-
T4-P	510 °C	45 min	4 h	polishing
T4-G	510 °C	45 min	4 h	grinding + polishing

An optical microscope (OM, ZEISS Axio Observer 3 materials, Oberkochen, Germany) was used to analyze the microstructure of samples. The phase constitutions of the samples were determined using X-ray diffraction (XRD, Bruker D8 ADVANCE, Billerica, MA, USA) with a Cu-Kα radiation source. The scanning angle ranged from 20° to 90 with a rate of 4°/min, and the accelerating voltage was 40 kV. The surface phases of the untreated and corroded samples were analyzed with X-ray photoelectron spectroscopy (XPS, PHI Quantera II, Japan). The XPS measurements were based on a classic X-ray optical scheme. Electrostatic focusing and magnetic screening were used to achieve an energy resolution of $\Delta E \leq 0.5$ eV for the Al Kα radiation (1487 eV). The XPS spectra were recorded with the 100 µm X-ray spot, and the X-ray power supplied to the sample was 23.6 W. All XPS measurements were performed with argon ion cleaning. Data were analyzed by Casa XPS software. The C-C binding energy was calibrated for the peak spectrum at 284.8 eV. The split peak fitting of the XPS peaks was performed based on a Shirley-type baseline. The microstructure of the near-surface region of the samples was observed using a transmission electron microscope (TEM, JEOL JEM-2100, Japan). The imaging was performed in a bright field (TEM/BF). The phase analysis was performed with the help of DM3 software based on selected area electron diffraction (SAED) patterns and high-resolution transmission electron microscopy (HRTEM). Energy dispersive X-ray spectroscopy (EDS) was used to determine the distribution of the surfaces' chemistry. The samples for TEM were cut out using the focused ion beam (FIB, FEI Quanta 200 FEG, Hillsboro, OR, USA) technique. The Vickers hardness test was performed on the surfaces of samples using a Vickers hardness tester (Struers Duramin-A300) with the test load of 50 g and the holding time of 15 s. The surfaces were measured multiple times at each test point.

A series of electrochemical tests, including dynamic potential polarization and electrochemical impedance, were performed with the assistance of an electrochemical workstation (VersaSTAT 3F, Berwyn, PA, USA) to assess the corrosion behavior of the nitriding sample in the cathode environment of Hank's solution. The composition (g/L) of Hanks solution is NaCl 8, KCl 0.4, $CaCl_2$ 0.14, $MgSO_4 \cdot 7H_2O$ 0.2, $Na_2HPO_4 \cdot H_2O$ 0.09, KH_2PO_4 0.06, $NaHCO_3$ 0.35, and $C_6H_{12}O_6$ 1.0. A three-electrode system was used for corrosion testing. Among them, samples, saturated calomel electrodes (SCE, saturated KCl), and platinum electrodes were used as working electrodes, reference electrodes, and counter electrodes, respectively. Prior to all experiments, the untreated Ti6Al4V and nitriding samples were immersed in Hank's solution at open circuit potential (OCP) for 1 h to achieve electrochemical stability. The potential polarization was tested from -1.0 V_{SCE} to 1.0 V_{SCE} with a scan rate of 0.5 mV/s. EIS measurements were performed at OCP with a frequency of 10^5 to 10^{-2} Hz and an amplitude of 10 mV.

3. Results

The XRD patterns of untreated Ti6Al4V and different nitriding time samples are shown in Figure 2a. The three main peaks were at 35.1°, 38.4°, and 40.2°, corresponding to the $(10\bar{1}0)$, (0002), and $(10\bar{1}1)$ phases of α-Ti (JCPDS 44-1294), respectively [40]. The α-Ti has a hexagonal close-packed (HCP) structure. The β-Ti was mainly in the (110) phase at 38.5° (JCPDS 44-1288), with a body-centered cubic (BCC) structure. All α-Ti and β-Ti peaks decreased with increasing nitriding time. The Ti_2N phase has been detected in the sample with a holding time of 0 h for T0 sample. According to the theory of nitrogen diffusion in titanium, nitrogen first entered the matrix as an interstitial solid solution. When the concentration reached a certain value, Ti_2N was generated at the surface. This process can be accelerated by using the hollow cathode plasma source. The (200) and (220) phases of TiN appeared at 42.6° and 61.8° for the T1 samples (JCPDS 38-1420), and the peaks increased with increasing nitriding time.

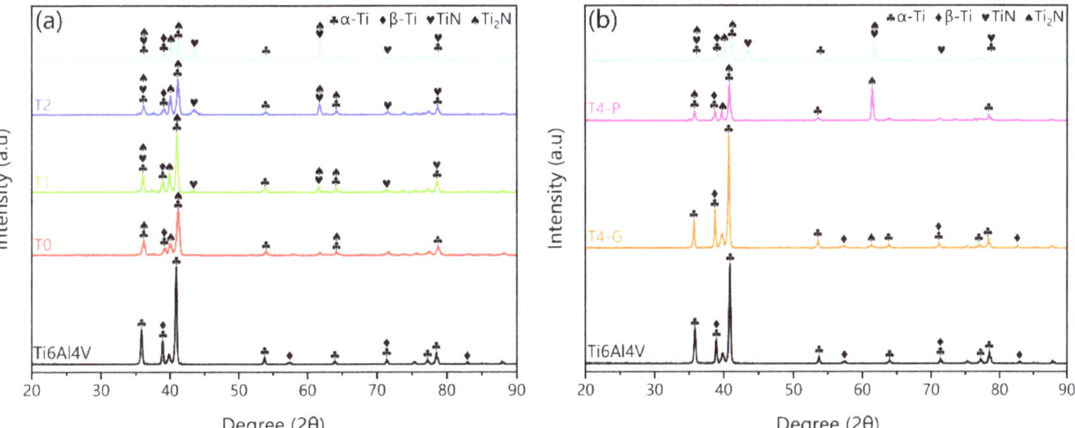

Figure 2. (**a**) The XRD patterns of the untreated Ti6Al4V, T0, T1, T2, and T4 samples; (**b**) The XRD patterns of the untreated Ti6Al4V, T4-G, T4-P, and T4 samples.

Figure 2b shows the XRD patterns of the Ti-N nitriding layer with different surface phases obtained by mechanical grinding and polishing methods. The T4-G sample has an overall shift of 0.5° to a small angle compared to the peak of the untreated Ti6Al4V sample, which is a common shift in solid solution XRD of nitrogen. The T4-P sample shows a significant increase in the (002) phase at 61° compared to the T0 sample in Figure 2a. This indicated that the newly generated Ti_2N phase on the surface of the T0 sample was different from the Ti_2N layer at the bottom of the TiN layer on the surface of the T4 sample by HCPSN treatment.

The cross-sectional optical micrograph of the sample is shown in Figure 3. The thickness of the compound layer increased significantly with the increase in nitriding time of the hollow cathode plasma source. The compound layer thicknesses of T0, T1, T2, and T4 samples were about 1.8, 2.5, 4.1, and 6.4 μm, respectively. According to the previous study [41], the thickness of the TiN layer is about 300~500 nm. Therefore, the corresponding thickness was removed by mechanical polishing on the basis of T4 samples, as shown in Figure 3e. The XRD pattern shows that the TiN phase is no longer present on the surface of this sample and is replaced by Ti_2N. Similarly, the T4-G samples were obtained by removing 6~7 μm on the surface of the T4 samples, as shown in Figure 3f.

The XPS images of the untreated and T2 samples are shown in Figure 4. The spectra of Ti2p, Al2p, N1s, and O1s were separated into peaks, respectively. The $Ti2p_{3/2}$ of the untreated Ti6Al4V sample had three peaks at binding energies of 453.2 eV, 456.9 eV, and 458.5 eV for Ti, Ti_2O_3, and TiO_2, as shown in Figure 4a. The N1s peak spectrum has no distinct peaks, as in Figure 4e. Figure 4g shows that the corresponding O1s peak spectrum has three characteristic peaks for TiO2, Ti_2O_3, and Al_2O_3 [42]. The $Ti2p_{3/2}$ in the surface layer of the T2 nitride sample was fitted to three representative peaks, respectively, and corresponding groups were TiN, TiON, and TiO_2 [43]. The presence of Al and Al_2O_3 was confirmed in Figure 4c [44]. Figure 4d indicated that there was no Al elemental present on the surface of the T2 sample and that Al migrated to the core of the sample, which needs to be discussed separately. The N1s peak spectrum has two correlation peaks at 396.1 eV and 397.3 eV, corresponding to TiON and TiN, as shown in Figure 4f [45]. TiO_2 and TiON were detected in the O1s peak spectrum on the surface of the nitriding samples, as shown in Figure 4h.

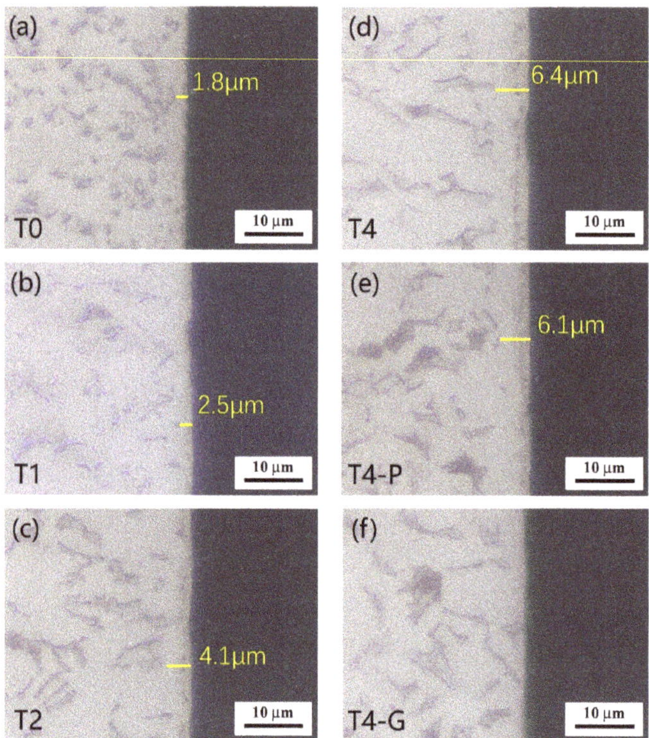

Figure 3. The cross-sectional optical micrographs of different samples.

The cross-sectional TEM image of the T2 sample is shown in Figure 5a. The EDS surface scan of the whole area is shown in Figure 5b–e. The Ti element content was high in the whole sample. The N element content decreased gradually from the outer layer to the inner part, while the opposite was true for the Al element. This was due to the migration of Al elements to the inner part of the sample. The HCPSN nitriding of the Ti6Al4V alloy pushed aluminum away from the near-surface and enriched the Ti_3Al, forming the Ti_3Al intermetallic phase [46]. The SEAD analysis was performed on Points 1, 2, and 3 in Figure 5a. The results are shown in Figure 5f–j. The typical nanocrystal diffraction ring is shown at Point 1. The (111), (200), (220), and (311) phases of TiN are shown from inside the center of the circle outward, respectively. The SEAD diagram at Point 2 is shown in Figure 5h. The composition of Point 2 was the (111), (−111), (200) phase of crystalline TiN on the [0, −1, 1] crystal band axis. Point 2 had the typical diffraction morphology of twin crystals. The (211), (200) phase of Ti_2N is shown in Figure 5j at Point 3. Activation of the nitridation reaction by hollow cathode assistance leads to more efficient nitridation [47]. The most superficial layer of the sample was nanocrystalline TiN, and the subsurface layer was the crystalline TiN layer. The lower layer of the TiN layer was the Ti_2N layer. The surface microstructure of hollow cathodic nitriding samples at 500 °C is similar to the result of conventional plasma nitriding at 850 and 900 °C studied by Czyrska-Filemonowicz et al. [48]. Furthermore, short-time nitriding assisted by hollow cathodes avoids surface defects of plasma nitriding (e.g., nano-whiskers of Ti and Al oxides [49]). The interface between the layers was obvious. Extended exposure to high temperatures during longer nitriding times increases grain growth [50]. The results of the point scan in Figure 5g–k confirmed this idea. The Wt. % of each component is shown in the table. These data were the average of multiple measurements. The standard deviation (S.D.) was also calculated.

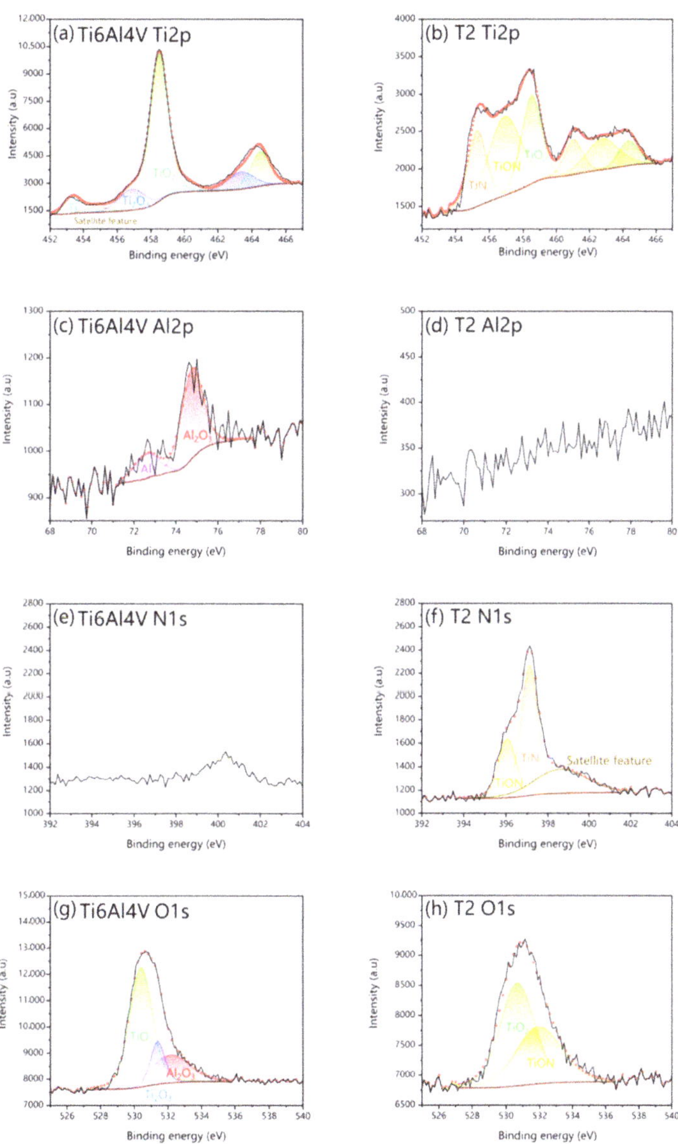

Figure 4. The XPS images of untreated Ti6Al4V: (**a**) Ti 2p, (**c**) Al 2p, (**e**) N 1s, (**g**) O 1s; and T2 samples: (**b**) Ti 2p, (**d**) Al 2p, (**f**) N 1s, (**h**) O 1s.

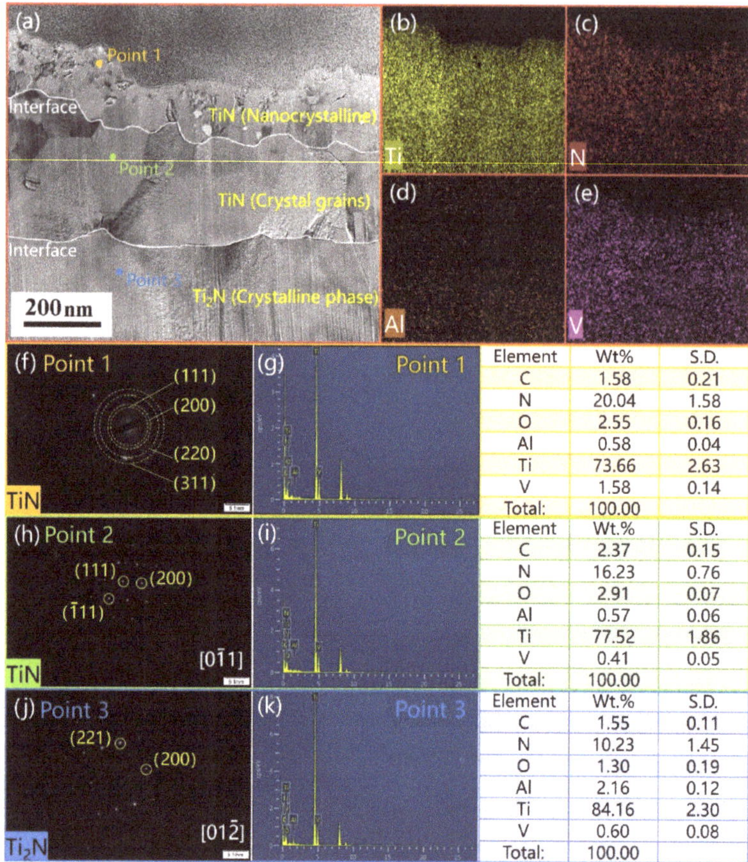

Element	Wt%	S.D.
C	1.58	0.21
N	20.04	1.58
O	2.55	0.16
Al	0.58	0.04
Ti	73.66	2.63
V	1.58	0.14
Total:	100.00	

Element	Wt.%	S.D.
C	2.37	0.15
N	16.23	0.76
O	2.91	0.07
Al	0.57	0.06
Ti	77.52	1.86
V	0.41	0.05
Total:	100.00	

Element	Wt.%	S.D.
C	1.55	0.11
N	10.23	1.45
O	1.30	0.19
Al	2.16	0.12
Ti	84.16	2.30
V	0.60	0.08
Total:	100.00	

Figure 5. (**a**) Cross-sectional TEM image of sample 510-2. The most superficial layer of the sample is nanocrystalline TiN, and the subsurface layer is the crystalline TiN layer. The lower layer of the TiN layer is the Ti_2N layer. The interface between the layers is obvious. EDS surface scan was performed for the whole area as shown in (**b**) Ti, (**c**) N, (**d**) Al, (**e**) V. The selected area electron diffraction (**f–j**), and EDS point scan (**g–k**) were performed for Points 1, 2, and 3. The tables show the results of the point scan. The data were averaged over multiple measurements, and the standard deviation (S.D.) was calculated.

The HRTEM images of the surface layer of nanocrystalline TiN are shown in Figure 6d. The most surface layer was the nanocrystalline layer of TiN (200). The amorphous zone surrounded the TiN nanocrystals. The shape of the crystals was irregular, and the diameter was about 50~100 nm. Hollow cathodes have high activity and a high density of plasma inside, and they are more likely to adsorb on the sample surface. The large number of incoming high-energy N^+ ions from the hollow cathode causes many surface defects, providing more nucleation sites with different orientations than in the case of conventional nitriding methods [51]. The dislocation density increases, thus promoting the refinement of high free energy grain size on the top surface and accelerating the diffusion of nitrogen [52]. The accelerated nitrogen diffusion kinetics leads to the formation of nanocrystalline TiN layers on the top surface and increases the nitrogen diffusion depth. This is attributed to the higher density of crystal defects in the nonequilibrium nanocrystalline TiN layer, which increases the surface energy storage of reactive nitrogen and its chemical reactivity, providing additional preferential nucleation sites for TiN [53].

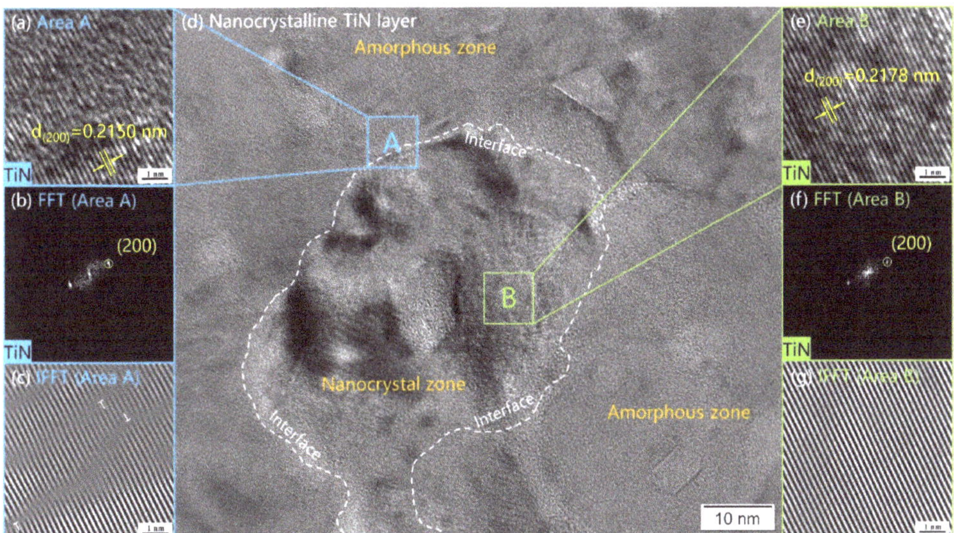

Figure 6. The HRTEM image (**d**) of surface layer nanocrystalline TiN. The most superficial layer was a nanocrystalline TiN layer. The TiN nanocrystals were surrounded by amorphous zone. The crystals were irregularly shaped and had a diameter of about 50~100 nm. The FFT (**b**) and IFFT (**c**) images of Area A (**a**). The FFT (**f**) and IFFT (**g**) images of Area B (**e**).

The HRTEM images of the subsurface layer are shown in Figure 7a. The enlarged HRTEM image of Area A is shown in Figure 7b, and the FFT image of Area A is shown in Figure 7c. Taking the interface in Figure 7a as the demarcation, the TiN matrix was on the right, and the twin crystal of TiN was on the left. In the [0–11] crystal zone, three crystal plane groups were detected with lattice spacings of 2.435, 2.406, and 2.154 Å. They are close to the (111) and (200) crystal planes of TiN (2.45Å and 2.12 Å, JCPDS 38-1420). There is a large amount of dislocation buildup inside the twins, as shown in Figure 7e. Dislocations affect the mechanical properties and corrosion resistance of nitrided samples. At the interface (Area C), the crystal plane undergoes a small angle shift, as shown in Figure 7f. Figure 7g showed a typical twin crystal diffraction dot pattern. The crystalline planes with lattice spacings of 2.506 Å, 2.478 Å, and 2.123 Å were observed in the TiN matrix, corresponding to (−111), (111), and (200) crystallographic planes (2.45Å and 2.12 Å, JCPDS 38-1420). The matrix and twin are symmetrical about the (−111) crystal plane. The superficial nanocrystalline TiN layer regulates the subsurface crystalline TiN by activating twinning and dislocation mechanisms. Grain boundaries, dislocations, layer faults, and twins act as diffusion shortcuts by providing convenient diffusion channels for interstitial nitrogen atoms [53].

The microstructure of the Ti_2N layer can be observed by HRTEM images, as in Figure 8. There are a large number of dislocations inside the Ti_2N layer. The lattice spacing was detected to be 2.469 Å, which is close to the lattice spacing of the (200) crystal plane of Ti_2N (2.47 Å, JCPDS 17-0386). The bottom boundary of TiN fluctuates from below along the top of the larger grains. At higher temperatures assisted by hollow cathodes, they also show that the growth of TiN generated in the nitriding treatment proceeds both toward the core of the sample, controlled by inward diffusion from the bottom of the TiN layer to form Ti_2N, and to a lesser extent toward the surface, controlled by outward diffusion of titanium [51]. On the other hand, the stresses caused by the difference in thermal contraction between TiN and α-Ti(N) after nitride cooling lead to late transformation to strain-induced Ti_2N [35].

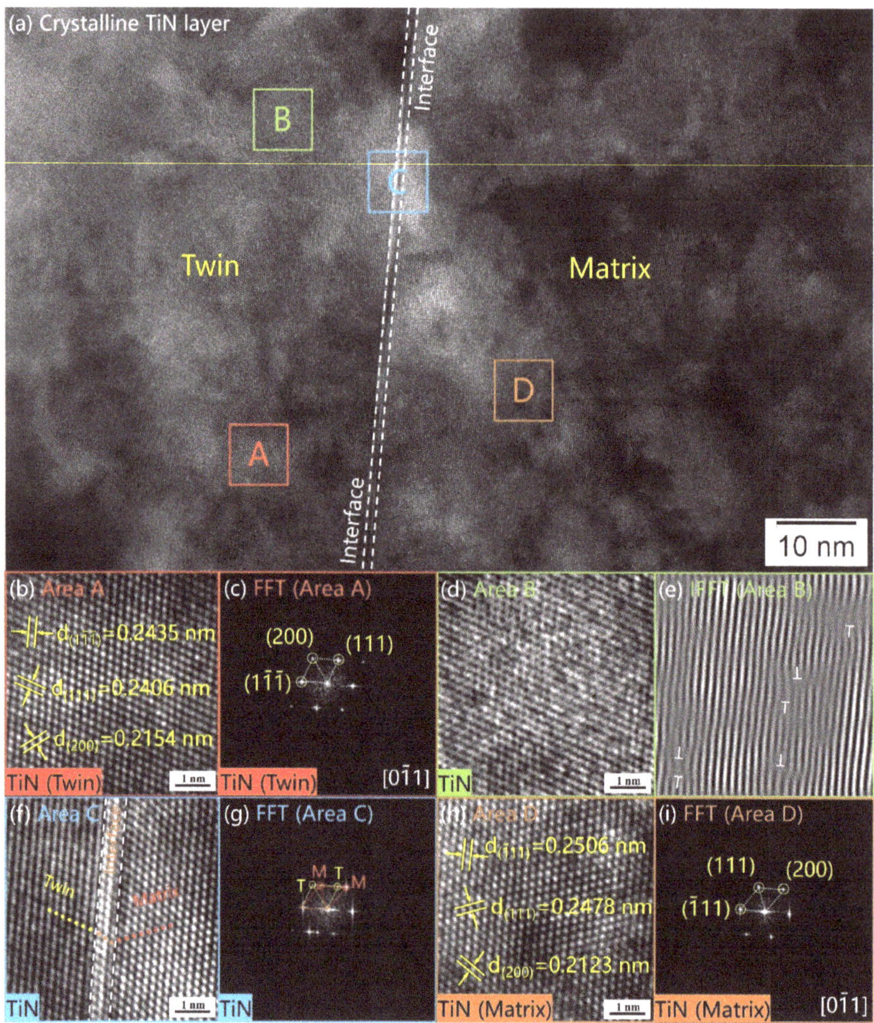

Figure 7. (**a**) The HRTEM image of crystalline TiN layer. With the interface in figure (**a**) as the boundary, the left side is the twin crystal, and the right side is the matrix; (**b**) The image of twin crystal in area A; (**c**) The FFT image of area A; (**d**) The image of area B; (**e**) The IFFT image of area B. There are a large number of dislocations distributed in area B; (**f**) The image of the interface in area C; (**g**) The FFT image of area C; (**h**) The image of the matrix in area D; (**i**) The FFT image of area D.

Figure 9 shows the cross-sectional hardness profile of nitriding, mechanically ground, and polished samples after nitriding. After nitriding, the surface hardness of the sample was increased. The surface hardness of T4 samples averaged 1340 $HV_{0.05}$, while T2 samples also had a surface hardness of 1032 $HV_{0.05}$. As the nitriding time increases, the surface hardness increases, and, at the same time, the error value of the hardness value increases. The error range of T4 samples was approximately 190 $HV_{0.05}$, while the error range of T0 samples was 48 $HV_{0.05}$. For the T4-G sample, the error range is only 26 $HV_{0.05}$. The surface phase of the T4-P sample is Ti_2N, with a hardness of 865 $HV_{0.05}$, which is higher than the surface hardness of the T1 sample. The diffusion layer thickness was around 25 μm for the T0 sample and between 50μm and 60μm for the rest of the samples. The grain size reduction due to hollow cathodic nitriding results in a fine grain strengthening effect [54]. The gradient structure of the nitriding layer nano-TiN, crystalline TiN, Ti_2N, and α-Ti(N) is the main reason for the increase in the hardness of the samples. The nano-structure of the TiN layer and the generation and accumulation of dislocations within the grains are effective ways to increase the surface strength of the samples [55]. The distribution of twins within the crystalline TiN results in a higher hardness of the TiN layer than the Ti_2N layer. The surface hardness is also related to the thickness of each layer, which eventually leads to the following arrangement of surface hardness from largest to smallest: T4 > T2 > T4-P > T1 > T0 > T4-G.

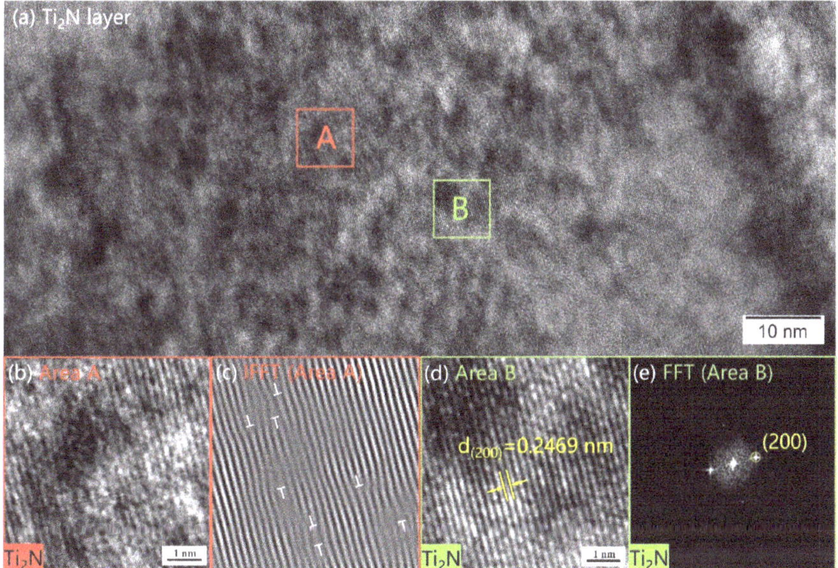

Figure 8. (**a**) The HRTEM image of the Ti_2N layer; (**b**) The image of area A; (**c**) The IFFT image of area A. Large dislocation distribution within Ti_2N layers; (**d**) The image of area B; (**e**) The FFT image of area B.

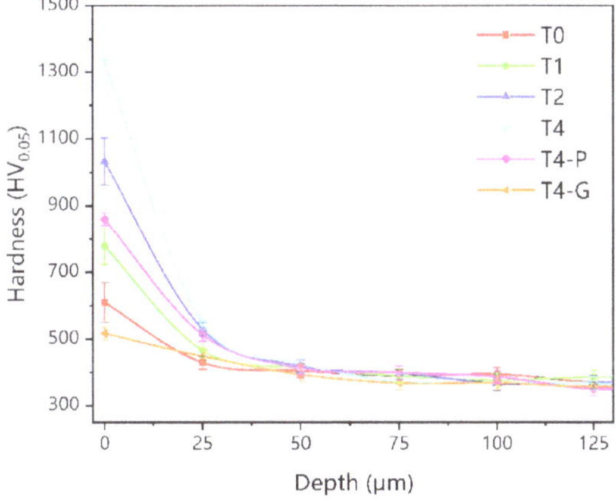

Figure 9. The cross-sectional hardness profile of different samples.

The untreated Ti6Al4V and nitriding samples were measured using the dynamic potential polarization method in Hank's solution. The results of the measured polarization curves are shown in Figure 10. The corrosion currents were obtained using the Tafel extrapolation method [56]. The corrosion rate is calculated by Equation (1) [57]. The a is the molecular weight of the electrode material; i_{corr} is the corrosion current; n is the number of equivalent exchange; F is the Faraday's constant; and ρ is the density of the electrode material:

$$\text{Corrosion rate } (\mu\text{m/year}) = 3.1536 \times 10^5 \frac{a i_{corr}}{nF\rho} \quad (1)$$

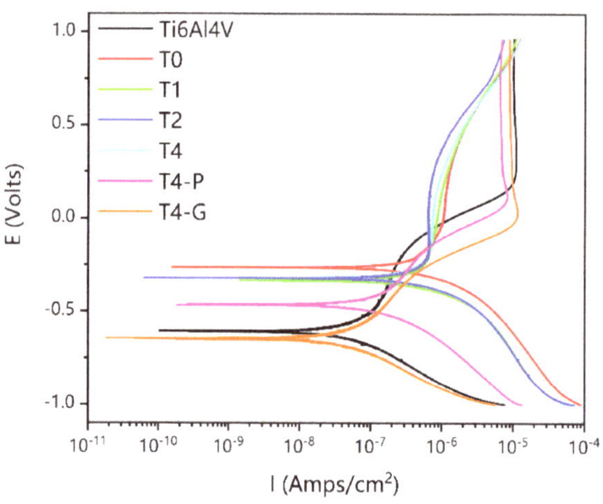

Figure 10. The images of the polarization curves in Hank's solution.

The I_{corr}, E_{corr}, and corrosion rates of different samples are shown in Table 3. The non-mechanically treated samples after nitriding had a more pronounced passivation zone at I = 10^{-6} A/cm^2. However, the untreated samples and the T4-P and T4-G samples had the passivation zone around I = 10^{-5} A/cm^2. The trends of the polarization curves of the T4-P and T4-G samples were similar. It is presumed that the surface of all these samples formed an oxide film of Ti. For the nitriding samples without mechanical treatment, the polarization curves were basically the same. The TiN layer on the surface, instead of the oxide layer, acted as a good barrier and alleviated the corrosion level of the substrate [58]. It can be further demonstrated that a small amount of TiN phase on the sample surface of T0 was not detected by XRD. In terms of self-corrosion currents, the untreated T2 and T4 samples have approximated I_{corr} values. However, the I_{corr} values of the mechanically treated samples were all elevated to different degrees. This is due to the loss of the corrosion inhibition mechanism of the TiN layer. In addition, the nitride of Ti and α-Ti(N) interstitial solid solution prevented the formation of the oxide layer to some extent. In nitriding samples at different times, I_{corr} and nitriding time were negatively correlated, which was directly related to the thickness of the nitriding layer. Both sufficient TiN layer and oxide layer can reduce the I_{corr} of the substrate, but when the nitride layer is too thin, or the barrier effect of the oxide layer is weakened due to the influence of N elements, the I_{corr} will be reduced. All E_{corr} values of the samples were ranked from smallest to largest as T4-G < Untreated Ti6Al4V < T4-P < T4 < T2 < T1 < T0. For E_{corr}, the nitride of Ti plays a dominant role. TiN is clearly superior to Ti_2N and TiO_2. A thinner nitriding layer increased the corrosion rate of the substrate, as shown in Table 3 for samples T0 and T1. The corrosion rate of the T2 sample was slightly lower than that of the Ti6Al4V untreated sample. Nitriding for 4 h, the thickness of the nitriding layer increased. However, the corrosion rate decreased. It is inferred that the deterioration of the surface layer is due to the effect of roughness. In comparison with the TiN phase and Ti_2N phase on the surface, the corrosion resistance decreased due to lower N concentration. The interstitial solid solution phase of surface α-Ti(N) generated an oxide barrier layer of Ti, so the corrosion resistance was improved to some extent.

Table 3. The I_{corr}, E_{corr}, and corrosion rate of different samples.

	I_{corr} (A/cm^2)	E_{corr} (V)	Corrosion Rate (μm/Year)
Ti6Al4V	7.565×10^{-8}	−0.608	23.8
T0	6.483×10^{-7}	−0.264	204
T1	4.540×10^{-7}	−0.318	143
T2	7.454×10^{-8}	−0.319	23.4
T4	7.645×10^{-8}	−0.320	24
T4-P	4.507×10^{-7}	−0.465	142
T4-G	1.497×10^{-7}	−0.646	47

The XPS of the untreated Ti6Al4V and T2 samples after corrosion testing confirmed a new generation of TiO, as shown in Figure 11a,b [59]. The TiO_2 barrier on the sample surface was severely damaged, and the TiO_2 characteristic peak area in the Ti2p peak spectrum was reduced. The dense TiN layer blocks the further occurrence of corrosion reactions. At the same time, some oxides (TiO, TiO_2) are formed on the surface of TiN [60]. Because of the good conductivity of TiO, TiO_2, and TiN, galvanic corrosion will occur between the nitriding layer and the substrate [61]. When the nitriding layer is not enough to block the corrosion medium, and the substrate is exposed to corrosion, the local corrosion is accelerated to a certain extent as an anode. Therefore, the corrosion rate of T0 and T1 samples is higher than that of untreated Ti6Al4V samples.

The EIS curves of different samples are shown in Figure 12. Figure 12a shows the Nyquist plot, and the impedance arc radius represents the corrosion resistance of the specimen. The corrosion resistance of each sample was good, and the impedance values were high, so the impedance arc did not form a complete half-arc, but only a section of the half-arc. Overall, for the nitriding samples of different times, the thicker the nitriding layer is, the better the corrosion resistance is. However, when the thickness of the nitriding layer is sufficient to block the adsorption of Cl^- ions, the surface roughness of the sample becomes the dominant factor. The sample surface showed a high roughness level in long-time plasma nitriding due to exposing the sample surface to continuous plasma bombardment by high temperature and high voltage in the furnace [62]. The samples in the study were subjected to intense bombardment by the plasma, and it is presumed that the surface roughness of the samples showed a high deterioration with time. Due to the effect of surface roughness, the sample impedance arc radius of T4 samples is effective with that of T2 samples. The Bode plots are shown in Figure 12b,c. The higher $|Z|$ values in the low-frequency region and wider theta peak widths in the mid-frequency region suggest better corrosion resistance [63]. For the surface phase, the effective TiN (sufficient thickness and low roughness) has the best corrosion resistance, followed by the oxides of Ti naturally occurring on the surface of untreated Ti6Al4V samples in the air. This is followed by the T4-G samples with the surface of the α-Ti(N) phase. Due to the high oxidation activity of Ti, it is presumed that a small amount of TiO_2 will form on the surface of T4-G samples in the air to protect the substrate from corrosion. The T4-P samples with the surface of Ti_2N phase have the worst corrosion resistance.

Based on the EIS curve, the simulated impedance circuit diagram is shown in Figure 13. The simple impedance circuit with a time constant was used for the untreated samples to simulate the oxide layer resistance, as shown in Figure 13a [22]. The surface phase of the mechanically ground and the polished sample was inferred from the polarization and EIS curves to be α-Ti(N) with an oxide layer blocked. A double-layer time-constant circuit was used to simulate, as shown in Figure 13b [64]. Figure 13c shows the fitted circuit diagram of the mechanically polished sample with surface Ti_2N phase. The unmechanically treated nitriding samples were fitted with a 3-time constant circuit, with R_1 representing the resistance of the top surface layer, R_2 representing the resistance of the subsurface layer, and R_{ct} representing the resistance of charge transfer resistance. The R_s represented the corrosion resistance of the seawater solution [65]. The constant phase element (CPE) was introduced to replace the ideal capacitor, as demonstrated in previous studies [41]. CPE_1, CPE_2, and CPE_{dl} represent the capacitance of each layer, respectively [58]. The values of the fitted circuit are shown in Table 4.

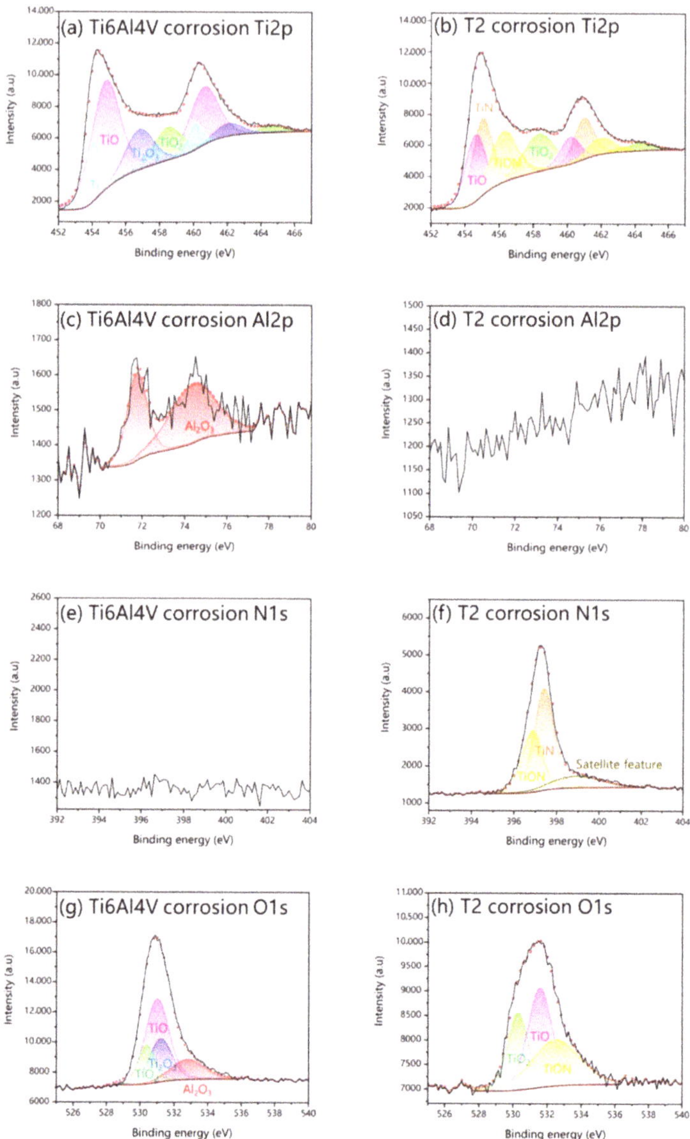

Figure 11. The XPS images after corrosion of untreated Ti6Al4V: (**a**) Ti 2p, (**c**) Al 2p, (**e**) N 1s, (**g**) O 1s; and T2 samples: (**b**) Ti 2p, (**d**) Al 2p, (**f**) N 1s, (**h**) O 1s.

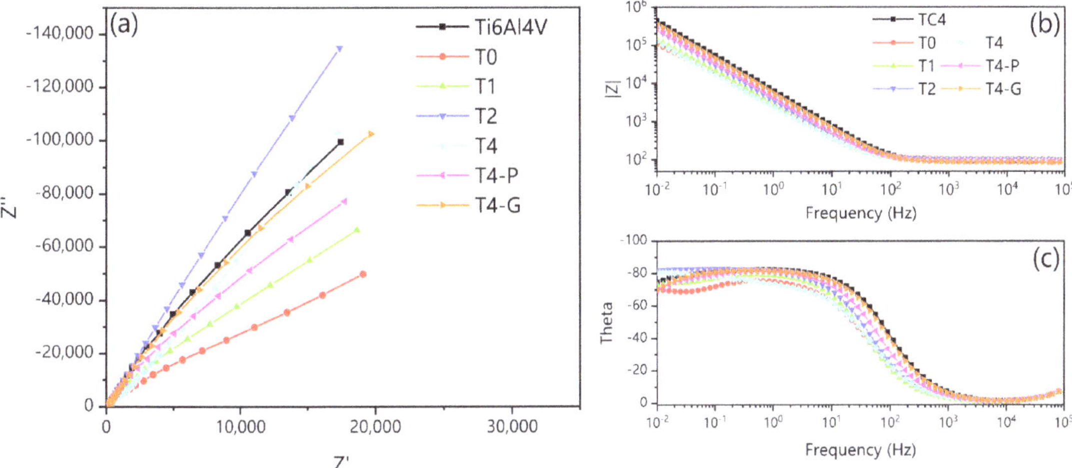

Figure 12. Images of EIS curves for untreated Ti6Al4V, nitriding, and mechanically treated samples: (**a**) Nyquist plots: The impedance arc radius represents the corrosion resistance of the specimen. Overall, for the nitriding samples of different times, the thicker the nitriding layer is, the better the corrosion resistance is; (**b**,**c**) Bode plots: the high-er |Z| values in the low-frequency region and wider theta peak widths in the mid-frequency region suggest better corrosion resistance.

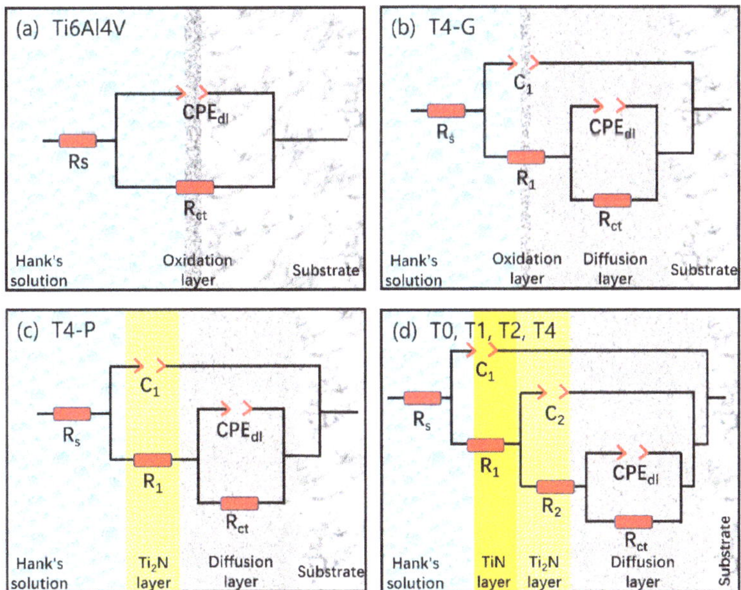

Figure 13. Simulated impedance circuit diagrams applied to different samples: (**a**) untreated Ti6Al4V samples. The simple impedance circuit with a time constant was used for the untreated samples to simulate the oxide layer resistance; (**b**) T4-G samples with surface mechanical grinding and polishing. A double-layer time-constant circuit was used to simulate; (**c**) T4-P samples with surface polishing; (**d**) T0, T1, T2, T4 nitriding samples. The unmechanically treated nitriding samples were fitted with a 3-time constant circuit.

Table 4. Fitted circuit values for untreated, mechanically ground and polished, and nitriding samples.

	R_s	R_1	CPE_1	R_2	CPE_2	R_{ct}	CPE_{dl}
Ti6Al4V	91.3					1.67×10^6	2.76×10^{-5}
T0	105.6	76.2	2.43×10^{-5}	446	1.13×10^{-5}	3.98×10^5	4.38×10^{-5}
T1	101.3	57.3	3.11×10^{-5}	635	1.17×10^{-5}	1.07×10^6	2.66×10^{-5}
T2	104	320.8	4.13×10^{-5}	12,417	5.58×10^{-6}	8.61×10^6	2.59×10^{-5}
T4	94.5	6493.0	7.43×10^{-5}	38,371	8.85×10^{-6}	3.10×10^6	7.11×10^{-5}
T4-P	93.0	350.9	3.00×10^{-5}			1.03×10^6	1.28×10^{-5}
T4-G	87.3	1805.0	3.32×10^{-5}			1.65×10^6	7.62×10^{-5}

4. Discussion

Among the different samples, the highest R_{ct} value was obtained for the T2 sample. In general, the high R_{ct} indicates better corrosion resistance, which acts as a physical barrier to retard the penetration of ions [66]. The R_{ct} values of the nitriding samples were generally higher than those of the untreated samples. The strong and relatively deep TiN nitride layer formed during this treatment at higher temperatures can be used as a tight corrosion protection barrier [34]. The T4-P samples with the Ti_2N surface after mechanical polishing lost the barrier of TiN on most surfaces of the nitriding layer with the lower R_{ct} value. The R_{ct} values of T4-G and untreated Ti6Al4V samples were close, and they both received protection from the surface TiO_2 oxide layer. H. Wang et al. found that N-TiO_2 coatings obtained by oxidative annealing after nitriding were superior to TiN and TiO_2. This finding may be attributed to the multilayer structure of N-TiO_2 (external oxide layer and internal diffusion zones of Ti and N), which can hinder the diffusion of reactive ions [67]. In this study, the TiO_2 in the outer layer of α-Ti(N) plays a similar role. However, the corrosion resistance is not as good as that of the T2 sample because it forms naturally at room temperature.

In electrochemical corrosion tests, oxygen absorption corrosion occurred mainly in neutral electrolytes. The chemical reactions occurring on the surface of untreated samples are shown as follows [66]:

$$2TiO_2 + 8Cl^- \rightarrow 2TiCl_4 + 2O_2 + 8e^- \quad (2)$$

Pitting corrosion occurred on the sample surface, and the TiO_2 layer barrier protected the substrate. The pitting pits gradually expanded and increased, and the corrosion solution reacted chemically with the substrate:

$$Ti + O_2 + 2H_2O \rightarrow Ti(OH)_4 \quad (3)$$

For the Ti-N nitriding layer, the reaction equation in the corrosion solution is usually [60]: A small part of the TiN reaction generates Ti oxide:

$$2TiN + 4O_2 \rightarrow 2TiO_2 + N_2 \quad (4)$$

$$Ti^{4+} + 2e^- \rightarrow Ti^{2+} \quad (5)$$

The reaction equation of TiN participating in corrosion is as follows:

$$2TiN + 8Cl^- \rightarrow TiCl_4 + N_2 + 8e^- \quad (6)$$

$$2TiN + 8H_2O \rightarrow 2Ti(OH)_4 + N_2 + 4H_2 \quad (7)$$

To conclude, a sufficient Ti-N nitriding layer acts as a barrier layer to block the intrusion of the corrosion solution. The T2 samples showed strong corrosion resistance. When the Ti-N nitriding layer is thin, the sample surface of TiN and the oxide of Ti work together. T0 and T1 samples undergo galvanic corrosion, deteriorating the corrosion resistance of the samples. The sample surface roughness also affects the corrosion resistance.

The corrosion resistance is weakened by the high roughness surface caused by the T4 sample being subjected to a long and intense bombardment sputtering inside the hollow cathode. Moreover, the T4-P and T4-G samples showed different corrosion resistance, both worse than the untreated Ti6Al4V. The surface of the T4-G sample with the α-Ti(N) phase generated a small amount of TiO_2 oxide layer to protect the substrate to some extent.

5. Conclusions

The microstructure, hardness, as well as electrochemical behaviors of the Ti-N nitriding layer of Ti6Al4V obtained by the HCPSN at 510 °C were investigated in detail. The main conclusions were drawn as follows:

1. XRD and TEM results show that the the Ti-N nitriding layer consists of a mixture of TiN, Ti_2N, and α-Ti(N) phases. The compound layer consists of the nanocrystalline TiN surface top layer, the crystalline TiN sub-surface layer, the Ti_2N interlayer, and the interstitial solid solution α-Ti(N) bottom layer. The compound layer thicknesses of the 0, 1, 2, and 4 h samples were 1.8, 2.5, 4.1, and 6.4 μm, respectively.
2. The thickness of the diffusion layer of the T0 sample was about 25 μm, and that of the remaining samples ranged from 50 μm to 60 μm. The surface hardness increased with the increase in nitriding time. The surface hardness of the T4 sample was 1340 $HV_{0.05}$ on average, and the surface phase of the T4-P sample was Ti_2N with a hardness of 865 $HV_{0.05}$. The surface hardness of the T4-G sample was 518 $HV_{0.05}$.
3. The polarization results in the human body solution showed that the the Ti-N nitriding layer showed a significant passivation zone, a significant reduction in corrosion rate, and a significant improvement in pitting resistance compared to the untreated Ti6Al4V alloy. The corrosion resistance of the nitriding layer with the Ti_2N phase on the surface deteriorated relatively. All Ecorr values of the samples were ranked from smallest to largest as T4-G < Untreated Ti6Al4V < T4-P < T4 < T2 < T1 < T0. The EIS analysis shows that the passivated film on the surface of the TiN nitride layer has higher charge transfer resistance and lower capacitance, which can effectively hinder the penetration and migration of reactive ions. Thus, the corrosion resistance is significantly improved.

Author Contributions: Conceptualization, Y.H. and Y.L.; methodology, L.Z.; software, M.S. and D.F.; validation, M.S., Z.Z. (Zhehao Zhang) and J.Y.; formal analysis, Y.L. and Y.H.; data curation, X.Y. and Z.Z. (Zelong Zhou); writing—original draft preparation, Y.L. and M.S.; writing—review and editing, Y.H.; supervision, Y.H.; project administration, Y.H. and Y.L.; funding acquisition, Y.H. and Y.L. All authors have read and agreed to the published version of the manuscript.

Funding: This research was funded by the National Natural Science Foundation of China, grant number 52175192.

Institutional Review Board Statement: Not applicable.

Informed Consent Statement: Not applicable.

Data Availability Statement: Not applicable.

Conflicts of Interest: The authors declare no conflict of interest.

References

1. Chen, Q.; Thouas, G.A. Metallic implant biomaterials. *Mater. Sci. Eng. R Rep.* **2015**, *87*, 1–57. [CrossRef]
2. Rodriguez-Contreras, A.; Punset, M.; Calero, J.A.; Gil, F.J.; Ruperez, E.; Manero, J.M. Powder metallurgy with space holder for porous titanium implants: A review. *J. Mater. Sci. Technol.* **2021**, *76*, 129–149. [CrossRef]
3. Balazic, M.; Kopac, J.; Jackson, M.J.; Ahmed, W. Titanium and titanium alloy applications in medicine. *Int. J. Nano Biomater.* **2007**, *1*, 3–34. [CrossRef]
4. Geetha, M.; Singh, A.K.; Asokamani, R.; Gogia, A.K. Ti based biomaterials, the ultimate choice for orthopaedic implants—A review. *Prog. Mater. Sci.* **2009**, *54*, 397–425. [CrossRef]
5. Noumbissi, S.; Scarano, A.; Gupta, S. A literature review study on atomic ions dissolution of titanium and its alloys in implant dentistry. *Materials* **2019**, *12*, 368. [CrossRef]

6. Liang, H.; Shi, B.; Fairchild, A.; Cale, T. Applications of plasma coatings in artificial joints: An overview. *Vacuum* **2004**, *73*, 317–326. [CrossRef]
7. Cordeiro, J.M.; Barão, V.A. Is there scientific evidence favoring the substitution of commercially pure titanium with titanium alloys for the manufacture of dental implants? *Mater. Sci. Eng. C* **2017**, *71*, 1201–1215. [CrossRef]
8. Wilson, J.A.-B.; Banfield, S.; Housden, J.; Olivero, C.; Chapon, P. On the response of Ti–6Al–4V and Ti–6Al–7Nb alloys to a Nitron-100 treatment. *Surf. Coat. Technol.* **2014**, *260*, 335–346. [CrossRef]
9. Fujita, K.; Ijiri, M.; Inoue, Y.; Kikuchi, S. Rapid Nitriding of Titanium Alloy with Fine Grains at Room Temperature. *Adv. Mater.* **2021**, *33*, 2008298. [CrossRef]
10. Hussein, M.A.; Ankah, N.K.; Kumar, A.M.; Azeem, M.A.; Saravanan, S.; Sorour, A.A.; Al Aqeeli, N. Mechanical, Biocorrosion, and Antibacterial Properties of Nanocrystalline TiN Coating for Orthopedic Applications. *Ceram. Int.* **2020**, *46*, 18573. [CrossRef]
11. Lin, N.; Li, D.; Zou, J.; Xie, R.; Wang, Z.; Tang, B. Surface Texture-Based Surface Treatments on Ti6Al4V Titanium Alloys for Tribological and Biological Applications: A Mini Review. *Materials* **2018**, *11*, 487. [CrossRef]
12. Travessa, D.N.; Guedes, G.V.B.; de Oliveira, A.C.; da Silva Sobrinho, A.S.; Roche, V.; Jorge, A.M., Jr. Corrosion performance of the biocompatible β-Ti12Mo6Zr2Fe alloy submitted to laser and plasma-nitriding surface modifications. *Corros. Sci.* **2022**, *209*, 110740. [CrossRef]
13. Guan, J.; Jiang, X.; Xiang, Q.; Yang, F.; Liu, J. Corrosion and tribocorrosion behavior of titanium surfaces designed by electromagnetic induction nitriding for biomedical applications. *Surf. Coat. Technol.* **2021**, *409*, 126844. [CrossRef]
14. Zhang, L.C.; Chen, L.Y.; Wang, L. Surface Modification of Titanium and Titanium Alloys: Technologies, Developments, and Future Interests. *Adv. Eng. Mater.* **2020**, *22*, 1901258. [CrossRef]
15. Zhang, L.; Shao, M.; Wang, Z.; Zhang, Z.; He, Y.; Yan, J.; Lu, J.; Qiu, J.; Li, Y. Comparison of tribological properties of nitrided Ti-N modified layer and deposited TiN coatings on TA2 pure titanium. *Tribol. Int.* **2022**, *174*, 107712. [CrossRef]
16. Bell, T.; Dong, H.; Sun, Y. Realising the Potential of Duplex Surface Engineering. *Tribol. Int.* **1998**, *31*, 127–137.
17. Martinez, A.; Flamini, D.; Saidman, S. Corrosion resistance improvement of Ti-6Al-4V alloy by anodization in the presence of inhibitor ions. *Trans. Nonferrous Met. Soc. China* **2022**, *32*, 1896–1909. [CrossRef]
18. Kamat, A.M.; Copley, S.M.; Segall, A.E.; Todd, J.A. Laser-Sustained Plasma (LSP) Nitriding of Titanium: A Review. *Coatings* **2019**, *9*, 283. [CrossRef]
19. Jing, Z.; Cao, Q.; Jun, H. Corrosion, wear and biocompatibility of hydroxyapatite bio-functionally graded coating on titanium alloy surface prepared by laser cladding. *Ceram. Int.* **2021**, *47*, 24641–24651. [CrossRef]
20. Gordin, D.; Busardo, D.; Cimpean, A.; Vasilescu, C.; Höche, D.; Drob, S.; Mitran, V.; Cornen, M.; Gloriant, T. Design of a nitrogen-implanted titanium-based superelastic alloy with optimized properties for biomedical applications. *Mater. Sci. Eng. C* **2013**, *33*, 4173–4182. [CrossRef] [PubMed]
21. Padilha Fontoura, C.; Ló Bertele, P.; Machado Rodrigues, M.; Elisa Dotta Maddalozzo, A.; Frassini, R.; Silvestrin Celi Garcia, C.; Tomaz Martins, S.; Crespo, J.d.S.; Figueroa, C.A.; Roesch-Ely, M.; et al. Comparative Study of Physicochemical Properties and Biocompatibility (L929 and MG63 Cells) of TiN Coatings Obtained by Plasma Nitriding and Thin Film Deposition. *ACS Biomater. Sci. Eng.* **2021**, *7*, 3683–3695. [CrossRef] [PubMed]
22. Bai, X.; Xu, L.; Shi, X.; Ren, J.; Xu, L.; Wang, Q.; Li, B.; Liu, Z.; Zheng, C.; Fu, Q. Hydrothermal oxidation improves corrosion and wear properties of multi-arc ion plated titanium nitride coating for biological application. *Vacuum* **2022**, *198*, 110871. [CrossRef]
23. Zhao, C.; Zhu, Y.; Yuan, Z.; Li, J. Structure and tribocorrosion behavior of Ti/TiN multilayer coatings in simulated body fluid by arc ion plating. *Surf. Coat. Technol.* **2020**, *403*, 126399. [CrossRef]
24. Sowińska, A.; Czarnowska, E.; Tarnowski, M.; Witkowska, J.; Wierzchoń, T. Structure and hemocompatibility of nanocrystalline titanium nitride produced under glow-discharge conditions. *Appl. Surf. Sci.* **2018**, *436*, 382–390. [CrossRef]
25. Sun, Y.; Haruman, E. Effect of carbon addition on low-temperature plasma nitriding characteristics of austenitic stainless steel. *Vacuum* **2006**, *81*, 114–119. [CrossRef]
26. Sun, Y. Tribocorrosive behaviour of low temperature plasma-nitrided PH stainless steel sliding against alumina under linear reciprocation with and without transverse oscillations. *Wear* **2016**, *362–363*, 105–113. [CrossRef]
27. Ahmadi, M.; Hosseini, S.; Hadavi, S. Comparison of auxiliary cathode and conventional plasma nitriding of gamma-TiAl alloy. *Vacuum* **2016**, *131*, 89–96. [CrossRef]
28. Yumusak, G.; Leyland, A.; Matthews, A. The effect of pre-deposited titanium-based PVD metallic thin films on the nitrogen diffusion efficiency and wear behaviour of nitrided Ti alloys. *Surf. Coat. Technol.* **2020**, *394*, 125545. [CrossRef]
29. Borisyuk, Y.V.; Oreshnikova, N.; Berdnikova, M.; Tumarkin, A.; Khodachenko, G.; Pisarev, A. Plasma nitriding of titanium alloy Ti5Al4V2Mo. *Phys. Procedia* **2015**, *71*, 105–109. [CrossRef]
30. Mohan, L.; Raja, M.D.; Uma, T.S.; Rajendran, N.; Anandan, C. In-Vitro Biocompatibility Studies of Plasma-Nitrided Titanium Alloy β-21S Using Fibroblast Cells. *J. Mater. Eng. Perform.* **2016**, *25*, 1508–1514. [CrossRef]
31. Venturini, L.F.; Artuso, F.B.; Limberger, I.d.F.; Javorsky, C.d.S. Differences in nitrided layer between classic active screen plasma nitriding and active screen plasma nitriding with hemispherical cathodic cage. *Int. Heat Treat. Surf. Eng.* **2012**, *6*, 19–23. [CrossRef]
32. Lin, K.; Li, X.; Dong, H.; Gu, D. Multistep active screen plasma co-alloying the treatment of metallic bipolar plates. *Surf. Eng.* **2020**, *36*, 539–546. [CrossRef]
33. Nishimoto, A.; Bell, T.E.; Bell, T. Feasibility Study of Active Screen Plasma Nitriding of Titanium Alloy. *Surf. Eng.* **2010**, *26*, 74. [CrossRef]

34. Morgiel, J.; Maj, Ł.; Szymkiewicz, K.; Pomorska, M.; Ozga, P.; Toboła, D.; Tarnowski, M.; Wierzchoń, T. Surface roughening of Ti-6Al-7Nb alloy plasma nitrided at cathode potential. *Appl. Surf. Sci.* **2022**, *574*, 151639. [CrossRef]
35. Szymkiewicz, K.; Morgiel, J.; Maj, Ł.; Pomorska, M.; Tarnowski, M.; Wierzchoń, T. TEM investigations of active screen plasma nitrided Ti6Al4V and Ti6Al7Nb alloys. *Surf. Coat. Technol.* **2020**, *383*, 125268. [CrossRef]
36. Li, Y.; He, Y.; Wang, W.; Mao, J.; Zhang, L.; Zhu, Y.; Ye, Q. Plasma Nitriding of AISI 304 Stainless Steel in Cathodic and Floating Electric Potential: Influence on Morphology, Chemical Characteristics and Tribological Behavior. *J. Mater. Eng. Perform.* **2018**, *27*, 948–960. [CrossRef]
37. Domínguez-Meister, S.; Ibáñez, I.; Dianova, A.; Brizuela, M.; Braceras, I. Nitriding of titanium by hollow cathode assisted active screen plasma and its electro-tribological properties. *Surf. Coat. Technol.* **2021**, *411*, 126998. [CrossRef]
38. Li, Y.; Bi, Y.; Zhang, M.; Zhang, S.; Gao, X.; Zhang, Z.; He, Y. Hollow cathodic plasma source nitriding of AISI 4140 steel. *Surf. Eng.* **2021**, *37*, 351–359. [CrossRef]
39. Shen, H.; Wang, L. Corrosion resistance and electrical conductivity of plasma nitrided titanium. *Int. J. Hydrogen Energy* **2021**, *46*, 11084–11091. [CrossRef]
40. Mohseni, H.; Nandwana, P.; Tsoi, A.; Banerjee, R.; Scharf, T. In situ nitrided titanium alloys: Microstructural evolution during solidification and wear. *Acta Mater.* **2015**, *83*, 61–74. [CrossRef]
41. Li, Y.; Wang, Z.; Shao, M.; Zhang, Z.; Wang, C.; Yan, J.; Lu, J.; Zhang, L.; Xie, B.; He, Y. Characterization and electrochemical behavior of a multilayer-structured Ti–N layer produced by plasma nitriding of electron beam melting TC4 alloy in Hank's solution. *Vacuum* **2023**, *208*, 111737. [CrossRef]
42. Niu, R.; Li, J.; Wang, Y.; Chen, J.; Xue, Q. Structure and high temperature tribological behavior of TiAlN/nitride duplex treated coatings on Ti6Al4V. *Surf. Coat. Technol.* **2017**, *309*, 232–241. [CrossRef]
43. Zhang, P.; Cheng, Q.; Yi, G.; Wang, W.; Liu, Y. The microstructures and mechanical properties of martensite Ti and TiN phases in a Ti6Al4V laser-assisted nitriding layer. *Mater. Charact.* **2021**, *178*, 111262. [CrossRef]
44. Li, Y.; Wang, L. Study of oxidized layer formed on aluminium alloy by plasma oxidation. *Thin Solid Film* **2009**, *517*, 3208–3210. [CrossRef]
45. McGee, R.C.; She, Y.; Nardi, A.; Goberman, D.; Dardas, Z. Enhancement of Ti-6Al-4V powder surface properties for cold spray deposition using fluidized Gas-Nitriding technique. *Mater. Lett.* **2021**, *290*, 129429. [CrossRef]
46. Morgiel, J.; Szymkiewicz, K.; Tarnowski, M.; Wierzchoń, T. TEM studies of low temperature cathode-plasma nitrided Ti6Al7Nb alloy. *Surf. Coat. Technol.* **2019**, *359*, 183–189. [CrossRef]
47. Fouquet, V.; Pichon, L.; Drouet, M.; Straboni, A. Plasma assisted nitridation of Ti-6Al-4V. *Appl. Surf. Sci.* **2004**, *221*, 248–258. [CrossRef]
48. Czyrska-Filemonowicz, A.; Buffat, P.; Łucki, M.; Moskalewicz, T.; Rakowski, W.; Lekki, J.; Wierzchoń, T. Transmission electron microscopy and atomic force microscopy characterisation of titanium-base alloys nitrided under glow discharge. *Acta Mater.* **2005**, *53*, 4367–4377. [CrossRef]
49. Szymkiewicz, K.; Morgiel, J.; Maj, Ł.; Pomorska, M. (Ti, Al) O_2 whiskers grown during glow discharge nitriding of Ti-6Al-7Nb alloy. *Materials* **2021**, *14*, 2658. [CrossRef]
50. Toboła, D.; Morgiel, J.; Maj, Ł. TEM analysis of surface layer of Ti-6Al-4V ELI alloy after slide burnishing and low-temperature gas nitriding. *Appl. Surf. Sci.* **2020**, *515*, 145942. [CrossRef]
51. Szymkiewicz, K.; Morgiel, J.; Maj, Ł.; Pomorska, M.; Tarnowski, M.; Tkachuk, O.; Pohrelyuk, I.; Wierzchoń, T. Effect of nitriding conditions of Ti6Al7Nb on microstructure of TiN surface layer. *J. Alloys Compd.* **2020**, *845*, 156320. [CrossRef]
52. Liu, J.; Suslov, S.; Vellore, A.; Ren, Z.; Amanov, A.; Pyun, Y.-S.; Martini, A.; Dong, Y.; Ye, C. Surface nanocrystallization by ultrasonic nano-crystal surface modification and its effect on gas nitriding of Ti6Al4V alloy. *Mater. Sci. Eng. A* **2018**, *736*, 335–343. [CrossRef]
53. Farokhzadeh, K.; Qian, J.; Edrisy, A. Effect of SPD surface layer on plasma nitriding of Ti–6Al–4V alloy. *Mater. Sci. Eng. A* **2014**, *589*, 199–208. [CrossRef]
54. Lin, L.; Tian, Y.; Yu, W.; Chen, S.; Chen, Y.; Chen, W. Corrosion and hardness characteristics of Ti/TiN-modified Ti6Al4V alloy in marine environment. *Ceram. Int.* **2022**, *48*, 34848–34854. [CrossRef]
55. Gao, C.; Wu, W.; Shi, J.; Xiao, Z.; Akbarzadeh, A. Simultaneous enhancement of strength, ductility, and hardness of TiN/AlSi10Mg nanocomposites via selective laser melting. *Addit. Manuf.* **2020**, *34*, 101378. [CrossRef]
56. Li, Q.; Wei, M.; Yang, J.; Zhao, Z.; Ma, J.; Liu, D.; Lan, Y. Effect of Ca addition on the microstructure, mechanical properties and corrosion rate of degradable Zn-1Mg alloys. *J. Alloys Compd.* **2021**, *887*, 161255. [CrossRef]
57. Karimi, S.; Nickchi, T.; Alfantazi, A. Effects of bovine serum albumin on the corrosion behaviour of AISI 316L, Co–28Cr–6Mo, and Ti–6Al–4V alloys in phosphate buffered saline solutions. *Corros. Sci.* **2011**, *53*, 3262–3272. [CrossRef]
58. Jin, J.; He, Z.; Zhao, X. Formation of a protective TiN layer by liquid phase plasma electrolytic nitridation on Ti–6Al–4V bipolar plates for PEMFC. *Int. J. Hydrogen Energy* **2020**, *45*, 12489–12500. [CrossRef]
59. Zhao, Y.; Wu, C.; Zhou, S.; Yang, J.; Li, W.; Zhang, L.-C. Selective laser melting of Ti-TiN composites: Formation mechanism and corrosion behaviour in H2SO4/HCl mixed solution. *J. Alloys Compd.* **2021**, *863*, 158721. [CrossRef]
60. Liu, C.; Revilla, R.I.; Li, X.; Jiang, Z.; Yang, S.; Cui, Z.; Zhang, D.; Terryn, H.; Li, X. New insights into the mechanism of localised corrosion induced by TiN-containing inclusions in high strength low alloy steel. *J. Mater. Sci. Technol.* **2022**, *124*, 141–149. [CrossRef]

61. Ye, Q.; Li, Y.; Zhang, M.; Zhang, S.; Bi, Y.; Gao, X.; He, Y. Electrochemical behavior of (Cr, W, Al, Ti, Si) N multilayer coating on nitrided AISI 316L steel in natural seawater. *Ceram. Int.* **2020**, *46*, 22404–22418. [CrossRef]
62. Tobola, D.; Morgiel, J.; Maj, Ł.; Pomorska, M.; Wytrwal-Sarna, M. Effect of tribo-layer developed during turning of Ti–6Al–4V ELI alloy on its low-temperature gas nitriding. *Appl. Surf. Sci.* **2022**, *602*, 154327. [CrossRef]
63. Jannat, S.; Rashtchi, H.; Atapour, M.; Golozar, M.A.; Elmkhah, H.; Zhiani, M. Preparation and performance of nanometric Ti/TiN multi-layer physical vapor deposited coating on 316L stainless steel as bipolar plate for proton exchange membrane fuel cells. *J. Power Sources* **2019**, *435*, 226818. [CrossRef]
64. Baltatu, M.S.; Vizureanu, P.; Sandu, A.V.; Florido-Suarez, N.; Saceleanu, M.V.; Mirza-Rosca, J.C. New titanium alloys, promising materials for medical devices. *Materials* **2021**, *14*, 5934. [CrossRef] [PubMed]
65. Movassagh-Alanagh, F.; Mahdavi, M. Improving wear and corrosion resistance of AISI 304 stainless steel by a multilayered nanocomposite Ti/TiN/TiSiN coating. *Surf. Interfaces* **2020**, *18*, 100428. [CrossRef]
66. Weicheng, K.; Zhou, Y.; Jun, H. Electrochemical performance and corrosion mechanism of Cr–DLC coating on nitrided Ti6Al4V alloy by magnetron sputtering. *Diam. Relat. Mater.* **2021**, *116*, 108398. [CrossRef]
67. Wang, H.; Zhang, R.; Yuan, Z.; Shu, X.; Liu, E.; Han, Z. A comparative study of the corrosion performance of titanium (Ti), titanium nitride (TiN), titanium dioxide (TiO_2) and nitrogen-doped titanium oxides (N–TiO_2), as coatings for biomedical applications. *Ceram. Int.* **2015**, *41*, 11844–11851. [CrossRef]

Review

Application of Electrophoretic Deposition as an Advanced Technique of Inhibited Polymer Films Formation on Metals from Environmentally Safe Aqueous Solutions of Inhibited Formulations

Natalia A. Shapagina * **and Vladimir V. Dushik**

Frumkin Institute of Physical Chemistry and Electrochemistry, Russian Academy of Sciences, Leninsky Prospect 31-4, 119071 Moscow, Russia
* Correspondence: fuchsia32@bk.ru

Abstract: The presented paper analyzes polymer films formed from aqueous solutions of organosilanes, corrosion inhibitors and their compositions. Methods of depositing inhibited films on metal samples, such as dipping and exposure of the sample in a modifying solution, as well as an alternative method, electrophoretic deposition (EPD), are discussed. Information is provided on the history of the EPD method, its essence, production process, areas of application of this technology, advantages over existing analogues, as well as its varieties. The article considers the promise of using the EPD method to form protective inhibited polymer films on metal surfaces from aqueous solutions of inhibitor formulations consisting of molecules of organosilanes and corrosion inhibitors.

Keywords: metals; electrophoretic deposition (EPD); cataphoretic (CPD) and anaphoretic deposition (APD); inhibited formulations (INFOR); organosilanes; corrosion inhibitors

Citation: Shapagina, N.A.; Dushik, V.V. Application of Electrophoretic Deposition as an Advanced Technique of Inhibited Polymer Films Formation on Metals from Environmentally Safe Aqueous Solutions of Inhibited Formulations. *Materials* **2023**, *16*, 19. https://doi.org/10.3390/ma16010019

Academic Editor: Yong Sun

Received: 11 November 2022
Revised: 12 December 2022
Accepted: 14 December 2022
Published: 20 December 2022

1. Introduction

Protection of metal constructions and facilities against corrosion is one of the main tasks in various industries [1–3]. One method to reduce of the corrosive activity of process media is the introduction of corrosion inhibitors. According to ISO 8044-2015 [4], corrosion inhibitors are chemical compounds or their compositions, the presence of which in sufficient concentration reduces the corrosion rate of metals without significantly changing in the concentration of corrosive reagents. Corrosion inhibitors can be applied in different forms: volatile [5], contact [6], chamber inhibitors [7], inhibited papers, sleeves, and films [8]. Despite the availability of an extensive range of existing corrosion inhibitors, there is the problem of expanding their assortment by creating new inhibitors or inhibited formulations (INFOR) with higher protective characteristics and lower cost [9–11]. When developing new inhibitors or their mixtures, it is necessary to consider the operating environment, the nature of the metal to be protected from corrosion damage, external influences (temperature, pressure, and other factors), etc. [11,12].

In this regard, in recent years, new methods of modifying the metal surface are being used to fight against corrosion, including those based on the use of inhibited formulations consisting of molecules of organosilanes and corrosion inhibitors [13–15].

2. Options for Protecting of Metal Surfaces from Corrosion Damage by Different Classes of Environmentally Friendly Organic Compounds

2.1. Corrosion Inhibitors

Since the slowdown of corrosion processes occurs due to a decrease in the active area of the metal surface and changes in the activation energy of electrode reactions that limit metal corrosion, we can divide inhibitors into three types, namely anodic (affecting the anodic dissolution of the metal surface, including their ability to cause passivation of metal), cathodic (reducing the rate of the cathodic process), and mixed (inhibiting both processes). The protective effect of corrosion inhibitors is based on a change in the state of the metal surface due to their adsorption or on the formation of hard-soluble compounds with metal ions [16–23]. Modern classification of corrosion inhibitors allows their systematization based on their chemical nature; there are oxidative, adsorptive, and co-ordination complex inhibitors, as well as polymeric inhibitors [11,22,24–26].

Oxidative corrosion inhibitors have protective properties against many structural metals in a wide pH range and can be used for their protection both in aqueous environments and when operating metals in aggressive atmospheric conditions. The efficient oxidative type inhibitors include salts of chromic acid; however, from an environmental point of view, they are highly toxic, so they had to be replaced by less dangerous ones. For example, molybdates and tungstates have a similar chemical structure to chromates but are not as efficient. In addition, their use is limited by their high cost [27–37]. An alternative to oxidative inhibitors is adsorptive inhibitors.

The protective effect of adsorption type inhibitors is based on the formation of protective layers firmly bonded to the metal surface, isolating the metal surface from the corrosive environment. This effect is due to the state of the metal surface and the charge of the adsorbing particles, as well as their ability to form chemical bonds with the metal or the products of its interaction with the components of the corrosive environment [24–26]. As a rule, cation-active inhibitors decelerate active anodic dissolution, i.e., they are effective in the region of potentials lower than the critical passivation potential, or they inhibit cathodic reactions. Anion-active corrosion inhibitors are more effective in preventing local (pitting) corrosion. Complex-forming corrosion inhibitors are very difficult to distinguish from adsorption-type inhibitors as they may not form thick films of complex compounds. Two main groups can be distinguished among the complex-forming inhibitors. The first group includes heterocyclic compounds capable of forming insoluble complexes in aqueous solutions; the second group consists of complexons and metal complexonates. In the first group, the widely used azoles (imidazoles, triazoles, thiazoles) which form on the protected metal surface thin insoluble films of complex compounds with cations of these metals stand out for their effectiveness. The best-known representatives of this class of inhibitors are primarily 1,2,3-benzotriazole (BTA) and its derivatives [38–43]. Recent studies show that BTA is widely used not only as a corrosion inhibitor for copper, but also for mild steel and even zinc [38,44–49]. In the second group, phosphorus-containing complexones, such as 1-hydroxyethane-1, 1-diphosphonic acid, nitrilotrimethylene phosphonic acid, and ethylenediamine-N,N,N′,N′-tetramethylene phosphonic acid, take the leading place [50–55].

Among polymeric inhibitors for neutral media, polyphosphates are the best known [56–58]. They are non-toxic, inexpensive, and can inhibit corrosion of steel at low concentrations. Their main disadvantages include the possibility of intensification of corrosion at high concentrations due to the formation of soluble complexes with cations of the protected metal. As early as in the 1970s, in addition to polyphosphates, water-soluble polymers containing $COOH^-$ and OH^- groups (mainly based on acrylic and maleic acids derivatives), characterized by high hydrolytic stability, started to be used for protection of water systems against scaling [24]. Polymers containing acidic groups are more often used as inhibitors of scaling [59], and cationic polymers for corrosion protection of metals [60]. Another group of corrosion inhibitors of this class is not the polymers themselves, but hydrophilic monomers capable of polymerizing upon adsorption on metal surfaces [59]. A consequence of such polymerization can be a decrease in the solubility of the adsorption layer, an increase in the protective effect, and irreversibility of adsorption.

Even though today a large number of corrosion inhibitors for the majority of structural metals have been effectively tested, there is still a need to expand the range of inhibitors and their compositions with higher protective properties and lower cost. In this vain, researchers aimed to find new corrosion inhibitors or inhibited formulations which do not lose their topicality.

2.2. Organosilicon Compounds. Organosilanes

Organosilicon compounds are a class of chemical compounds that contain a bond ≡Si—C≡ in their molecules. The main difference of organosilicon compounds from others is largely due to the low bond strength of bond ≡Si–Si≡ compared to bond ≡C–C≡, and conversely, significantly higher bond strength of bond ≡Si–O–Si≡ than bond ≡C–O–C≡. The values of the bond energies of silicon atoms with each other and with oxygen (213.7 444.1 kJ/mol, respectively) compared to those of the bond energies of carbon atoms with each other and with oxygen (347.7 358.2 kJ/mol, respectively) confirms this statement. Typically, organosilicon compounds are divided into the following groups: organohalogensilanes, organosilanes, organosiloxanes, and heterocyclic compounds [61–64]. Let us consider in more detail the class of organosilanes.

Organosilanes are environmentally friendly substances that are not found in nature; they are mainly synthesized from silicon dioxide [65]. The general formula of organosilanes is shown in Figure 1 [62,65,66].

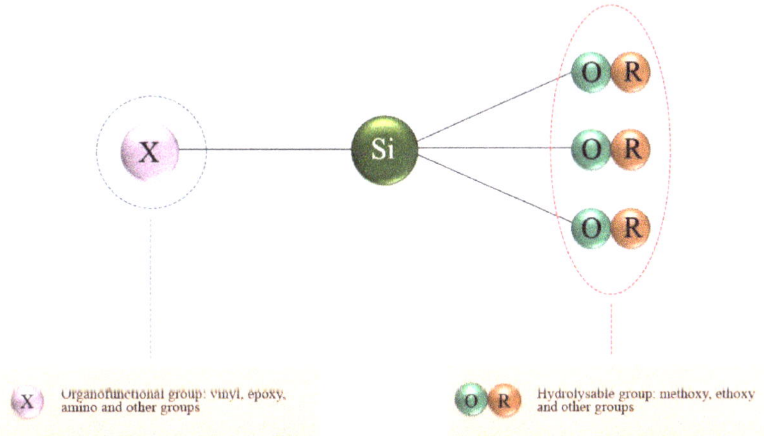

Figure 1. The general formula of organosilanes.

Organosilanes are more prone to condensation reactions, resulting in polysiloxane structure formations that differ markedly in thermal stability from their carbon counterparts [65,66]. Such compounds contain thermally stable siloxane bonds where elements with positive and negative polarization alternate. The presence of a polar substituent in the hydrocarbon radical bonded to a silicon atom leads to an increase in polarity of the polymer molecule and, as a result, an increase in adhesion, mechanical strength, and other properties. In this regard, organosilanes are widely used in the paint industry as adhesion promoters or crosslinking agents that form strong bonds with the overlying layers of coatings and as surface hydrophobisers [65–72]. Thus, the addition of a small amount of organosilanes in the form of 0.1 ÷ 0.5% aqueous solutions improves the adhesion of the polymer matrix to glass fiber [62,73]. The hydrophobicity of the surface after treatment with organosilanes depends on the orientation effect in the organosiloxane layer formed on the surface. The best results are achieved when the siloxane bond is oriented toward the surface and the hydrocarbon radical is oriented from the surface into the external environment [62,74,75].

According to their chemical structure, organosilanes can be divided into two groups: monosilanes (single Si atom) and bis-silanes (two Si atoms). Monosilanes are used as organosilane crosslinking agents, while bis-silanes are used to form crosslinks in silane crosslinking agents [76–78]. In the presence of water, in general, the following transformations occur with organosilanes as shown in Figure 2 [62,65,66,74,79,80].

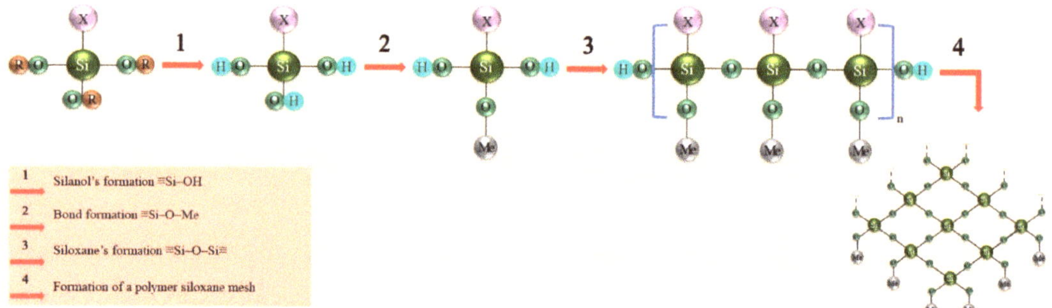

Figure 2. The reactions that result in the formation of a polymer siloxane film on a metal surface.

In the first stage, hydrolysis of organosilanes to form silanol occurs in aqueous solution. If a metal sample is placed in an aqueous organosilane solution, the silanol molecules will diffuse to the metal's oxide-hydroxide surface (stage 2), displacing the adsorbed water molecules from the surface and starting to interact with the hydroxyl groups of the metal surface. As a result of this interaction, hydrogen bonds are formed and enter a condensation reaction with the formation of ≡Si–O–Me bonds on the metal surface layer. In parallel with stages 1 and 2, the adsorbed silanol molecules enter a polycondensation reaction, forming ≡Si–O–Si≡ bridging siloxane bonds resistant to hydrolysis (stage 3). In the last stage, under the influence of temperature, water evaporation and "solidification/cross-linking" of the siloxane structures take place, forming a polymeric siloxane film on the metal surface.

The hydrolysis reaction of organosilanes proceeds spontaneously, without catalysts, but the presence of acids or bases accelerates their hydrolysis [73,79]. It was shown in [80] that the rate of hydrolysis slows down at neutral pH (6.7 ÷ 7.0), and at pH (2.0 ÷ 4.0) the rate of hydrolysis increases by 1000 times. The completeness of hydrolysis is affected by the size of the hydroxyl group in the organosilane molecule; the smaller it is, the higher the rate of hydrolysis. For example, the hydrolysis rate of methoxysilane is 6–10 times higher than that of ethoxysilane [73,80]. In addition, as was shown in [81,82], besides the addition of acids, the hydrolysis reaction rate can be increased if the water–organosilane solution is subjected to ultrasound treatment. Thus, by adjusting the pH of the solution, the rate of hydrolysis, the formation of siloxane bonds to the metal surface, and the polymerization and orientation of the organosilane molecules can be controlled [70,73].

There are several ways to apply organosilanes on different substrates:

- By dissolving of organosilane in a solvent mixture of ethanol and water at pH 4.0 to 5.0 (the sample is immersed in the solution and then removed for drying) [77,79–82];
- Vapor phase deposition (in a closed chamber, a tank with organosilane is heated at reduced pressure, forming its vapor, which condenses on a metal surface) [83];
- Spin-on deposition (organosilane solution is deposited on a low-speed rotating substrate, followed by washing) [83];
- Spray application from aqueous or alcoholic solutions followed by air drying [83].

Some papers [84,85] reported that ornagosilanes such as 3-aminopropyltriethoxysilane, 3-glycidoxypropyltrimethoxysilane, ureidopropyltriethoxysilane, vinyltrimethoxysilane, and functional silanes with the addition of crosslinking silanes such as bis [trimethoxysilylpropyl] amine are quite resistant to corrosive environments. Although organosilanes inhibit corrosive processes on metals, their protective effect is much lower compared to corrosion inhibitors [86,87].

2.3. *Inhibited Formulations (INFOR) Consisting of Organosilane Molecules and Corrosion Inhibitors*

As noted earlier, organosilane films are not always able to protect metals from corrosion damage. The closest analogues of such films are anticorrosive chromate coatings [88,89]; however, their use is limited due to their toxicity. The main disadvantage of siloxane films before chromate coatings is the lack of the so-called "self-healing" ability of siloxanes. Chromate in an aqueous medium is able to diffuse to the damaged area and incorporate into the film, providing recovery of the damaged areas. Numerous studies have been aimed at eliminating this disadvantage of organosilanes by introducing chromium-free corrosion inhibitors into the siloxane film, which contribute to the "healing" of film defects [89].

The use of aqueous solutions of organosilanes with corrosion inhibitors is of interest. Treatment of metal samples in such compositions can lead to a high protective anti-corrosion effect. Films that are formed on metal from INFOR solutions are promising and environmentally safe. Since the compounds used are consumed in small concentrations, the size of the protected product is not changed and the problem associated with its de-conservation is removed [14,15]. Previous studies have shown that INFOR can be used to protect metals in various aggressive environments: in solutions, in atmospheric conditions, etc. Thus, in [90] it was shown that in an aqueous chloride-containing solution, when using the inhibitor composition consisting of organosilane molecules (vinyltrimethoxysilane or diaminsilane) and corrosion inhibitor (1,2,3-benzotriazole), siloxanoazole fragments are formed on the metal surface which provide additional cross-linking of surface adsorbed molecules, thereby increasing the degree of polymerization and, consequently, the density of the surface layers. Moreover, similar inhibitor compositions can be used to form films on metal surfaces to protect metal in various corrosive atmospheres [14,15,90,91]. In this case, the use of carboxylic or phosphonic acids leads to an additional interaction with siloxane groups, which contributes to the formation of a thicker film, and therefore provides improved corrosion resistance [91]. Summarized information from this section is presented in Table 1.

Table 1. Summary of protection of metals by organic compounds.

Criteria	Corrosion Inhibitors	Organosilanes	INFOR
Type of protective action	Oxidative [11,22,27–37] Adsorptive [24–26,30–37] Complex forming [38–49] Polymeric [56–60]	Film forming (isolaing) [79–87]	Isolating [88–91]
Healing effect	High for chromates [24,25,88,89]	Moderate [61–67]	High [14,15,90,91]
Application form	Volatile [4,5,11–13] Contact [6,16–38,40,41,48,49] Chamber [7,9,10,39,42] Inhibited papers, sleeves [8]	Ethanol-water solution, applied by immersion [77,79–82], vapor phase, spin-on, spray [83]	Water solution with organosilane and contact inhibitor [14,15,90,91]

3. Methods of Forming Protective Films and Coatings on Metal Surfaces from INFOR Aqueous Solutions

The conventional technique of coating and film deposition is the method of dipping the sample into a solution containing inhibitor composition [11,12,92]. However, this method has disadvantages, the most important one being that the protective properties of the film or coating vary with the exposure time of the sample in the solution (Figure 3) [91].

Figure 2 shows that if the conventional method of a polymer film formation on the metal surface is used, its performance properties (continuity, uniformity, adhesion, and protective properties, etc.) are determined by the duration of exposure of the sample in the INFOR modifying solution. Thus, to form a quality film on the surface of steel and copper, it is necessary to spend at least 9 h [91]. Therefore, it is obviously necessary to optimize the process of obtaining a quality protective film on a metal sample. In this case, electrodeposition can be an alternative method of film formation.

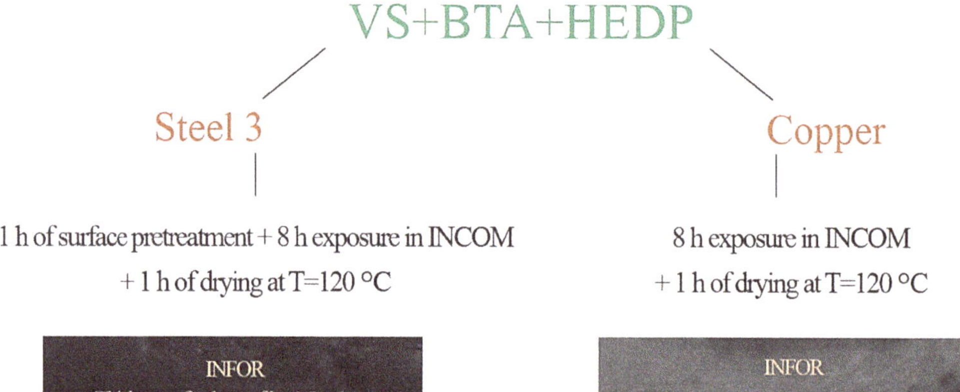

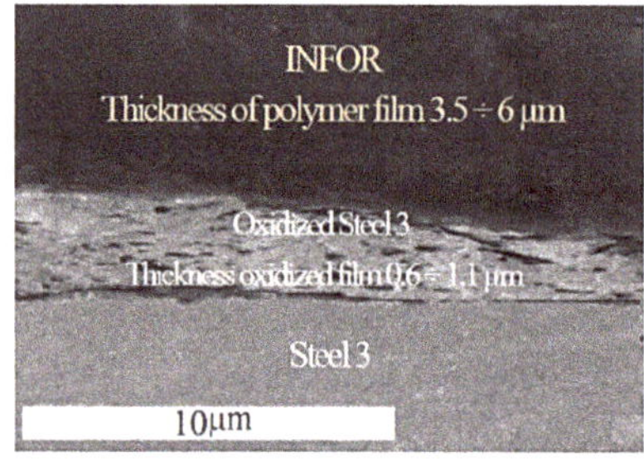

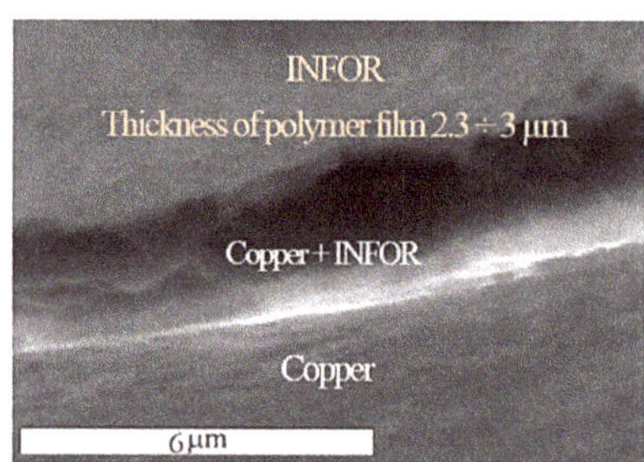

Figure 3. SEM images of cross-sections of polymer films on steel 3 and copper.

3.1. Electrophoretic Deposition (EPD): History of the Method, Its Essence, Advantages, Production Process, and EPD Varieties

The principle of moving particles of matter under the influence of electricity was discovered in 1809 by professors P. I. Strakhov and F. F. Reiss of Moscow University. It was called electrophoresis [93–95]. Gradually the study of this phenomenon gained momentum; already in 1917, the first patent for the use of electrophoretic painting was received by General Electric. Beginning in the 1920s, the process began to be used to apply latex rubber [96]. In the 1930s, some of the first patents were obtained describing basic neutralized water-dispersible resins specifically designed for EPD coatings on metals. In the late 1950s, the Ford Motor Company engineering team actively began to develop a methodology for the EPD coating process for cars. The first commercial anodic coating system for automobiles went live in 1963. The first patent for the cathodic product was issued in 1965, and already in 1975, the created technology resulted in the rapid application of cathodic EPD in the automotive industry [95]. In the USSR, this method became popular in the 1980s and found its application in many plants of the automotive, aviation, and machine-building industries [97,98]. Today, the electrophoretic deposition technology accounts for about 70% of the coating of various products. A major part of EPD use is in the automotive industry. It is probably one of the effective methods that can significantly increase the service life of metal structures and products [99,100].

The term EPD includes a wide range of industrial operations: cathodic and anodic electrodeposition as well as electrophoretic coating or electrophoretic painting. During EPD, colloidal particles suspended in a liquid medium migrate under the action of an electric field (electrophoresis) and are deposited on the surface of an oppositely charged electrode [101]. All colloidal particles that can be used to form stable suspensions and that can carry a charge are useful for electrophoretic deposition. Such materials include metals, polymers, pigments, dyes, and ceramics. There are two modes of EPD, namely constant voltage and constant current. The first mode forms thinner coatings than the second one [102–106].

EPD processing has several advantages [99,101,107–109]:

- The coatings/films applied to the product are continuous and uniform in thickness;
- Flms/coatings can be formed on products with complex geometry;
- EPD-formed coatings/films have better corrosion and mechanical properties, which ensure a longer service life of the treated product;
- Less time is spent per unit compared to immersion/aging samples in modifying solutions;
- The technology is applicable to a wide class of materials (metals, ceramics, polymers, etc.);
- The process is automated as a rule and does not require large amounts of human resources and special requirements to the operating personnel, which significantly reduces the cost of the films/coatings produced by EPD technology;
- Generally, an aqueous solvent is used, reducing the risk of fire in comparison to the solvent-based films/coatings they replace;
- Modern electrophoretic materials (varnishes, paints, and other products) are largely more environmentally friendly than materials of other film/coating technologies.

Despite the obvious advantages of EPD, this method has a variety of disadvantages:

- Limited choice of solution compositions because of electrical conductivity and solubility of the components used;
- This method allows the application of only a single-layer film/coating;
- It is necessary to use expensive equipment, e.g., high-power current sources and drying cabinets of large volume, which leads to an increase in industrial area.

In general terms, the EPD production process can be schematically represented as shown in Figure 4, [109–112].

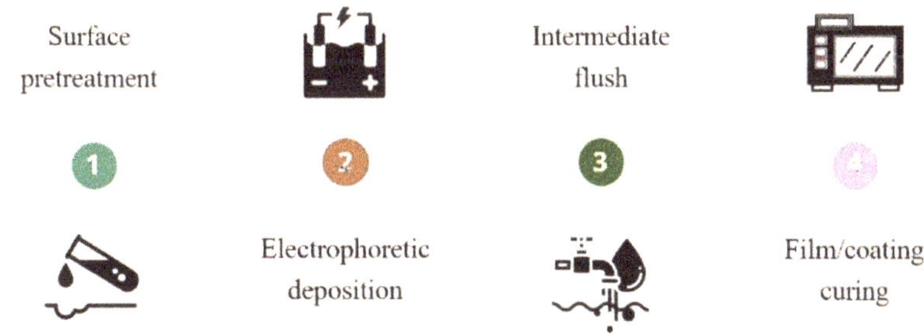

Figure 4. The production process of electrophoretic deposition.

The first step is surface preparation. This is usually a process of machining, cleaning/degreasing the metal, and applying pretreatments such as oxidizing or phosphating [109,110]. In the second step, the EPD process begins. The sample is immersed in an electrolyte bath/cell and an electric current is applied through the EPD bath using electrodes. Typically, when electrophoretic films/coatings are deposited, the voltage ranges from 25 to 400 V DC. After deposition, the sample is washed in water to remove the excess undeposited film/coating (step 3). An ultrafilter may be used during the washing process, on which excess deposition material accumulates and then can be returned to the deposition bath; this ensures high material efficiency and reduces the amount of wasting. In the last step, the sample is subjected to a heat treatment, which allows the polymer film or coating to cure. As a result of the thermal treatment, the film, which was porous due to the gas released during the EPD process, spreads out, acquiring a smooth, uniform, and defectless structure [111,112].

In addition, as previously mentioned, EPD is either cathodic or anodic (Figure 5), [99,106,113].

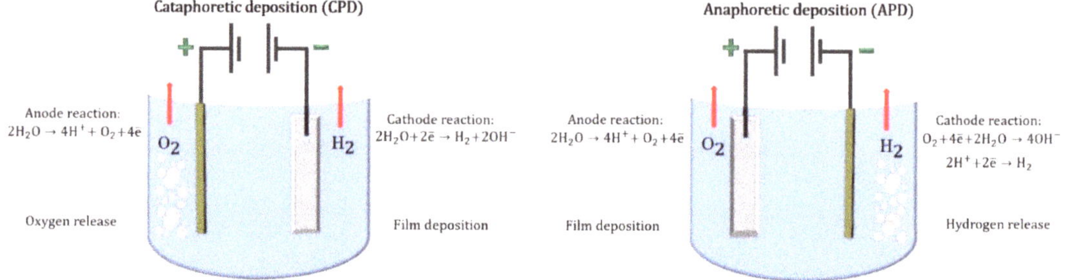

Figure 5. The processes taking place during cataphoretic and anaphoretic deposition.

Figure 4 shows that the CPD and APD processes are similar. The difference lies in the polarity of the charge on the surface of the processed product (cathode or anode): in CPD, the surface has a negative electrical charge, and the counter electrodes have a positive charge; in APD, the sample surface is charged with a positive electrical charge, and the counter electrodes have a negative charge [109,113]. In APD, the deposited material contains acidic salts that play the role of charge-carrying groups. These negatively charged anions react with positively charged hydrogen ions (protons), which are formed at the anode as a result of water electrolysis with the conversion of the original acid. A fully protonated acid carries no charge. It is less soluble in water and can precipitate out of the water onto the anode. During CPD, the precipitated material contains basics as charge-carrying groups. If the basic salt was formed by the protonating of the base, such a base will react with the hydroxyl ions produced by the electrolysis of the water to form a neutrally charged base and water. Both types of processes have their advantages and disadvantages. Let us consider the main positives for each method [99,108,109,111,114,115]. The key features of these processes are presented in Table 2.

Table 2. The key features of the APD and CPD processes.

Method	Feature
CPD	The films/coatings produced by this method have higher protective properties. However, this effect may be due to the cross-linking chemistry of the raw material (polymer) used rather than to the electrode on which the film/coating is deposited; The product can be designed with less current density due to the higher throw power of the medium; The oxidation process takes place at the anode, so staining and other problems that could result from the oxidation of the metal substrate are eliminated.
APD	Compared to CPD, APD is less expensive; Less sensitivity to changes in substrate quality; The substrate is not exposed to strong alkaline attack which can dissolve phosphate, oxide, and other coatings used as substrate pretreatment; The anodic process avoids hydrogen embrittlement, which can occur during the cathodic process, due to hydrogen ion discharge.

Thus, when choosing one or another method of electrodeposition, one should consider the nature of the substrate, its pretreatment, the pH of the electrolyte used, as well as the cost of film/coating formation by the chosen method. Consideration of the above factors will greatly facilitate the task of choosing the electrodeposition method for the formation of a high-quality film/coating on the substrate.

3.2. Formation of Protective Inhibited Polymer Films on Metals using EPD from INFOR Aqueous Solutions

A review of the current scientific progress on the subject showed that the electrodeposition method, in suspensions containing organosilane, is used either for coatings already pretreated with organosilane or for sol-gel coatings on metal (Figure 6) [116–123].

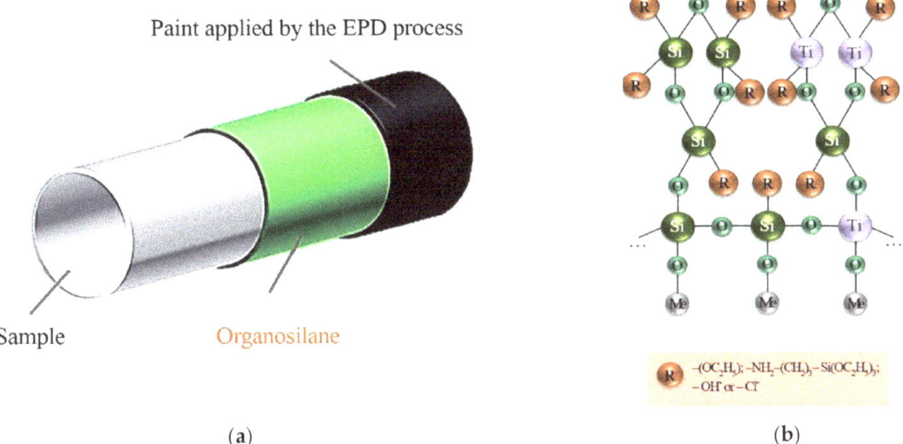

Figure 6. Examples of the use of organosilanes in EPD: (**a**) preliminary silanization of the metal sample surface; (**b**) application of organosilanes in colloidal sol-gels.

In the case of pre-silanization, the organosilane is used as an intermediate layer between the metal substrate and the main coating to improve the adhesion properties of the main coating (Figure 6a) [116–119]. In the case of sol-gels, solutions usually contain one or more organometallic compounds such as zirconium, aluminum, or titanium compounds, one or more organosilanes, and acids, bases, glycols, etc., are used as catalysts (Figure 6b). (Figure 6b) [120–123]. The addition of organosilanes leads to the formation of denser particles and the sols themselves become more viscous, which, as a result of further electrodeposition, results in virtually defect-free coatings [123,124].

As a result, we can conclude that the very idea of using the EPD method to form polymer inhibited films on metals from aqueous INFOR solutions is brand new. The originality of this proposal lies in the electrolyte used and the film formation method. The electrolyte is an aqueous suspension containing organosilane and corrosion inhibitors in which the main film-forming component is organosilane due to its ability to polycondense. EPD is used to accelerate the deposition of the film on the sample, as well as to orient the siloxane bonds to the metal surface. Since the proposed technology is a "new trend" for such INFORs, there are few works in this area. For example, work [125] reports that using CPD on the surface of tungsten from an aqueous INFOR solution was able to form a polymer film on the surface of tungsten. The paper proposes INFOR composition with acceptable concentrations of the substances used and suggests optimal modes of formation of a quality inhibited polymer film on the surface of tungsten. For example, increasing the duration of cataphoresis contributes to increasing the number of deposited siloxane groups, which complicates the process of further polycondensation (Figure 7) [125].

The figure shows if that the duration of CPD less than 5 min, it leads to the formation of an "island film" on the sample. Thus, it has been experimentally established that, for the formation of a continuous film from aqueous INFOR solutions on the surface of tungsten, the optimal CPD time is 5 min. During this time, the necessary number of siloxane groups is chemisorbed for further formation of a solid film.

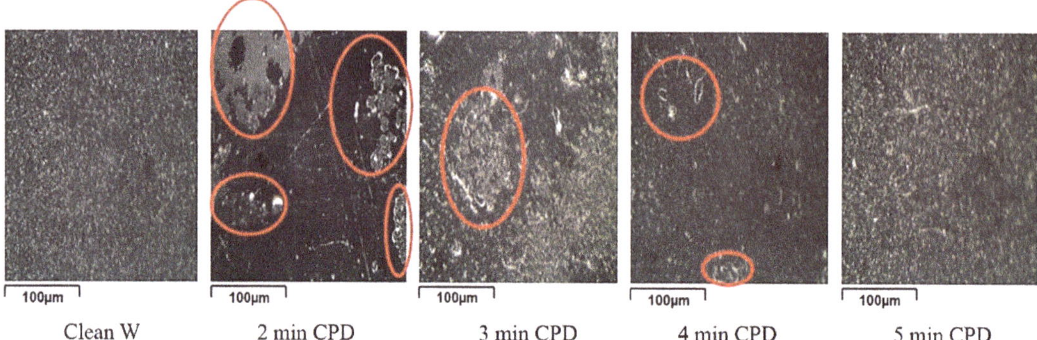

Figure 7. Appearance of tungsten samples subjected to CPD in INFOR aqueous solution depending on the duration of CPD. Uncovered areas of the sample are marked with the red circle.

3.3. The Main Similar Methods of Forming Protective Inhibited Polymer Films on Metals from Aqueous Solutions of INFOR

The proposed EPD technology for the formation of polymer films on metals from aqueous solutions of INFOR can find application in the following areas:

- Corrosion protection of metals (preservation of metal products, protection of metals from atmospheric corrosion, etc.) [11,13,126];
- Metal surface pretreatment with a primer for its subsequent painting with paint and varnish materials [127,128];
- Decorative films [129,130].

Based on the field of application, and the proposed method of polymer films application, it is possible to distinguish the following range of products that can compete with the discussed technology. First of all, these are cataphoresis varnishes, inhibited polymer films/sleeves, as well as water-borne organosoluble paint coatings. The EPD method has several advantages over the listed counterparts; further, we will analyze its positive aspects.

3.3.1. Cataphoresis Varnishes

As a rule, such formulations have a complex, multicomponent, and expensive chemical composition. Usually, the films produced from such suspensions are colorless. However, if there is a need to obtain a color film, then after the application of the polymer film to the substrate, the sample must be further dyed in the pigment toner, which, in turn, leads to an increase in the cost of the product. After the polymer film/coating is applied to the sample, it must be heat cured at T = 120 ÷ 180 °C for 20 ÷ 40 min [103,104,129,131–135]. Compared to cataphoresis lacquers, the technology of applying inhibited films of aqueous solutions of INFOR using EPD has the following advantages:

- The use of INFOR will lead to a simplification of the electrolyte composition;
- Speeding up the drying process of metal products by 25% since, according to preliminary experimental data, it takes about 10 ÷ 15 min for the thermal curing of films;
- The cost of an aqueous suspension is significantly lower.

3.3.2. Inhibited Polymer Films/Sleeves

Usually, such films contain a volatile corrosion inhibitor (VCI). The principle of their action is as follows: the product is covered with a film which releases inhibitor vapors, filling all the space inside the package and creating a protective gas environment around the parts or structures. On the metal surface, the VCIs condense and form a monomolecular film that prevents corrosion. Since these compounds are in a gaseous state, they easily penetrate any crevices and cavities, providing protection in the most difficult-to-reach places. This is a major advantage when using inhibited materials containing volatile compounds. Protective action is carried out at a distance of up to 70 cm from the film [8,136–139].

The main disadvantages of such protection: an auxiliary barrier is required to prevent VCI from escaping from the volume of the protective sleeve/film, and the protected product must be hermetically sealed. This additionally requires the use of special adhesive tape or heat welding [8,139]. Such aspects also increase the cost of the offered products. Compared to inhibited polymer films/sleeves, the films made from INFOR solutions by the EPD method have the following advantages:

- The proposed technology does not require additional packaging material;
- Economical consumption of the protective material;
- Reduction of production labor costs by 2 times.

3.3.3. Water-Borne, Organosoluble Paint Coatings

They are easily available and relatively cheap paintwork materials (paints). Depending on the condition of the polymer binder, waterborne paints are subdivided into water-dispersible and water-soluble. Water-dispersible paintwork materials are suspensions of pigments and fillers in aqueous dispersions of film-forming substances of the synthetic polymers type with the addition of emulsifiers, dispersants, and other auxiliaries. Water-soluble paints, according to the type of film-forming substance, are subdivided into copolymervinyl acetate (the basis is an aqueous dispersion of vinyl acetate copolymers with dibutyl maleate or ethylene); polyvinyl acetate (the basis is a polyvinyl acetate dispersion); butadiene styrene (the basis is latex, which is a copolymer of butadiene with styrene); polyacrylic (the basis is an acrylic dispersion copolymer), etc. [140–144].

Organic solvent paints are paints based organic solvents which evaporate in the drying process. Such paints are divided into two groups: oil and alkyd paints. For oil paints, the binder is oil which dries in the oxidation process. They are linseed oil, linseed oil varnish, oil-saturated alkyd resin, or a mixture of different oils. They are characterized by a fairly long drying process. Alkyd resin is the binder of alkyd materials. As a rule, it is received by boiling vegetable oils, linseed, tall oil, soya oil, etc., together with alcohol and organic acids or acid anhydrides. Like oils, alkyd resins dry out as a result of oxidation [145–150].

The main disadvantage of the considered paint-and-lacquer coatings is that they rather quickly sorb moisture, which results in the peeling of the coating from the metall. In addition, some representatives of these materials can be fire-hazardous and toxic [151–153].

In comparison with water-soluble and organosoluble paint coatings, the offered method of the formation of polymer films from INFOR solutions possesses the following positive features:

- INFOR components are safe;
- The polymer inhibited films have a more solid structure that should lead to an increase in the adhesive strength of the film/coating to the metal;
- No long preparation of the surface is required;
- Formed films can be used as a primer for the following painting of the product.

Summarized information on the comparison of the described methods is presented in Table 3.

Table 3. Comparison of the key features of the protective film formation methods.

Method/Technique	Conveniences	Limitations
Organosilanes films by dipping in INFOR	No complex equipment is required Environmentally safe	Metal surface needs to be pre-treated Long process of coating formation
[14,15,90,91,125]		
Organosilanes films by EPD of INFOR	Accelerated coating formation process Complex shapes can be coated Environmentally safe	Requires more expensive equipment
[125–130]		
Cataphoresis varnishes	Uniform coating Complex shapes can be coated Relatively high wear resistance	Complex solution composition Requires more expensive Equipment
[103,104,131–135]		
Inhibited sleeves/films	Easy to apply Relatively cheap	Need to be sealed due to a danger of inhibitor volatilization
[8,136–139]		
Paint coatings	Proven process of coating formation Relatively cheap	Sorb moisture, toxic Relatively high consumption of paint material
[140–153]		

4. Conclusions

Thus, we can conclude that, for obtaining polymer films/coatings on substrates of different materials, the EPD method is an alternative to traditional methods. This method has several advantages: it optimizes the process of obtaining polymer films/coatings; it is possible to coat samples of different geometric shapes; the formed films/coatings have better performance properties; the risk of ignition of used materials is reduced; environmentally safe substances are used as raw materials. The EPD option is chosen based on the following factors: the nature of the substrate; the method of its pretreatment; the pH of the electrolyte used; and the cost of forming the film/coating by the chosen method. The use of electrophoretic deposition technology is a relatively new and promising method of forming corrosion-resistant polymer films on metal samples from aqueous solutions of inhibitor compositions consisting of molecules of organosilanes and corrosion inhibitors. Cataphoresis varnishes should be considered as direct analogues of this method; however, unlike polymeric inhibited films formed from aqueous INFOR solutions, cataphoresis varnishes do not always have satisfactory adhesion strength and good protective capacity.

Author Contributions: Conceptualization, N.A.S.; investigation, N.A.S.; writing—original draft preparation, N.A.S. and V.V.D.; writing—review and editing, N.A.S. and V.V.D.; funding acquisition, V.V.D. All authors have read and agreed to the published version of the manuscript.

Funding: This research was funded by the Ministry of Science and Higher Education of the Russian Federation (122011300078-1).

Institutional Review Board Statement: Not applicable.

Informed Consent Statement: Not applicable.

Data Availability Statement: Not applicable.

Acknowledgments: The authors would like to thank the Centre for Collective Use of Scientific Equipment of the Frumkin Institute of Physical chemistry and Electrochemistry at the Russian academy of Sciences (IPCE RAS, Russia) for some of the measurements taken using the Centre equipment.

Conflicts of Interest: The authors declare no conflict of interest.

References

1. *GOST 5272–68*; Corrosion of Metals. Terms. In Term 8. Application: Zh; IPK Publishing Standards: Moscow, Russia, 1999; p. 15.
2. *GOST 9.014–75*; Unified System of Corrosion and Ageing Protection. Temporary Corrosion Protection of Products. General Requirements. IPK Publishing Standards: Moscow, Russia, 2005; p. 43.
3. *GOST 9.054–75*; Unified System of Corrosion and Ageing Protection. Anticorrosive Oils, Greases and Inhibited Film-Forming Petroleum Compounds. Accelerated Test Methods of Protective Property. IPK Publishing Standards: Moscow, Russia, 2006; p. 11.
4. *ISO 8044:2020*; Corrosion of Metals and Alloys—Vocabulary. ISO copyright office: Geneva, Switzerland, 2020; p. 29.
5. Goncharova, O.A.; Luchkin, A.Y.; Kuznetsov, Y.I.; Andreev, N.N. Vapor-Phase Protection of Zinc from Atmospheric Corrosion by Low-Volatile Corrosion Inhibitors. *Prot. Met. Phys. Chem. Surf.* **2019**, *55*, 1299–1303. [CrossRef]
6. Askari, M.; Aliofkhazraei, M.; Jafari, R.; Hamghalam, P.; Hajizadeh, A. Downhole corrosion inhibitors for oil and gas production–A review. *Appl. Surf. Sci. Adv.* **2021**, *6*, 1–23. [CrossRef]
7. Goncharova, O.A.; Kuznetsov, D.S.; Andreev, N.N.; Kuznetsov, Y.I.; Andreeva, N.P. Chamber Inhibitors of Corrosion of AMg6 Aluminum Alloy. *Prot. Met. Phys. Chem. Surf.* **2020**, *56*, 1293–1298. [CrossRef]
8. Andreev, N.N.; Goncharova, O.A.; Vesely, S.S. Volatile inhibitors of atmospheric corrosion. IV. Evolution of vapor-phase protection in the light of patent literature. *Int. J. Corros. Scale Inhib.* **2013**, *2*, 162–193. [CrossRef]
9. Luchkin, Y.I.; Goncharova, O.A.; Andreev, N.N. Mixture inhibitors. Mutual influence of components. *Korroz. Mater. Zashchita.* **2021**, 27–32. [CrossRef]
10. Goncharova, O.A.; Luchkin, Y.I.; Kuznetsov, Y.I.; Andreev, N.N.; Andreeva, N.P.; Vesely, S.S. Octadecylamine, 1,2,3-benzotriazole and a mixture thereof as chamber inhibitors of steel corrosion. *Int. J. Corros. Scale Inhib.* **2018**, *7*, 203–212. [CrossRef]
11. Kuznetsov, Y.I. *Organic Inhibitors of Corrosion of Metals*; Springer Science + Business Media: Boston, MA, USA, 1996; p. 283.
12. Kuznetsov, Y.I. Organic corrosion inhibitors: Where are we now? A review. Part IV. Passivation and the role of mono- and diphosphonates. *Int. J. Corros. Scale Inhib.* **2017**, *6*, 384–427. [CrossRef]
13. Kuznetsov, Y.I.; Andreev, N.N.; Marshakov, A.I. Physicochemical Aspects of Metal Corrosion Inhibition. *Russ. J. Phys. Chem. A.* **2020**, *94*, 505–515. [CrossRef]
14. Makarychev, Y.; Gladkikh, N.; Arkhipushkin, I.; Kuznetsov, Y. Corrosion Inhibition of Low-Carbon Steel by Hydrophobic Organosilicon Dispersions. *Metals* **2021**, *11*, 1269. [CrossRef]
15. Gladkikh, N.; Makarychev, Y.; Petrunin, M.; Maleeva, M.; Maksaeva, L.; Marshakov, A. Synergistic effect of silanes and azole for enhanced corrosion protection of carbon steel by polymeric coatings. *Prog. Org. Coat.* **2020**, *138*, 105386. [CrossRef]
16. Kuznetsov, Y.I.; Red'kina, G.V. Thin Protective Coatings on Metals Formed by Organic Corrosion Inhibitors in Neutral Media. *Coatings* **2022**, *12*, 149. [CrossRef]
17. Abd El Wanees, S.; Bahgat Radwan, A.; Alsharif, M.A.; Abd El Haleem, S.M. Initiation and inhibition of pitting corrosion on reinforcing steel under natural corrosion conditions. *Mater. Chem. Phys.* **2017**, *190*, 79–95. [CrossRef]
18. Redkina, G.V.; Kuznetsov, Y.I.; Andreeva, N.P.; Arkhipushkin, I.A.; Kazansky, L.P. Features of zinc passivation by sodiumdodecylphosphonate in a neutral aqueous solution. *Corros. Sci.* **2020**, *168*, 108554. [CrossRef]
19. Franco, J.P.; Ribeiro, J. 1-Hydroxyethylidene-1,1-diphosphonic Acid (HEDP) as a Corrosion Inhibitor of AISI 304 Stainless Steel in a Medium Containing Chloride and Sulfide Ions in the Presence of Different Metallic Cations. *Adv. Chem. Eng. Sci.* **2020**, *10*, 225–257. [CrossRef]
20. *Corrosion Inhibitors: Principles, Mechanisms and Applications*; Hart, E. (Ed.) Nova Science Publishers Incorporated: Hauppauge, NY, USA, 2016; p. 173.
21. Aramaki, K. Effects of organic inhibitors on corrosion of zinc in an aerated 0.5 M NaCl solution. *Corros. Sci.* **2001**, *43*, 1985–2000. [CrossRef]
22. Kozlova, L.S.; Sibileva, S.V.; Chesnokov, D.V.; Kutyrev, A.E. Corrosion inhibitors (review). *Aviatsionnyye Mater. I Tekhnologii* **2015**, *35*, 67–75. [CrossRef]
23. Asmara, Y.P.; Kurniawan, T. Corrosion Prediction for Corrosion Rate of Carbon Steel in Oil and Gas Environment: A Review. *Indones. J. Sci. Technol.* **2018**, *3*, 64–74. [CrossRef]
24. Kuznetsov, Y.I. Physicochemical aspects of metal corrosion inhibition in aqueous solutions. *Russ. Chem. Rev.* **2004**, *73*, 75–87. [CrossRef]
25. Kuznetsov, Y.I.; Kazansky, L.P. Physicochemical aspects by azoles as corrosion inhibitors. *Russ. Chem. Rev.* **2008**, *77*, 219–232. [CrossRef]
26. Kuznetsov, Y.I. Role of the complexation concept in the present views on the initiation and inhibition of metal pitting. *Prot. Met.* **2001**, *37*, 485–490.
27. Nguyen, T.D.; Nguyen, S.A.; Tran, B.A.; Vu, K.O.; Tran, D.L.; Phan, T.T.; Scharnagl, N.; Zheludkevich, M.L.; Hang, T.X.T. Molybdate intercalated hydrotalcite/graphene oxide composite as corrosion inhibitor for carbon steel. *Surf. Coat. Technol.* **2020**, *399*, 1–12. [CrossRef]
28. Osipenko, M.A.; Kharytonau, D.S.; Kasach, A.A.; Ryl, J.; Adamiec, J.; Kurilo, I.I. Inhibitive effect of sodium molybdate on corrosion of AZ31 magnesium alloy in chloride solutions. *Electrochim. Acta* **2022**, *410*, 140175. [CrossRef]
29. Coelho, L.B.; Fava, E.B.; Kooijman, A.M.; Gonzalez-Garcia, Y.; Olivier, M.-G. Molybdate as corrosion inhibitor for hot dip galvanised steel scribed to the substrate: A study based on global and localised electrochemical approaches. *Corros. Sci.* **2020**, *175*, 108893. [CrossRef]
30. Moutarlier, V.; Gigandet, M.P.; Pagetti, J.; Ricq, L. Molybdate/sulfuric acid anodising of 2024-aluminium alloy: Influence of inhibitor concentration on film growth and on corrosion resistance. *Surf. Coat. Technol.* **2003**, *173*, 87–95. [CrossRef]

31. Wang, Y.-Q.; Kong, G.; Che, C.-S.; Zhang, B. Inhibitive effect of sodium molybdate on the corrosion behavior of galvanized steel in simulated concrete pore solution. *Constr. Build. Mater.* **2018**, *162*, 383–392. [CrossRef]
32. Alentejano, C.R.; Aoki, I.V. Localized corrosion inhibition of 304 stainless steel in pure water by oxyanions tungstate and molybdate. *Electrochim. Acta* **2004**, *49*, 2779–2785. [CrossRef]
33. Anil Kumar, G.N.; Shruthi, D.L. The nature of the chemical bond in sodium tungstate based on ab initio, DFT and QTAIM topological analysis of electron density. *Mater. Today* **2021**, *44*, 3127–3132. [CrossRef]
34. Mu, G.; Li, X.; Qu, Q.; Zhou, J. Molybdate and tungstate as corrosion inhibitors for cold rolling steel in hydrochloric acid solution. *Corros. Sci.* **2006**, *48*, 445–459. [CrossRef]
35. Qu, Q.; Li, L.; Bai, W.; Jiang, S.; Ding, Z. Sodium tungstate as a corrosion inhibitor of cold rolled steel in peracetic acid solution. *Corros. Sci.* **2009**, *51*, 2423–2428. [CrossRef]
36. Tamborim Takeuchi, S.M.; Azambuja, D.S.; Saliba-Silva, A.M.; Costa, I. Corrosion protection of NdFeB magnets by phosphating with tungstate incorporation. *Surf. Coat. Technol.* **2006**, *200*, 6826–6831. [CrossRef]
37. Tsai, C.-Y.; Liu, J.-S.; Chen, P.-L.; Lin, C.-S. A roll coating tungstate passivation treatment for hot-dip galvanized sheet steel. *Surf. Coat. Technol.* **2011**, *205*, 5124–5129. [CrossRef]
38. Petrunin, M.; Maksaeva, L.; Gladkikh, N.; Makarychev, Y.; Maleeva, M.; Yurasova, T.; Nazarov, A. Thin Benzotriazole films for Inhibition of Carbon Steel Corrosion in Neutral Electrolytes. *Coatings* **2020**, *10*, 362. [CrossRef]
39. Luchkin, A.Y.; Goncharova, O.A.; Arkhipushkin, I.A.; Andreev, N.N.; Kuznetsov, Y.I. The effect of oxide and adsorption layers formed in 5-Chlorobenzotriazole vapors on the corrosion resistance of copper. *J. Taiwan Inst. Chem. Eng.* **2020**, *117*, 231–241. [CrossRef]
40. Arkhipushkin, I.A.; Agafonkina, M.O.; Kazansky, L.P.; Kuznetsov, Y.I.; Shikhaliev, K.S. Characterization of adsorption of 5-carboxy-3-amino-1,2,4-triazole towards copper corrosion prevention in neutral media. *Electrochim. Acta* **2019**, *308*, 392–399. [CrossRef]
41. Kuznetsov, Y.I.; Shikhaliev, K.S.; Agafonkina, M.O.; Andreeva, N.P.; Arkhipushkin, I.A.; Potapov, A.Y.; Kazansky, L.P. Effect of substituents in 5-R-3-amino-1,2,4-triazoles on the chemisorption on copper surface in neutral media. *Corros. Eng. Sci. Technol.* **2021**, *56*, 60–70. [CrossRef]
42. Luchkin, A.Y.; Goncharova, O.A.; Andreev, N.N.; Arkhipushkin, I.A.; Kazanskii, L.P.; Kuznetsov, Y.I. 5-Chloro-1,2,3-benzotriazole as a Chamber Corrosion Inhibitor for the MA8 Magnesium Alloy. *Prot. Met. Phys. Chem. Surf.* **2021**, *57*, 1319–1327. [CrossRef]
43. Kazansky, L.P.; Pronin, Y.E.; Arkhipushkin, I.A. XPS study of adsorption of 2-mercaptobenzothiazole on a brass surface. *Corros. Sci.* **2014**, *89*, 21–29. [CrossRef]
44. Abd El Haleem, S.M.; Abd El Wanees, S.; Bahgat, A. Environmental factors affecting the corrosion behaviour of reinforcing steel. VI. Benzotriazole and its derivatives as corrosion inhibitors of steel. *Corros. Sci.* **2014**, *87*, 321–333. [CrossRef]
45. Mennucci, M.M.; Banczek, E.P.; Rodrigues, P.R.P.; Costa, I. Evaluation of benzotriazole as corrosion inhibitor for carbon steel in simulated pore solution. *Cem. Concr. Compos.* **2009**, *31*, 418–424. [CrossRef]
46. Niu, L.; Cao, C.N.; Lin, H.C.; Song, G.L. Inhibitive effect of benzotriazole on the stress corrosion cracking of 18cr-9ni-ti stainless steel in acidic chloride solution. *Corros. Sci.* **1998**, *40*, 1109–1117. [CrossRef]
47. Onyeachu, I.B.; Solomon, M.M. Benzotriazole derivative as an effective corrosion inhibitor for low carbon steel in 1 M HCl and 1 M HCl + 3.5 wt% NaCl solutions. *J. Mol. Liq.* **2020**, *313*, 113536. [CrossRef]
48. Rodriguez, J.; Mouanga, M.; Roobroeck, A.; Cossement, D.; Mirisola, A.; Olivier, M.G. Study of the inhibition ability of benzotriazole on the Zn-Mg coated steel corrosion in chloride electrolyte. *Corros. Sci.* **2018**, *132*, 56–67. [CrossRef]
49. Wint, N.; Griffiths, C.M.; Richards, C.J.; Williams, G.; Mcmurray, H.N. The role of benzotriazole modified zinc phosphate in preventing corrosion-driven organic coating disbondment on galvanised steel. *Corros. Sci.* **2020**, *174*, 108839. [CrossRef]
50. Srinivasa Rao, S.; Roopas Kiran, S.; Chaitanya Kumar, K.; Diwakar, B.S. Electrochemical behaviour of interface of carbon steel/solution containing three-component formulations. *Mater. Today* **2019**, *18*, 2003–2011. [CrossRef]
51. Zhu, J.; Wang, S.; Li, H.; Qian, J.; Lv, L.; Pan, B. Degradation of phosphonates in Co(II)/peroxymonosulfate process: Performance and mechanism. *Water Res.* **2021**, *202*, 117397. [CrossRef]
52. Moschona, A.; Plesu, N.; Mezei, G.; Thomas, A.G.; Demadis, K.D. Corrosion protection of carbon steel by tetraphosphonates of systematically different molecular size. *Corros. Sci.* **2018**, *145*, 135–150. [CrossRef]
53. Labjar, N.; Lebrini, M.; Bentiss, F.; Chihib, N.-E.; Hajjaji, S.E.; Jama, C. Corrosion inhibition of carbon steel and antibacterial properties of aminotris-(methylenephosphonic) acid. *Mater. Chem. Phys.* **2010**, *119*, 330–336. [CrossRef]
54. Somov, N.V.; Chausov, F.F.; Kazantseva, I.S.; Vorob'yov, V.L.; Shumilova, M.A.; Maratkanova, A.N. Cerium(III) chelate complex with monoprotonated nitrilo-tris(methylenephosphonic) acid: Structure and chemical bonding. *J. Mol. Struct.* **2022**, *1270*, 133935. [CrossRef]
55. Srinivasa Rao, S.; Chaitanya Kumar, K.; Roopas Kiran, S.; Diwakar, B.S. Protective behaviour of two phosphonate-based inhibitor systems containing lactobionic acid in corrosion control of carbon steel. *Mater. Today* **2022**, *49*, 588–592. [CrossRef]
56. Hejjaj, C.; Ait Aghzzaf, A.; Bouali, I.; Hakkou, R.; Dahbi, M.; Fischer, C.B. Layered aluminum tri-polyphosphate as intercalation host for 6-aminohexanoic acid – Synthesis, characterization and application as corrosion protection inhibitor for low carbon steel. *Corros. Sci.* **2021**, *181*, 109239. [CrossRef]
57. Lebrini, M.; Bentiss, F.; Chihib, N.-E.; Jama, C.; Hornez, J.P.; Lagrenée, M. Polyphosphate derivatives of guanidine and urea copolymer: Inhibiting corrosion effect of Armco iron in acid solution and antibacterial activity. *Corros. Sci.* **2008**, *50*, 2914–2918. [CrossRef]
58. Naderi, R.; Attar, M.M. Electrochemical assessing corrosion inhibiting effects of zinc aluminum polyphosphate (ZAPP) as a modified zinc phosphate pigment. *Electrochim. Acta* **2008**, *53*, 5692–5696. [CrossRef]
59. Soror, T. Scale and Corrosion Prevention in Cooling Water Systems Part I: Calcium Carbonate. *Open Corros. J.* **2009**, *2*, 45–50. [CrossRef]

60. Farhat, T.; Schlenoff, J. Corrosion Control Using Polyelectrolyte Multilayers. *ESL* **2002**, *5*, B13–B15. [CrossRef]
61. Larson, G.L. Silicon- the silicon-carbon bond: Annual survey for the year 1987. *J. Organomet. Chem.* **1989**, *374*, 1–347. [CrossRef]
62. Plueddenmann, E.P. *Silane Coupling Agents*, 2nd ed.; Plenum Press: New York, NY, USA, 1991; pp. 79–152.
63. Subramanian, V.; van Ooij, W.J. Effect of the amine functional group on corrosion rate of iron coated with films of organofunctional silanes. *Corrosion* **1998**, *54*, 204–215. [CrossRef]
64. Bažant, V.; Chvalovský, V.; Rathouský, J. *Organosilicon Compounds*; Publishing House of the Czechoslovak Academy of Sciences: New York, NY, USA; Academic Press: Prague, Czech Republic, 1965; p. 587.
65. Moriguchi, K.; Utagava, S. *Silane: Chemistry, Applications, and Performance*; Nova Science Publishers Incorporated: New York, NY, USA, 2013; p. 176.
66. Petrunin, M.A.; Gladkikh, N.A.; Maleeva, M.A.; Maksaeva, L.B.; Yurasova, T.A. The use of organosilanes to inhibit metal corrosion. A review. *Int. J. Corros. Scale Inhib.* **2019**, *8*, 882–907. [CrossRef]
67. Maleeva, M.A.; Ignatenko, V.E.; Shapagin, A.V.; Sherbina, A.A.; Maksaeva, L.B.; Marshakov, A.I.; Petrunin, M.A. Modification of bituminous coatings to prevent stress corrosion cracking of carbon steel. *Int. J. Corros. Scale Inhib.* **2015**, *4*, 226–234. [CrossRef]
68. Kuznetsov, Y.I.; Semiletov, A.M.; Chirkunov, A.A.; Arkhipushkin, I.A.; Kazanskii, L.P.; Andreeva, N.P. Protecting Aluminum from Atmospheric Corrosion via Surface Hydrophobization with Stearic Acid and Trialkoxysilanes. *Russ. J. Phys. Chem. A.* **2018**, *92*, 621–629. [CrossRef]
69. Semiletov, A.M.; Kuznetsov, Y.I.; Chirkunov, A.A. Surface Modification of Aluminum Alloys by Two-Stage Passivation in Solutions of Vinyltrimethoxysilane and Organic Inhibitors. *Prot. Met. Phys. Chem. Surf.* **2019**, *55*, 1311–1316. [CrossRef]
70. Semiletov, A.M.; Chirkunov, A.A.; Kuznetsov, Y.I. Protection of aluminum alloy AD31 from corrosion by adsorption layers of trialkoxysilanes and stearic acid. *Mater. Corros.* **2020**, *71*, 77–85. [CrossRef]
71. Ngo, D.T.; Sooknoi, T.; Resasco, D.E. Improving stability of cyclopentanone aldol condensation MgO-based catalysts by surface hydrophobization with organosilanes. *Appl. Catal. B* **2018**, *237*, 835–843. [CrossRef]
72. Rahimipour, S.; Rafiei, B.; Salahinejad, E. Organosilane-functionalized hydrothermal-derived coatings on titanium alloys for hydrophobization and corrosion protection. *Prog. Org. Coat.* **2020**, *142*, 105594. [CrossRef]
73. Ishida, H.; Koenig, J.L. Vibrational Assignments of Organosilanetriols. I. Vinylsilanetriol and Vinylsilanetriol-d3 in Aqueous Solutions. *Appl. Spectrosc.* **1978**, *32*, 462–469. [CrossRef]
74. Osterholtz, F.D.; Pohl, E.R. Kinetics of the hydrolysis and condensation of organofunctional alkoxysilanes—A review. *J. Adhes. Sci. Technol.* **1992**, *6*, 127–149. [CrossRef]
75. Metwalli, E.; Haines, D.; Becker, O.; Conzone, S.; Pantano, C.G. Surface characterizations of mono-, di-, and tri-aminosilane treated glass substrates. *J. Colloid Interface Sci.* **2006**, *298*, 825–831. [CrossRef]
76. Volkis, V.; Averbuj, C.; Eisen, M.S. Reactivity of group 4 benzamidinate complexes towards mono- and bis-substituted silanes and 1,5-hexadiene. *J. Organomet. Chem.* **2007**, *692*, 1940–1950. [CrossRef]
77. Uneyama, K. Functionalized fluoroalkyl and alkenyl silanes: Preparations, reactions, and synthetic applications. *J. Fluorine Chem.* **2008**, *129*, 550–576. [CrossRef]
78. Singh, G.; Sushma; Singh, A.; Priyanka; Chowdhary, K.; Singh, J.; Esteban, M.A.; Espinosa-Ruíz, C.; González-Silvera, D. Designing of chalcone functionalized 1,2,3-triazole allied bis-organosilanes as potent antioxidants and optical sensor for recognition of Sn^{2+} and Hg^{2+} ions. *J. Organomet. Chem.* **2021**, *953*, 122049. [CrossRef]
79. Arkles, B.; Steinmetz, J.R.; Zazyczny, J.; Mehta, P. Factors contributing to the stability of alkoxysilanes in aqueous solution. *J. Adhes. Sci. Technol.* **1992**, *6*, 193–206. [CrossRef]
80. Pohl, E.R.; Chaves, A. Sterically hindered silanes for waterborne systems: A model study of silane hydrolysis. In *Silanes and Other Coupling Agents*; CRC Press: Boca Raton, FL, USA, 2004; Volume 2, pp. 3–9.
81. Silvestro, L.; dos Santos Lima, G.T.; Ruviaro, A.S.; de Matos, P.R.; Mezalira, D.Z.; Gleize, P.J.P. Evaluation of different organosilanes on multi-walled carbon nanotubes functionalization for application in cementitious composites. *J. Build. Eng.* **2022**, *51*, 104292. [CrossRef]
82. Wang, L.; Sanders, J.E.; Gardner, D.G.; Han, Y. In-situ modification of cellulose nanofibrils by organosilanes during spray drying. *Ind. Crops Prod.* **2016**, *93*, 129–135. [CrossRef]
83. Arkles, B. *Silane Coupling Agents Connecting Across Boundaries*; Gelets, Inc.: Morrisville, PA, USA, 2006; p. 302.
84. Fedel, M.; Olivier, M.; Poelman, M.; Deflorian, F.; Rossi, S.; Druart, M.E. Corrosion protection properties of silane pre–treated powder coated galvanized steel. *Prog. Org. Coat.* **2009**, *66*, 118–128. [CrossRef]
85. Puomi, P.; Fagerholm, H.M. Performance of silane treated primed hot–dip galvanised steel. *Anti–Corros. Method M.* **2001**, *48*, 7–17. [CrossRef]
86. Avdeev, Y.G.; Tyurina, M.V.; Kuznetsov, Y.I. Protection of low-carbon steel in phosphoric acid solutionsby mixtures of a substituted triazole with sulfur-containing compounds. *Int. J. Corros. Scale Inhib.* **2014**, *3*, 246–253. [CrossRef]
87. Haasnoot, J.G. Mononuclear, oligonuclear and polynuclear metal coordination compounds with 1,2,4-triazole derivatives as ligands. *Coord. Chem. Rev.* **2000**, *200–202*, 131–185. [CrossRef]
88. Lu, G.; Zangari, G. Investigations of the effect of chromate conversion coatings on the corrosion resistance of Ni-based alloys. *Electrochim. Acta* **2004**, *49*, 1461–1473. [CrossRef]
89. Mekhalif, Z.; Delhalle, J. Investigation of the protective action of chromate coatings on hot-dip galvanized steel: Role of wetting agents. *Corros. Sci.* **2005**, *47*, 547–566. [CrossRef]

90. Gladkikh, N.; Makarychev, Y.; Maleeva, M.; Petrunin, M.; Maksaeva, L.; Rybkina, A.; Marshakov, A.; Kuznetsov, Y. Synthesis of thin organic layers containing silane coupling agents and azole on the surface of mild steel. Synergism of inhibitors for corrosion protection of underground pipelines. *Prog. Org. Coat.* **2019**, *132*, 481–489. [CrossRef]
91. Gladkikh, N.; Makarychev, Y.; Chirkunov, A.; Shapagin, A.; Petrunin, M.; Maksaeva, L.; Maleeva, M.; Yurasova, T.; Marshakov, A. Formation of polymer-like anticorrosive films based on organosilanes with benzotriazole, carboxylic and phosphonic acids. Protection of copper and steel against atmospheric corrosion. *Prog. Org. Coat.* **2020**, *141*, 105544. [CrossRef]
92. Ulman, A. Formation and structure of self-assembled monolayers. *Chem. Rev.* **1996**, *96*, 1533–1534. [CrossRef]
93. Reiss, F.F. *Zametka o Novom Deystvii Gal'vanicheskogo Elektrichestva (1808). Izbrannyye Trudy po Elektrichestvu*; State Publishing House of Technical Theoretical Literature: Moscow, Russia, 1956; pp. 159–168.
94. Hamaker, H.C. Formation of a deposit by electrophoresis. *Trans. Faraday Soc.* **1940**, *35*, 279–287. [CrossRef]
95. Wei, M.; Ruys, A.J.; Milthorpe, B.K.; Sorrell, C.C. Solution ripening of hydroxyapatite nanoparticles: Effects on electrophoretic deposition. *J. Biomed. Mater. Res.* **1999**, *45*, 11–19. [CrossRef]
96. Zaitseva, E.A. Deutsche an der Moskauer Universität im 19. Jahrhundert: Ferdinand Friedrich v.Reuss (1778–1852). In *Deutsch-Russische Beziehungen in Medizin und Naturwissenschaften, D.v.Engelhardt u. I.Kästner (Hgg)*; Shaker Verlag: Aachen, Germany, 2001; pp. s.209–s.226.
97. Dukhin, S.S.; Deryagin, B.V. *Elektroforez*; Publishing House Nauka: Moscow, Russia, 1974; 332p.
98. Deynega, Y.F. Printsipy formirovaniya kompozitsionnykh elektroforezo- elektrokhimicheskikh pokrytiy. *Ukr. Khim. Zhurnal.* **1980**, *46*, 1016–1023.
99. Besra, L.; Liu, M. A review on fundamentals and applications of electrophoretic deposition (EPD). *Prog. Mater. Sci.* **2007**, *52*, 1–61. [CrossRef]
100. Gurrappa, I.; Binder, L. Electrodeposition of nanostructured coatings and their characterization—A review. *Sci. Technol. Adv. Mater.* **2008**, *9*, 043001. [CrossRef]
101. Van der Biest, O.O.; Vandeperre, L.J. Electrophoretic deposition of materials. *Annu. Rev. Mater. Sci.* **1999**, *29*, 327–352. [CrossRef]
102. Boccaccini, A.R.; Zhitomirsky, I. Application of electrophoretic and electrolytic deposition techniques in ceramics processing. *Curr. Opin. Solid State Mater. Sci.* **2002**, *6*, 251–260. [CrossRef]
103. Biest Van Der, O.; Put, S.; Anné, G.; Vleugels, J. Electrophoretic deposition for coatings and free standing objects. *J. Mater. Sci.* **2004**, *39*, 779–785. [CrossRef]
104. Hanaor, D.; Michelazzi, M.; Veronesi, P.; Leonelli, C.; Romagnoli, M.; Sorrell, C. Anodic aqueous electrophoretic deposition of titanium dioxide using carboxylic acids as dispersing agents. *J. Eur. Ceram. Soc.* **2011**, *31*, 1041–1047. [CrossRef]
105. Farrokhi-Rad, M.; Shahrabi, T. Effect of suspension medium on the electrophoretic deposition of hydroxyapatite nanoparticles and properties of obtained coatings. *Ceram. Int.* **2014**, *40*, 3031–3039. [CrossRef]
106. Chávez-Valdez, A.; Boccaccini, A.R. Innovations in electrophoretic deposition: Alternating current and pulsed direct current methods. *Electrochim. Acta* **2012**, *65*, 70–89. [CrossRef]
107. Babaei, N.; Yeganeh, H.; Gharibi, R. Anticorrosive and self-healing waterborne poly(urethane-triazole) coatings made through a combination of click polymerization and cathodic electrophoretic deposition. *Eur. Polym. J.* **2019**, *112*, 636–647. [CrossRef]
108. Fukada, Y.; Nagarajan, N.; Mekky, W.; Bao, Y.; Kim, H.S.; Nicholson, P.S. Electrophoretic deposition – mechanisms, myths and materials. *J. Mater. Sci.* **2004**, *39*, 787–801. [CrossRef]
109. Almeida, E.; Alves, I.; Brites, C.; Fedrizzi, L. Cataphoretic and autophoretic automotive primers: A comparative study. *Prog. Org. Coat.* **2003**, *46*, 8–20. [CrossRef]
110. Freeman, D.B. *Phosphating and Metal Pretreatment*, 1st ed.; Woodhead-Faulkner in Association with Pyrene Chemical Services Ltd.: Cambridge, UK, 1986; 229p.
111. Brenner, A. *Electrodeposition of Alloys: Principles and Practice*, 1st ed.; Academic Press: New York, NY, USA, 1963; 734p.
112. West, J.M. *Electrodeposition and Corrosion Processes*, 2nd ed.; Van Nostrand: London, UK, 1965; 189p.
113. Boccaccini, A.R.; Dickerson, J.H. Electrophoretic deposition: Fundamentals and applications. *J. Phys. Chem. B.* **2013**, *117*, 1501. [CrossRef] [PubMed]
114. Brown, D.R.; Salt, F.W. The mechanism of electrophoretic deposition. *J. Appl. Chem.* **1965**, *15*, 40–48. [CrossRef]
115. Ferrari, B.; Moreno, R. EPD kinetics: A review. *J. Eur. Ceram. Soc.* **2010**, *30*, 1069–1078. [CrossRef]
116. Chng, E.; Watson, A.; Suresh, V.; Fujiwara, T.; Bumgardner, J.; Gopalakrishnan, R. Adhesion of electrosprayed chitosan coatings using silane surface chemistry. *Thin Solid Film.* **2019**, *692*, 137454. [CrossRef]
117. Abele, L.; Jäger, A.K.; Schulz, W.; Ruck, S.; Riegel, H.; Sörgel, T.; Albrecht, J. Superoleophobic surfaces via functionalization of electrophoretic deposited SiO_2 spheres on smart aluminum substrates. *Appl. Surf. Sci.* **2019**, *490*, 56–60. [CrossRef]
118. Wua, L.; Zhang, J.T.; Hua, J.; Zhang, J.Q. Improved corrosion performance of electrophoretic coatings by silane addition. *Corros. Sci.* **2012**, *56*, 58–66. [CrossRef]
119. Zhu, R.; Zhang, J.; Chang, C.; Gao, S.; Ni, N. Effect of silane and zirconia on the thermal property of cathodic electrophoretic coating on AZ31 magnesium alloy. *J. Magnes. Alloys* **2013**, *1*, 235–241. [CrossRef]
120. Castro, Y.; Ferrari, B.; Moreno, R.; Duran, A. Coatings produced by electrophoretic deposition from nano-particulate silica sol–gel suspensions. *Surf. Coat. Technol.* **2004**, *182*, 199–203. [CrossRef]
121. Oltean, G.; Valvo, M.; Nyholm, L.; Edström, K. On the electrophoretic and sol–gel deposition of active materials on aluminium rod current collectors for three-dimensional Li-ion micro-batteries. *Thin Solid Film.* **2014**, *562*, 63–69. [CrossRef]

122. Yu, M.; Xue, B., Liu. Electrophoretic deposition of hybrid coatings on aluminum alloy by combining 3-aminopropyltrimethoxysilan to silicon–zirconium sol solutions for corrosion protection. *Thin Solid Film.* **2015**, *590*, 33–39. [CrossRef]
123. Hayati, Z.; Hoomehr, B.; Khalesi, F.; Raeissi, K. Synthesis and electrophoretic deposition of TiO_2-SiO_2 composite nanoparticles on stainless steel substrate. *J. Alloys Compd.* **2023**, *931*, 167619. [CrossRef]
124. Castro, Y.; Aparicio, M.; Moreno, R.; Duran, A. Silica-zirconia sol–gel coatings obtained by different synthesis routes. *J. Sol-Gel Sci. Technol.* **2005**, *35*, 41–50. [CrossRef]
125. Gladkikh, N.A.; Dushik, V.V.; Shaporenkov, A.A.; Shapagin, A.V.; Makarychev, Y.B.; Gordeev, A.V.; Marshakov, A.I. Water Suspension Containing Organosilan, Corrosion Inhibitor and Polycondensation Promoter and Method for Producing Protective Films on Surface of Tungsten and Coatings on Its Basis from Water Suspension Containing Organosilan, Corrosion Inhibitor and Polycondensation. RU2744336C1, 5 March 2021.
126. Morcillo, M.; Díaz, I.; Cano, H.; Chico, B.; Fuente, D. Atmospheric corrosion of weathering steels. Overview for engineers. Part II: Testing, inspection, maintenance. *Constr. Build Mater.* **2019**, *222*, 750–765. [CrossRef]
127. Bahadori, A. Chapter 1–Surface Preparation for Coating, Painting, and Lining. In *Essentials of Coating, Painting, and Lining for the Oil, Gas and Petrochemical Industries*; Bahadori, A., Ed.; Gulf Professional Publishing: Boston, MA, USA, 2015; pp. 1–105.
128. Bayliss, D.A. 12–Paint Coatings for the Plant Engineer. In *Plant Engineer's Handbook*; Mobley, R.K., Ed.; Butterworth-Heinemann: Woburn, MA, USA, 2001; pp. 147–160.
129. Ramdé, T.; Ecco, L.G.; Rossi, S. Visual appearance durability as function of natural and accelerated ageing of electrophoretic styrene-acrylic coatings: Influence of yellow pigment concentration. *Prog. Org. Coat.* **2017**, *103*, 23–32. [CrossRef]
130. Sharifalhoseini, Z.; Entezari, M.H.; Davoodi, A.; Shahidi, M. Surface modification of mild steel before acrylic resin coating by hybrid ZnO/GO nanostructures to improve the corrosion protection. *J. Ind. Eng. Chem.* **2020**, *83*, 333–342. [CrossRef]
131. Zhu, L.; Claude-Montigny, B.; Gattrell, M. Insulating method using cataphoretic paint for tungsten tips for electrochemical scanning tunnelling microscopy (ECSTM). *Appl. Surf. Sci.* **2005**, *252*, 1833–1845. [CrossRef]
132. Ranjbar, Z.; Rastegar, S. Influence of co-solvent content on electro-deposition behavior of acrylic lattices of different glass transition temperatures. *Prog. Org. Coat.* **2006**, *57*, 365–370. [CrossRef]
133. Padash, F.; Dorff, B.; Liu, W.; Ellwood, K.; Okerberg, B.; Zawacky, S.R.; Harb, J.N. Characterization of initial film formation during cathodic electrodeposition of coatings. *Prog. Org. Coat.* **2019**, *133*, 395–405. [CrossRef]
134. Leca, M.; Micutz, M.; Serban, R. Stable aqueous dispersions of some cataphoretically applicable film-forming resins. *Prog. Org. Coat.* **1997**, *30*, 241–245. [CrossRef]
135. San José García, G.; Fobbe, H. The influence of reduced pressures on the film formation of cathodic electrodeposition paints. *Prog. Org. Coat.* **2017**, *104*, 110–117. [CrossRef]
136. Vertyachikh, I.M.; Voronezhtsev, J.I.; Goldade, V.A.; Pinchuk, L.S.; Rechits, G.V.; Liberman, S.Y. Method of making sleeve inhibited polyethylene film. WO1986007004 23 May 1985.
137. Damiano, F. Corrosion Inhibiting Protective Sleeves. US20050118375A1, 26 February 2013.
138. Vido, M. Packaging Material for Metal. CA2390278C, 19 April 2011.
139. Allen, W.M. Corrosion Inhibiting Protective Foam Packaging. US20090111901A1, 29 October 2007.
140. Biegańska, B.; Zubielewicz, M.; Śmieszek, E. Anticorrosive water-borne paints. *Prog. Org. Coat.* **1987**, *15*, 33–56. [CrossRef]
141. Kobayashi, T. Pigment dispersion in water-reducible paints. *Prog. Org. Coat.* **1996**, *28*, 79–87. [CrossRef]
142. Larson, R.G.; Van Dyk, A.K.; Chatterjee, T.; Ginzburg, V.V. Associative thickeners for waterborne paints: Structure, characterization, rheology, and modeling. *Prog. Polym. Sci.* **2022**, *129*, 101546. [CrossRef]
143. Malshe, V.C. Paints: Water-Based. In *Reference Module in Chemistry, Molecular Sciences and Chemical Engineering*; Elsevier: Amsterdam, The Netherlands, 2019; pp. 1–157.
144. Martinez, M.; Gámez, E.; Bellotti, N.; Deyá, C. Alkyd based water-reducible anticorrosive paints and their antifungal potential. *Prog. Org. Coat.* **2021**, *152*, 106069. [CrossRef]
145. Araujo, W.S.; Margarit, I.C.P.; Mattos, O.R.; Fragata, F.L.; de Lima-Neto, P. Corrosion aspects of alkyd paints modified with linseed and soy oils. *Electrochim. Acta* **2010**, *55*, 6204–6211. [CrossRef]
146. Duce, C.; Bernazzani, L.; Bramanti, E.; Spepi, A.; Colombini, M.P.; Tiné, M.R. Alkyd artists' paints: Do pigments affect the stability of the resin? A TG and DSC study on fast-drying oil colours. *Polym. Degrad. Stab.* **2014**, *105*, 48–58. [CrossRef]
147. İşeri-Çağlar, D.; Baştürk, E.; Oktay, B.; Kahraman, M.V. Preparation and evaluation of linseed oil based alkyd paints. *Prog. Org. Coat.* **2014**, *77*, 81–86. [CrossRef]
148. Udell, N.A.; Hodgkins, R.E.; Berrie, B.H.; Meldrum, T. Physical and chemical properties of traditional and water-mixable oil paints assessed using single-sided NMR. *Microchem. J.* **2017**, *133*, 31–36. [CrossRef]
149. Otabor, G.O.; Ifijen, I.H.; Mohammed, F.U.; Aigbodion, A.I.; Ikhuoria, E.U. Alkyd resin from rubber seed oil/linseed oil blend: A comparative study of the physiochemical properties. *Heliyon* **2019**, *5*, e01621. [CrossRef]
150. Fujisawa, N.; Bronken, I.A.T.; Freeman, A.A.; Łukomski, M. Nanoindentation of softening modern oil paints. *Int. J. Solids Struct.* **2022**, 112009. [CrossRef]
151. Jeffs, R.A.; Jones, W. Additives for paint. In *Paint and Surface Coatings*, 2nd ed.; Lambourne, R., Strivens, T.A., Eds.; Woodhead Publishing: Sawston, UK, 1999; pp. 185–197.

152. Lambourne, R. Solvents, thinners, and diluents. In *Paint and Surface Coating*, 2nd ed.; Lambourne, R., Strivens, T.A., Eds.; Woodhead Publishing: Sawston, UK, 1999; pp. 166–184.
153. Hayward, G.R. Health and safety in the coatings industry. In *Paint and Surface Coatings*, 2nd ed.; Lambourne, R., Strivens, T.A., Eds.; Woodhead Publishing: Sawston, UK, 1999; pp. 725–766.

MDPI

Article

N⁺-Implantation on Nb Coating as Protective Layer for Metal Bipolar Plate in PEMFCs and Their Electrochemical Characteristics

Yu-Sung Kim [1], Jin-Young Choi [1], Cheong-Ha Kim [1], In-Sik Lee [1], Shinhee Jun [2], Daeil Kim [3], Byung-Chul Cha [1,*] and Dae-Wook Kim [1,*]

[1] Advanced Manufacturing Process R & D Group, Ulsan Regional Division, Korea Institute of Industrial Technology (KITECH), 55, Jongga-ro, Jung-gu, Ulsan 44313, Republic of Korea
[2] DAE-IL Co., 8, Bonggyenonggong-gil, Ulju-gun, Ulsan 44914, Republic of Korea
[3] School of Materials Science & Engineering, University of Ulsan, 55-12, Techno Saneop-ro, Nam-gu, Ulsan 44776, Republic of Korea
* Correspondence: bccha76@kitech.re.kr (B.-C.C.); dwkim@kitech.re.kr (D.-W.K.)

Abstract: Nitrogen ions were implanted into the coated Nb layer by plasma immersion ion implantation to improve resistance to corrosion of a metal bipolar plate. Due to nitrogen implantation, the corrosion behavior of the Nb layer was enhanced. The electron microscope observation reveals that the microstructure of the Nb layer became denser and had fewer defects with increasing implantation energy. As a result, the densified structure effectively prevented direct contact with the corrosive electrolyte. In addition, at a higher implantation rate (6.40×10^{17} N_2/cm^2), a thin amorphous layer was formed on the surface, and the implanted nitrogen ions reacted at neighboring Nb sites, resulting in the localized formation of nitrides. Such phase and structural changes contributed to further improve corrosion resistance. In particular, the implanted Nb layer at bias voltage of 10 kV exhibited a current density more than one order of magnitude smaller with a two times faster stabilization than the as-deposited Nb layer under the PEMFC operating conditions.

Keywords: plasma immersion ion implantation; niobium; bipolar plate; corrosion resistance

Citation: Kim, Y.-S.; Choi, J.-Y.; Kim, C.-H.; Lee, I.-S.; Jun, S.; Kim, D.; Cha, B. C.; Kim, D. W. N⁺ Implantation on Nb Coating as Protective Layer for Metal Bipolar Plate in PEMFCs and Their Electrochemical Characteristics. *Materials* **2022**, *15*, 8612. https://doi.org/10.3390/ma15238612

Academic Editor: Yong Sun

Received: 25 October 2022
Accepted: 29 November 2022
Published: 2 December 2022

1. Introduction

To solve global warming from climate change, many countries have shown their support to the regulation of greenhouse gases from automobiles and vessels. In this regard, green vehicle technologies have been intensively researched to reduce fossil fuel consumption. Polymer electrolyte membrane fuel cells (PEMFCs) represents one of the promising candidate technologies in electric vehicles due to their higher power density, lower operating temperature, and comparatively reduced noise.

A unit cell of PEMFC stacks is composed of a membrane electrode assembly (MEA) including membrane, catalysts layers, and gas diffusion layers located between pairs of bipolar plates. Among these, the bipolar plate is one of the most important components due to its multifunctional role. The bipolar plate serves as the electrical connector between serially connected unit cells in the stack, and should be separate from fuel and oxidant.

Its other role is to provide structural supporter to the stacks, as well as water management for protecting membranes. Additionally, it should be tolerant to an aggressive acidic environment resulting from corrosive operating conditions. Therefore, the bipolar plate should provide excellent electric conductivity, gas and liquid impermeability, high mechanical properties and workability, and corrosion resistance in acidic media [1,2].

Graphite is commonly used materials in the bipolar plate because of its high electric conductivity, chemical stability, and corrosion resistance. However, graphite has a porous and brittle structure, making it difficult for machines to work on the graphite bipolar plate, which increases the production cost and time [3,4]. Thus, an alternative material with high mechanical strength and workability is needed for an effective mass production of bipolar plates to be made possible. Stainless steels have an excellent workability and mechanical properties and can be an alternative to bipolar plate. Due to higher corrosion resistance in stainless steels, 316L austenitic stainless steel (SS316L) has attracted significant attention and is widely researched [5–7]. However, when SS316L is exposed to highly corrosive acid environments for a long period, the metal ions from the exposed surface are dissolved, contaminating the electrolytes and catalysts and degrading performance of the PEMFC stacks.

Various surface modification techniques, including nitriding, cladding, and coating, have been applied to protect the surface of bipolar plates [8–12]. The nitriding process is known to be a useful method to improve the mechanical properties and corrosion resistivity of steels. However, in the case of the austenitic stainless steels, a higher treatment temperature is required to terminate the native oxide layer on the surface. This high temperature leads to the formation of CrN precipitate, which causes chromium deficiency. As a result, corrosion resistance is degraded. In another method, the cladding could be a useful modification due to this being a simple process without the need for vacuum. However, during the post annealing process, a higher temperature leads to the formation of a brittle intermetallic layer between the substrate and the cladding layer, resulting in the fracture of the intermetallic layer. Hence, a modification process that operates under low temperature is needed to prevent the above problems. The coating process, in particular physical vapor deposition, could be promising due to its ready controlling of the thickness and uniformity of the protecting layer without a high treatment temperature [11–13]. As a protecting material, pure niobium (Nb) possesses excellent chemical stability and corrosion resistance in an acidic environment with higher electric conductivity. Recently, we demonstrated the effect of a Nb layer on the corrosion resistance of bipolar plates through tuning microstructures. Our findings showed that the densified Nb layer with fewer defects effectively prevented direct contact between electrolyte and substrate, and exhibited good corrosion resistance in the acidic environment [14]. However, although the Nb layer was densely coated on the substrate, it could not completely eliminate inside defects including pinholes and voids, which could be a pathway for the corrosive electrolyte to penetrate. Therefore, for overcoming the above problem, additional modification of coating layer is required.

Recently, the plasma immersion ion implantation (PIII) method has beeb used as an effective way to modify the physical and chemical properties of substrate without apparent phase transformation and precipitation, i.e., improvement with maintaining its bulk properties [15,16]. A PIII process is performed at a low temperature, and physically implants ions into the substrate by adjusting the ion dose and implant ion energy. Such properties may give an advantage compared with other thermal processes. In particular, the effect of energetic ion bombardment could modify surface properties, and densify the film structure. We believe these effects could positively affect the corrosion behavior of the coated layer. Moreover, the implanted ions could contribute to further improvement in corrosion behavior. As a dopant ion, the implanted nitrogen ions (N^+) lead to improved corrosion behavior and mechanical properties [17–21]. Hence, we expected that the implantation of N^+ ions on the Nb layer would modify its structural properties and reinforce its corrosion resistance ability. In the present study, the microstructural features of the N^+-implanted Nb layer and the interaction of implanted ions with the neighboring Nb sites were investigated. The experimental results revealed the effects of N^+-implantation on the corrosion behavior in the simulated PEMFC environment.

2. Materials and Methods

A SS316L sheet with a thickness of 0.1 mm was used as the base material, and its chemical composition is shown in Table 1. In this study, the specimen was first cleaned by argon plasma etching for 30 min at 0.66 Pa. Next, a Nb layer of ca. 1.5 μm thickness was deposited on the SS316L sheet using pulsed DC magnetron sputtering in accordance with a previous study [14], as summarized in Table 2. The base pressure in the chamber was maintained at 0.001 Pa. Subsequently, N^+ ions were implanted on the Nb-coated SS316L using PIII technique using a 60 kV class pulsed power supply, and the implantation conditions are summarized in Table 2.

Table 1. Chemical composition of 316L stainless steel.

Elements	Cr	Ni	Mn	Si	C	Mo	S	N	Ti	B	Cu	Fe
Wt%	16.62	10.1	1.35	0.43	0.018	2.06	0.028	0.046	0.01	0.01	0.34	balance

Table 2. Experimental conditions for Nb sputtering and N^+ implantation.

Nb sputtering	Pressure (Pa)	0.666
	Gas flow (sccm)	10 (Ar)
	Target power (W)	600
	Bias voltage (V)	700
N^+ implantation	Pressure (Pa)	0.14
	Gas flow (sccm)	5 (N_2)
	Pulse width (μs)	3
	Bias voltage (kV)	5, 10

In this study, both experimental and computational approaches were performed to investigate the distribution of the implanted ions. The Monte Carlo simulation program (TRIM-2013.00) was used to predict the distribution of ions. As the target material, Nb was set at 8.57 g/cm^3 of density, and the total nitrogen ions for calculation were set at 99,999 ions. A secondary ion mass spectrometry (SIMS, IMS-6f, CAMECA, Paris, France) analysis was performed with 15 keV impact energy and 34 nA current to investigate the distribution of practically implanted ions. The phase and structure of the modified Nb layer were identified using X ray diffraction (XRD) analysis. For this, an X-ray diffractometer (Rigaku, Tokyo, Japan, D/Max 2500-V/PC) was performed in the 2θ range of 30° to 140° with step size of 0.2°. The partial phase transformation was investigated by X-ray photoelectron spectroscopy (XPS) with a monochromic Al-Kα source (NEXSA at 15 kV, 15 mA; ThermoFisher, Waltham, MO, USA). A field emission scanning electron microscope (FE-SEM, Tokyo, Japan, JEOL JSM-5900LV) was used to examine the variation in microstructure due to the implantation of ions. The electrochemical properties were investigated using a three-electrode system by a potentiostat (Wonatech, Seoul, Republic of Korea, WPG-100). The saturated calomel electrode was used as a reference electrode, and a platinum mesh was used as a counter electrode. The PEMFC environment was simulated using 0.5 M sulfuric acid (H_2SO_4) aqueous solution as an electrolyte at 70 °C with saturating air as the cathode and hydrogen as the anode. The potentiodynamic curves were determined in the range of −1.5 to +1.5 V with a scan rate of 5 mV/s. In addition, a potentiostatic test was performed at the constant voltages of +0.6 V and −0.1 V with a scan rate of 1 mV/s for cathode and anode conditions, respectively, to evaluate durability in the long-term operation.

3. Results and Discussion

3.1. Range and Distribution of Implanted Ions

When ions bombard on to a target surface, since the ions randomly settle on the inside of the target, the implanted ions with the same energy or incidence angle cannot be sure to reach the same site. Thus, even though ions implant under the identical conditions, they might possibly distribute onto different site. The distributions of implanted N^+ ions in the coated Nb layers were predicted using computational approaches, which were compared with the results from experimental approaches obtained by a secondary ion mass spectrometry (SIMS) analysis.

As shown in the simulated results (Figure 1), the depth and width of implantation tended to increase with increasing applied bias voltages. According to distribution theory for implanted ions, higher ion energy induces a wider range of ion distribution [22,23]. However, as the bias voltages increased, the peak nitrogen concentration migrated from the surface into the inside of the Nb layer with an ever increasing in implantation depth. Thus, although the total ion range and concentration increased, the nitrogen concentration at the surface decreased. The obtained results imply that a higher bias voltage induces a relatively deficient concentration of N^+ ions at the surface, resulting in an insufficient surface modification of the Nb layer. Comparing the computational and the experimental results (Figure 2, Table 3), the implantation depth and the peak nitrogen concentration presented a similar tendency. Although 10 kV of bias voltage led to the peak nitrogen concentration migrating into the inside, the number of nitrogen ions at the surface presented a higher value than the 5 kV applied to the sample. Therefore, the obtained results suggest that if 10 kV was applied to the sample, there would be sufficiently implanted N^+ ions at both surface and inside of the Nb layer.

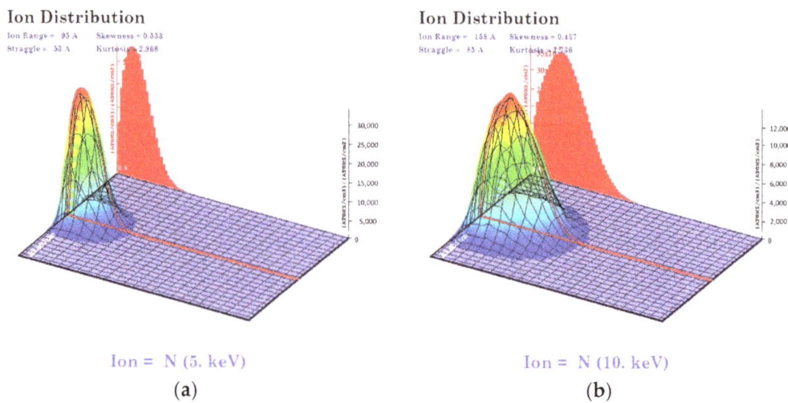

Figure 1. Monte Carlo simulation program (TRIM) simulations for implanted nitrogen profiles in the Nb target at the bias voltage (**a**) 5 kV and (**b**) 10 kV.

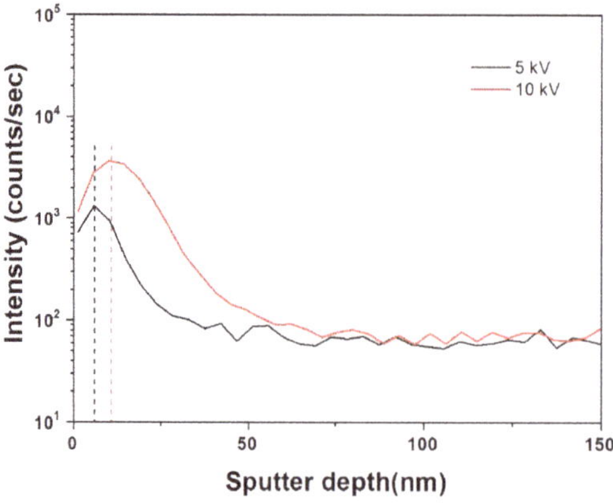

Figure 2. Depth profiles of nitrogen contents in the N^+-implanted samples by secondary ion mass spectrometry (SIMS).

Table 3. Implanted ion range and concentration of TRIM simulation and SIMS analysis.

Bias (kV)	Ion Range (Å)		Implantation Dose (Ions/cm^2)	Peak Nitrogen Concentration (Atoms/cm^3)
	TRIM	SIMS		
5	95	57	2.78×10^{16}	1.61×10^{22}
10	158	140	6.40×10^{17}	2.37×10^{23}

3.2. *Phase Characterization*

The XRD patterns were compared to investigate the effect of implantation of ions on the crystal phase, as shown in Figure 3. The parameters are summarized in detail in Table 4. After the implantation of N^+, the Nb peaks were slightly shifted to a lower angle and became broader. These results might indicate the implantation of N^+ ions into the interstitial sites of the Nb lattices whereby the Nb lattices were expanded and the d-spacing increased. In addition, the grain refinement and partial amorphization might contribute to peak broadening [24]. The crystallite size was derived from the XRD patterns using the Williamson–Hall method. As can be seen in Table 4, the crystallite size tended to decreased with N^+-implantation due to energetic ion bombardment. As previously reported, the Nb structure and physical properties were drastically changed depending on the implantation rate [25–27]. When the implantation rate was below 3×10^{16} N_2/cm^2, small defects including lattice strain or dislocation loop occurred. Subsequently, when the implantation rate exceeded 6×10^{16} N_2/cm^2, structural changes to the amorphous layer were initiated. Eventually, when the implantation rate reached 1×10^{17} N_2/cm^2, continuous amorphous layers were formed. In other words, the Nb layer changed its structure from crystalline to amorphous or a highly disordered phase when the implantation rate exceeded a certain level [27]. Based on the literature, since the implantation rate of the sample when 10 kV bias voltage (6.40×10^{17} N_2/cm^2) was applied exceeds the anticipated transition point to an amorphous phase, the amorphous layer should have formed on the surface. However, the formed amorphous layer was not very thick, and presented crystalline diffraction patterns with small, broadening peaks. In the case of 5 kV bias voltage, the amorphous layer was not expected to be formed since the implantation rate was below a certain level. Meanwhile, no additional peaks were observed in the XRD patterns such as Nb nitride. We therefore further performed an XPS analysis to investigate partial phase transformation. As shown in the N1s spectra (Figure 4), NbN, NbN_x, and NbN_xO_y peaks were detected, indicating that the implanted N^+ ions randomly and partially formed nitride in the Nb matrix. The results also indicate that the intensity of the nitride peak was stronger at higher bias voltage. The detected NbN_xO_y peak might arise from the formation of native oxide due to exposure in the air.

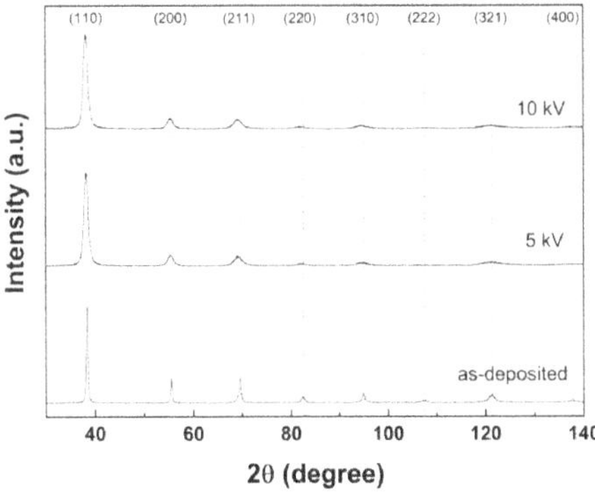

Figure 3. XRD patterns of the as-deposited Nb layer and N^+-implanted samples.

Table 4. Conducted parameters of as-deposited Nb and N^+-implanted samples from XRD patterns.

Bias (kV)	2θ (Degree)	*d*-Spacing (Å)	FWHM (Radian)	Crystallite Size (nm)
As-deposited Nb	38.40	2.342	0.020	17.00
5 kV	38.12	2.358	0.024	12.28
10 kV	38.16	2.356	0.025	11.76

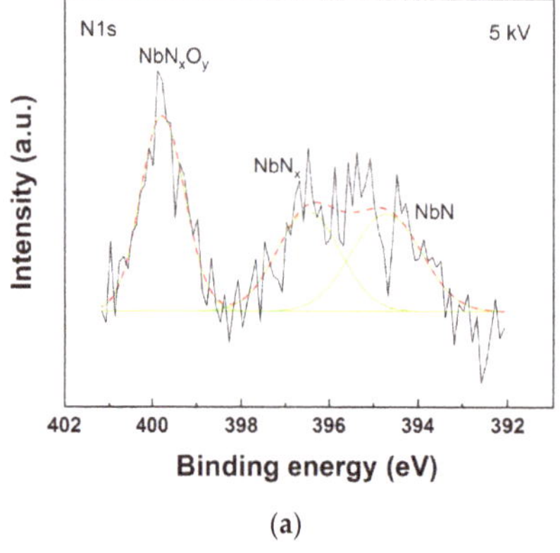

(a)

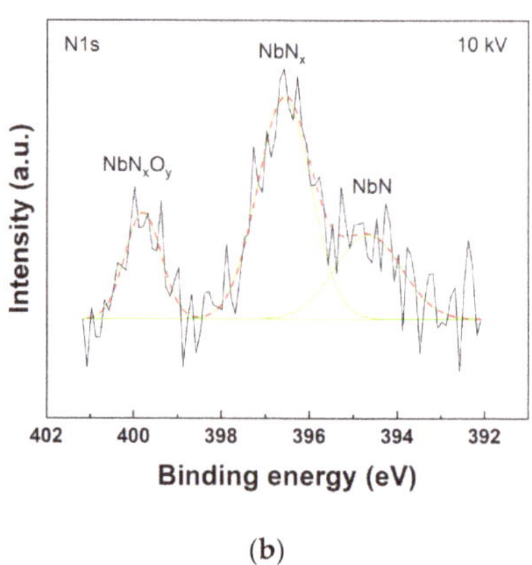

(b)

Figure 4. N1s XPS core level spectra of implanted N^+ at (**a**) 5 kV and (**b**) 10 kV.

The phase characterization results suggested that a thin amorphous layer was formed on the surface and Nb nitride was partially formed in the Nb layer. The amorphous alloy is known to be a corrosion inhibitor, and often shows higher corrosion resistance than crystalline alloys at the same composition [28–31]. Furthermore, Nb nitride possesses good corrosion resistance even under aggressive environmental conditions, which have been widely studied for preventing metal corrosion [32–35]. Therefore, partial phase changes in the Nb layer might positively affect the corrosion behavior.

3.3. Microstructure

The surface and cross-section of the as-deposited and N^+-implanted Nb layers were observed by using electron micrographs, as shown in Figure 5. The FE-SEM images of the surface showed that N^+-implanted Nb layers present truncated grain morphology due to ion bombardment which was absent in the as-deposited Nb layer. When 10 kV bias voltage was applied, a relatively nonuniform and blurred grain boundary was formed. The obtained results are consistent with the results of the surface roughness in Figure 6. Due to higher bias voltage, the crystalline structure is truncated and becomes smaller, resulting in a decrease in surface roughness. In the case of the cross-section, the as-deposited Nb layer presented a columnar structure with voids between columns; by contrast, in the N^+-implanted Nb layers the interface between the columns became smoother and most of the voids were filled up. The obtained results demonstrated that the Nb layer was densified due to the bombardment of high energy ions.

To clarify the densification of the Nb layer, the defects on the surface and the inside of layers including voids and pinholes were examined. The critical passivation current density (CPCD) method was used to evaluate density and distribution of defects. The CPCD method electrochemically evaluates the defect area ratio (R_{da}) using a current density at the active–passive transition region that is proportional to the exposed area of the substrate [36]. The calculated R_{da} pattern was as follows: as-deposited Nb (0.61 %) > 5 kV (0.57 %) > 10 kV (0.32 %). The R_{da} decreased with N^+-implantation, and higher bias voltage led to further decrease. Therefore, a higher ion energy could contribute to densification of the structure, and it having less defects, which is consistent with microstructural observation. We assume that these microstructural changes would contribute to improving the corrosion behavior. The densified microstructure could prevent direct contact with the corrosive electrolyte, and decreased surface roughness could reduce surface area an exposed to the electrolyte.

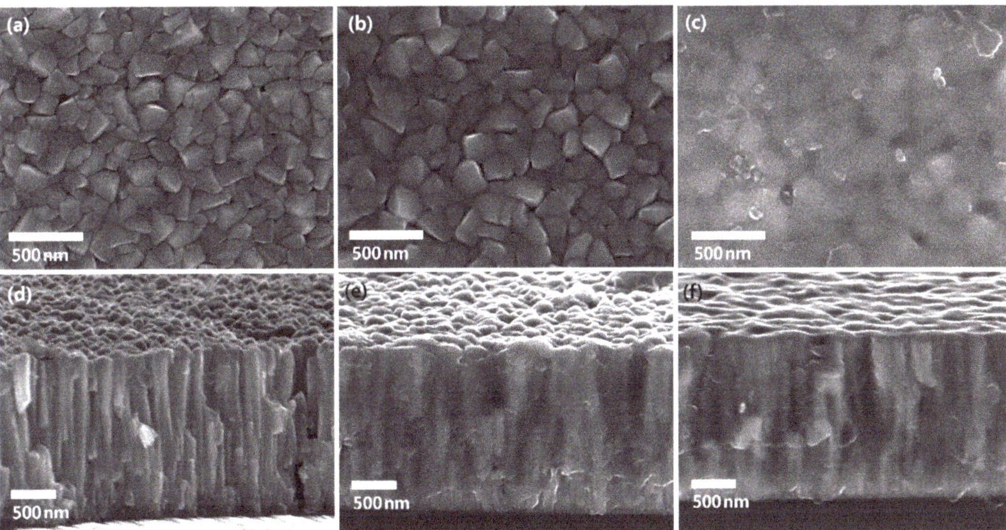

Figure 5. Surface and cross-section images: (**a**,**d**) as-deposited, respectively; (**b**,**e**) at 5 kV, respectively; and (**c**,**f**) 10 kV, respectively.

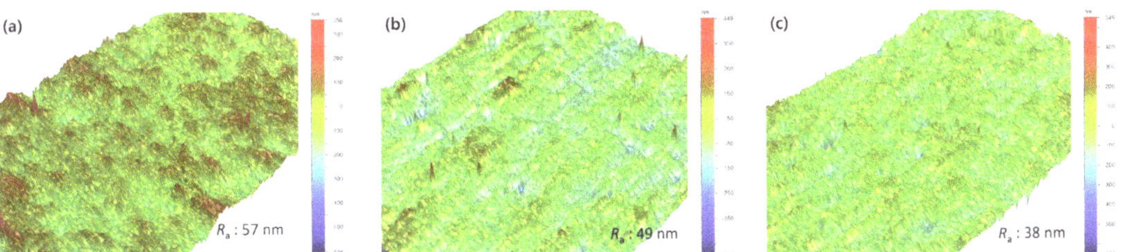

Figure 6. 3D surface profiling images of (**a**) as-deposited Nb layer, (**b**) 5 kV, and (**c**) 10 kV.

3.4. Electrochemical Properties

The potentiodynamic polarization test was performed under simulated PEMFC operating conditions (Figure 7, Table 5). The simulated cathode and anode conditions were described in the method section. As shown in Figure 7, the N^+-implanted Nb layers exhibited a wide passive region compared to the as-deposited Nb layer and maintained passive state at both the cathode (0.6 V) and anode (−0.1 V) operating potentials. The corrosion current density (i_{corr}) was obtained from the Tafel extrapolation and linear polarization, and was slightly decreased at the N^+-implanted Nb layers. The $i_{corr,\ Tafel}$ was determined from extrapolation between the E_{corr} with the anodic and cathodic Tafel slopes in the obtained polarization curve. The $i_{corr,\ Rp}$ was calculated by the Stern–Geary equation using polarization resistance and tafel slopes. Both the obtained $i_{corr,\ Tafel}$ and $i_{corr,\ Rp}$ showed similar values and the same trend. Likewise, the passivation current density (i_{pass}) was smaller than that of the as-deposited Nb layer. The lower value of i_{pass} implies a facile and fast transition to the passivation state. The polarization resistance (R_p) was determined from linear polarization. In comparison with the as-deposited layer, the R_p of the N^+-implanted Nb layers was increased. Moreover, when 10 kV bias voltage was applied, the R_p was more than twice that of the as-deposited Nb layer. Since the R_p implies tolerance to an oxidation reaction when an external potential is applied, it can be used to estimate corrosion resistance. Thus, the N^+-implantation might contribute to greater protection against corrosion and enhanced tolerance to an aggressive oxidation environment. Although the corrosion potential was slightly decreased with N^+-implantation, the corrosion and passivation current density and polarization resistance were improved with increasing bias voltage. In other words, even though the corrosion occurred a little earlier, the passive film was quickly formed and reached a passive state due to the N^+-implantation effects.

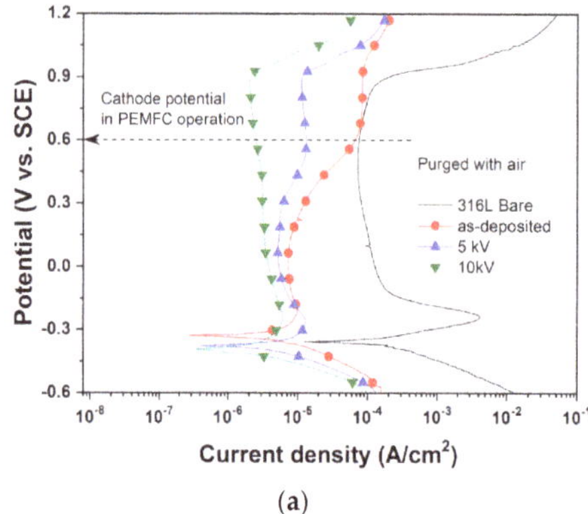

(a)

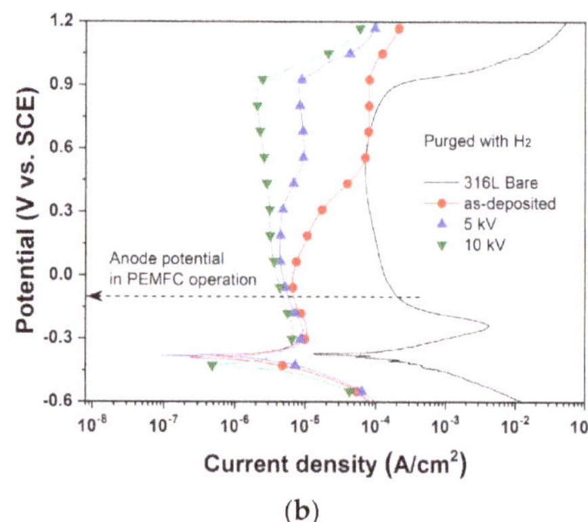

(b)

Figure 7. Potentiodynamic polarization curves of as-deposited Nb layer and N^+-implanted samples with purged (**a**) air and (**b**) H_2.

Table 5. Corrosion parameters extracted from potentiodynamic polarization curves.

Sample	E_{corr} (V)	$i_{corr,\ Tafel}$ (A/cm^2)	$i_{corr,\ Rp}$ (A/cm^2)	$i_{0.6V\ vs.\ SCE}$ (A/cm^2)	$i_{-0.1V\ vs.\ SCE}$ (A/cm^2)	i_{pass} (A/cm^2)	R_p (Ω cm^2)
316L bare	−0.3374	1.33×10^{-4}	1.67×10^{-4}	4.95×10^{-5}	1.31×10^{-4}	3.64×10^{-5}	61
As-deposited	−0.3278	1.26×10^{-6}	2.08×10^{-6}	4.78×10^{-5}	7.88×10^{-6}	7.69×10^{-6}	5010
5 kV	−0.3688	1.19×10^{-6}	1.43×10^{-6}	1.31×10^{-5}	5.67×10^{-6}	5.08×10^{-6}	6688
10 kV	−0.3746	1.08×10^{-6}	1.23×10^{-6}	2.52×10^{-6}	4.66×10^{-6}	1.99×10^{-6}	11,457

Meanwhile, for the bipolar plate application, the corrosion behavior at the operating potential is highly significant. The N^+-implanted Nb layers presented a smaller current density at the cathode ($i_{0.6\ V}$) and anode ($i_{-0.1\ V}$) operating potential when compared with that of the as-deposited Nb layer and bare 316L SS. In particular, the 10 kV applied sample showed the lowest value of current density in the cathode potential, that is one and two orders of magnitude smaller current density than that of the as-deposited Nb layer or bare 316L SS, respectively. Thus, these results suggest that the implanted N^+ ions effectively protect the Nb layer and the substrate under an aggressive PEMFC operating environment.

The bipolar plate is immersed in the corrosive electrolyte at both cathode and anode potentials during PEMFC operation. Thus, a durable corrosion resistance at the electrode potential is necessary. A potentiostatic polarization test performed under the simulated PEMFC operating conditions (Figure 8) showed that the higher initial current density of all the samples at the cathode (Figure 8a) rapidly decreased and stabilized to a lower value. However, a difference in the stabilization time was observed. The N^+-implanted Nb layers stabilized within 2000 s, while the as-deposited Nb layer took more than two times longer for the current density to stabilize. This difference might correspond to the different redox reaction rates. A relatively fast redox reaction of the N^+-implanted Nb layers resulted in the fast formation of a passive film on the surface [37]. Moreover, in the stabilized current density, the N^+-implanted Nb layers exhibited a current density one order of magnitude smaller than that of the as-deposited Nb layer. In the case of the anode (Figure 7b), the polarization curves revealed a similar tendency to the cathode. Consequently, N^+-implantation induced a more stable and lower current density in both cathode and anode, thereby satisfying the target set by the US department of energy (DOE).

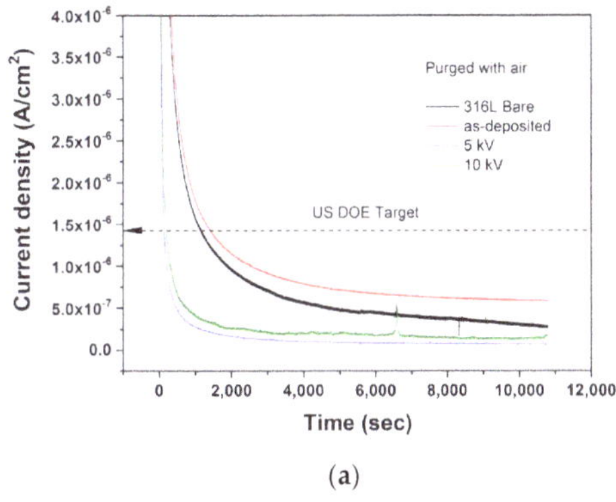

(a)

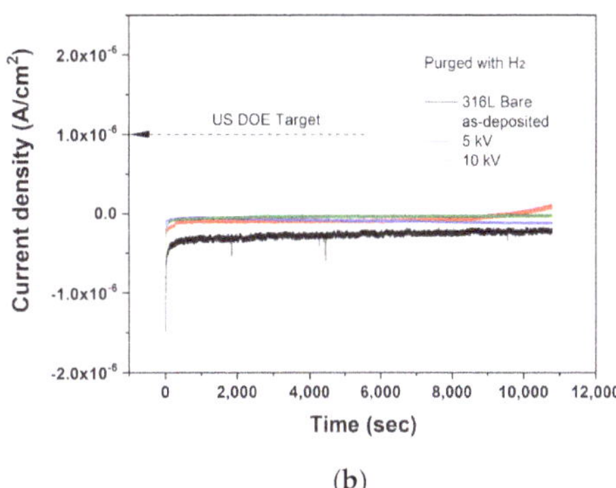

(b)

Figure 8. Potentiostatic polarization curves of the as-deposited Nb layer and N^+-implanted samples with purged (**a**) air (constant 0.6 V) and (**b**) H_2 (constant −0.1 V).

The above results confirmed that the N^+-implantation effectively modified the Nb layer, thereby improving electrochemical properties under the simulated PEMFC operating conditions. We regard the reason for such results likely to arise from the following effects. First, the bombardment of higher energy ions induced a densified Nb layer with fewer defects, which prevented direct contact between substrate and corrosive electrolyte. The synergetic effect between partially formed Nb nitride and the Nb matrix might contribute to additional improvement of electrochemical properties. In addition, tuning of the surface roughness by ion sputtering led to an increased electron work function and reduced the exposed area at the surface, thereby improving corrosion behavior [38–40]. It should also be considered that a thinly formed amorphous layer on the surface might contribute to surface passivation. Furthermore, in the present work, we investigated effective methods for modification of the protective layer. We found that this—in combination with reinforcement of substrate materials—will help further improve corrosion behavior of bipolar plate materials.

4. Conclusions

The deposited pure Nb layers were modified by the implantation of N^+ ions, and the FE-SEM observation demonstrated that the higher ion energy could densify the Nb layer. The reduced defects inside the Nb layer effectively prevented the penetration of the corrosive electrolyte, resulting in improved corrosion resistance. Moreover, partially formed nitride in the Nb matrix further contributed to the improvement in corrosion resistance. In the polarization test, N^+-implantation at 10 kV resulted in a current density one order of magnitude smaller than that of the as-deposited Nb layer. With respect to durability, N^+-implanted Nb layers exhibited a current density one order of magnitude smaller with rapid surface passivation. Therefore, the N^+-implantation effectively modified the deposited Nb layer to form an improved protective film for the metal bipolar plate. We believe that these results will provide new possibilities for surface modification and improvement of fuel cell performance, as well as other potential applications.

Author Contributions: Conceptualization, S.J., B.-C.C. and D.-W.K.; analysis, C.-H.K., I.-S.L. and J.-Y.C.; validation, B.-C.C. and D.K.; investigation, Y.-S.K., I.-S.L. and J.-Y.C.; writing—manuscript,

Y.-S.K. and D.-W.K.; writing—review and editing, S.J., D.K. and D.-W.K. All authors have read and agreed to the published version of the manuscript.

Funding: This study has been conducted with the support of the Korea Institute of Industrial Technology as "Development of Core Manufacturing Technology for green Energy Ship (Kitech JA-22-0008)". This research was supported by the Korea Research Fellowship program funded by the Ministry of Science and ICT through the National Research Foundation of Korea (2019H1D3A1A01071089).

Institutional Review Board Statement: Not applicable.

Informed Consent Statement: Not applicable.

Data Availability Statement: The data can be provided by authors on request.

Conflicts of Interest: The authors declare no conflict of interest.

References

1. Tang, A.; Crisci, L.; Bonville, L.; Jankovic, J. An overview of bipolar plates in proton exchange membrane fuel cells. *J. Renew. Sustain. Energy* **2021**, *13*, 022701. [CrossRef]
2. Pourrahmani, H.; Siavashi, M.; Yavarinasab, A.; Matian, M.; Chitgar, N.; Wang, L.; Herle, J.V. A Review on the Long-Term Performance of Proton Exchange Membrane Fuel Cells: From Degradation Modeling to the Effects of Bipolar Plates, Sealings, and Contaminants. *Energies* **2022**, *15*, 5081. [CrossRef]
3. Carmo, M.; Fritz, D.L.; Mergel, J.; Stolten, D. A comprehensive review on PEM water electrolysis. *Int. J. Hydrog. Energy* **2013**, *38*, 4901–4934. [CrossRef]
4. Lettenmeier, P.; Wang, R.; Abouatallah, R.; Saruhan, B.; Freitag, O.; Gazdzicki, P.; Morawietz, T.; Hiesgen, R.; Gago, A.S.; Friedrich, K.A. Low-cost and durable bipolar plates for proton exchange membrane electrolyzers. *Sci. Rep.* **2017**, *7*, 44035. [CrossRef]
5. Jin, J.; Zhang, J.; Hu, M.; Li, X. Investigation of high potential corrosion protection with titanium carbonitride coating on 316L stainless steel bipolar plates. *Corros. Sci.* **2021**, *191*, 109757. [CrossRef]
6. Rojas, N.; Sanchez-Molina, M.; Sevilla, G.; Amores, E.; Almandoz, E.; Esparza, J.; Vivas, M.R.C.; Colominas, C. Coated stainless steels evaluation for bipolar plates in PEM water electrolysis conditions. *Int. J. Hydrog. Energy* **2021**, *46*, 25929–25943. [CrossRef]
7. Novalin, T.; Eriksson, B.; Proch, S.; Bexell, U.; Moffatt, C.; Westlinder, J.; Lagergren, C.; Lindbergh, G.; Lindström, R.W. Concepts for preventing metal dissolution from stainless-steel bipolar plates in PEM fuel cells. *Energy Convers. Manag.* **2022**, *253*, 115153. [CrossRef]
8. Shen, H.; Wang, L.; Sun, J. Characteristics and properties of CrN compound layer produced by plasma nitriding of Cr-electroplated of AISI 304 stainless steel. *Surf. Coat. Technol.* **2020**, *385*, 125450. [CrossRef]
9. Yi, P.; Dong, C.; Zhang, T.; Xiao, K.; Ji, Y.; Wu, J.; Li, X. Effect of plasma electrolytic nitriding on the corrosion behavior and interfacial contact resistance of titanium in the cathode environment of proton-exchange membrane fuel cells. *J. Power Sources* **2019**, *418*, 42–49. [CrossRef]
10. Avram, D.N.; Davidescu, C.M.; Dan, M.L.; Mirza-Rosca, J.C.; Hulka, I.; Stanciu, E.M.; Pascu, A. Corrosion resistance of NiCr (Ti) coatings for metallic bipolar plates. *Mater. Today Proc.* **2022**, *in press*. [CrossRef]
11. Mani, S.P.; Agilan, P.; Kalaiarasan, M.; Ravichandran, K.; Rajendran, N.; Meng, Y. Effect of multilayer CrN/CrAlN coating on the corrosion and contact resistance behavior of 316L SS bipolar plate for high temperature proton exchange membrane fuel cell. *J. Mater. Sci. Technol.* **2022**, *97*, 134–146. [CrossRef]
12. Atapour, M.; Rajaei, V.; Trasatti, S.; Casaletto, M.P.; Chiarello, G.L. Thin niobium and niobium nitride PVD coatings on AISI 304 stainless steel as bipolar plates for PEMFCs. *Coatings* **2020**, *10*, 889. [CrossRef]
13. Alexeeva, O.K.; Fateev, V.N. Application of the magnetron sputtering for nanostructured electrocatalysts synthesis. *Int. J. Hydrog. Energy* **2016**, *41*, 3373–3386. [CrossRef]
14. Kim, Y.S.; Lee, I.S.; Choi, J.Y.; Jun, S.; Kim, D.; Cha, B.C.; Kim, D.W. Corrosion behavior of niobium-coated 316L stainless steels as metal bipolar plates for polymer electrolyte membrane fuel cells. *Materials* **2021**, *14*, 4972. [CrossRef]
15. Gupta, D. Plasma immersion ion implantation (PIII) process physics and technology. *Int. J. Adv. Technol.* **2011**, *2*, 471–490.
16. Mändl, S.; Manova, D. Modification of metals by plasma immersion ion implantation. *Surf. Coat. Technol.* **2019**, *365*, 83–93. [CrossRef]
17. Ramezani, A.H.; Hoseinzadeh, S.; Ebrahiminejad, Z.; Hantehzadeh, M.R.; Shafiee, M. The study of mechanical and statistical properties of nitrogen ion-implanted Tantalum bulk. *Optik* **2021**, *225*, 165628. [CrossRef]
18. Huang, H.H.; Shiau, D.K.; Chen, C.S.; Chang, J.H.; Wang, S.; Pan, H.; Wu, M.F. Nitrogen plasma immersion ion implantation treatment to enhance corrosion resistance, bone cell growth, and antibacterial adhesion of Ti-6Al-4V alloy in dental applications. *Surf. Coat. Technol.* **2019**, *365*, 179–188. [CrossRef]
19. Sharma, P.; Dhawan, A.; Sharma, S.K. Influence of nitrogen ion implantation on corrosion behavior of $Zr_{55}Cu_{30}Ni_5Al_{10}$ amorphous alloy. *J. Non. Cryst. Solids* **2019**, *511*, 186–193. [CrossRef]
20. Hapsari, S.; Sujitno, T.; Ahmadi, H.; Aziz, R.A. Analysis of nitrogen ion implantation on the corrosion resistance and mechanical properties of aluminum alloy 7075. *J. Phys. Conf. Ser.* **2020**, *1436*, 012075. [CrossRef]

21. Vlcak, P.; Fojt, J.; Weiss, Z.; Kopeček, J.; Perina, V. The effect of nitrogen saturation on the corrosion behaviour of Ti-35Nb-7Zr-5Ta beta titanium alloy nitrided by ion implantation. *Surf. Coat. Technol.* **2019**, *358*, 144–152. [CrossRef]
22. Okuda, S.; Kimura, T.; Akimune, H. Depth profiles of implanted H and He in metal Mo determined with backscattered protons. *Jpn. J. Appl. Phys.* **1979**, *18*, 465. [CrossRef]
23. Winterbon, K.B. Ion implantation distributions in inhomogeneous materials. *Appl. Phys. Lett.* **1977**, *31*, 649–651. [CrossRef]
24. Fewell, M.P.; Mitchell, D.R.G.; Priest, J.M.; Short, K.T.; Collins, G.A. The nature of expanded austenite. *Surf. Coat. Technol.* **2000**, *131*, 300–306. [CrossRef]
25. Camerlingo, C.; Scardi, P.; Tosello, C.; Vaglio, R. Disorder effects in ion-implanted niobium thin films. *Phys. Rev. B* **1985**, *31*, 3121–3123. [CrossRef]
26. Torche, M.; Schmerber, G.; Guemmaz, M.; Mosser, A.; Parlebas, J.C. Non-stoichiometric niobium nitrides: Structure and properties. *Thin Solid Films* **2003**, *436*, 208–212. [CrossRef]
27. Gamo, K.; Goshi, H.; Takai, M.; Iwaki, M.; Masuda, K.; Namba, S. Control of Tc for niobium by N ion implantation. *Jpn. J. Appl. Phys.* **1977**, *16*, 1853. [CrossRef]
28. Xie, C.; Milošev, I.; Renner, F.U.; Kokalj, A.; Bruna, P.; Crespo, D. Corrosion resistance of crystalline and amorphous CuZr alloys in NaCl aqueous environment and effect of corrosion inhibitors. *J. Alloys Compd.* **2021**, *879*, 160464. [CrossRef]
29. Zander, D.; Köster, U. Corrosion of amorphous and nanocrystalline Zr-based alloys. *Mater. Sci. Eng. A* **2004**, *375*, 53–59. [CrossRef]
30. Masumoto, T.; Hashimoto, K. Chemical properties of amorphous metals. *Annu. Rev. Mater. Sci.* **1978**, *8*, 215–233. [CrossRef]
31. Gebert, A.; Buchholz, K.; Leonhard, A.; Mummert, K.; Eckert, J.; Schultz, L. Investigations on the electrochemical behavior of Zr-based bulk metallic glasses. *Mater. Sci. Eng. A* **1999**, *267*, 294–300. [CrossRef]
32. Jin, W.; Wu, G.; Li, P.; Chu, P.K. Improved corrosion resistance of Mg-Y-RE alloy coated with niobium nitride. *Thin Solid Films* **2014**, *572*, 85–90. [CrossRef]
33. Olaya, J.J.; Rodil, S.E.; Muhl, S. Comparative study of niobium nitride coatings deposited by unbalanced and balanced magnetron sputtering. *Thin Solid Films* **2008**, *516*, 8319–8326. [CrossRef]
34. Chen, M.; Ding, J.C.; Kwon, S.H.; Wang, Q.; Zhang, S. Corrosion resistance and conductivity of NbN-coated 316L stainless steel bipolar plates for proton exchange membrane fuel cells. *Corros. Sci.* **2022**, *196*, 110042. [CrossRef]
35. Fonseca, R.M.; Soares, R.B.; Carvalho, R.G.; Tentardini, E.K.; Lins, V.F.C.; Castro, M.M.R. Corrosion behavior of magnetron sputtered NbN and Nb1-xAlxN coatings on AISI 316L stainless steel. *Surf. Coat. Technol.* **2019**, *378*, 124987. [CrossRef]
36. Uchida, H.; Yamashita, M. Effect of preparation conditions on pinhole defect of TiN films by ion mixing and vapor deposition. *Vacuum* **2022**, *65*, 555–561. [CrossRef]
37. Feng, K.; Shen, Y.; Liu, D.; Chu, P.K.; Cai, X. Ni–Cr Co-implanted 316L stainless steel as bipolar plate in polymer electrolyte membrane fuel cells. *Int. J. Hydrog. Energy* **2010**, *35*, 690–700. [CrossRef]
38. Li, W.; Li, D.Y. Variations of work function and corrosion behaviors of deformed copper surfaces. *Appl. Surf. Sci.* **2005**, *240*, 388–395. [CrossRef]
39. Xue, M.; Xie, J.; Li, W.; Wang, F.; Ou, J.; Yang, C.; Li, C.; Zhong, Z.; Jiang, Z. Changes in surface morphology and work function caused by corrosion in aluminum alloys. *J. Phys. Chem. Solids* **2012**, *73*, 781–787. [CrossRef]
40. Li, W.; Li, D.Y. Influence of surface morphology on corrosion and electronic behavior. *Acta Mater.* **2006**, *54*, 445–452. [CrossRef]

 materials

Article

The Influence of Heat Treatment on Corrosion Resistance and Microhardness of Hot-Dip Zinc Coating Deposited on Steel Bolts

Dariusz Jędrzejczyk [1,*] and Elżbieta Szatkowska [2]

1 Department of Mechanical Engineering Fundamentals, University of Bielsko-Biala, Willowa 2, 43-309 Bielsko-Biała, Poland
2 BOSMAL Automotive Research and Development Institute Ltd., Sarni Stok 93, 43-300 Bielsko-Biala, Poland
* Correspondence: djedrzejczyk@ath.bielsko.pl

Abstract: The analyzed topic refers to the corrosion resistance and changes in microhardness of the heat-treated (HT) hot-dip zinc coating deposited on bolts. The research aimed to evaluate the influence of the HT on the increase of the coating hardness and changes in anticorrosion properties. Hot-dip zinc coating was deposited in industrial conditions (acc. EN ISO 10684) on chosen bolts (M12x60). The achieved results were assessed based on corrosion resistance tests in neutral salt spray (salt chamber) and microhardness measurements. Tests were conducted in accordance with the adopted fractional plan, generated in the DOE module of the Statistica software. Using the conjugate gradient method optimal parameters of HT were determined. The conducted tests proved that the controlled heat treatment may increase the hardness of the hot-dip zinc coating without a significant deterioration in its basic protective function (corrosion resistance). The observed changes in the hardness and corrosion resistance of the zinc coating are a consequence of changes in its structure.

Keywords: heat treatment; corrosion resistance; hot-dip zinc galvanizing; coating hardness

Citation: Jędrzejczyk, D.; Szatkowska, E. The Influence of Heat Treatment on Corrosion Resistance and Microhardness of Hot-Dip Zinc Coating Deposited on Steel Bolts. *Materials* **2022**, *15*, 5887. https://doi.org/10.3390/ma15175887

Academic Editor: Yong Sun

Received: 30 July 2022
Accepted: 24 August 2022
Published: 26 August 2022

1. Introduction

Zinc is the most commonly used element in the production of anticorrosion coatings [1]. Processes of Zn coating deposition are relatively simple and do not require advanced devices, complicated technologies or large financial resources [2]. The report of the International Lead and Zinc Study Group regarding years 2016–2020 [3] states that global zinc production amounted to about 12 million tons yearly and more than 6 million tons were used as anticorrosion protection for steel. In practice, four galvanizing methods are used. The oldest method used on an industrial scale for corrosion protection of steel parts is immersion galvanizing, commonly known as "hot-dip galvanizing". By contrast, the latest technology using zinc to protect steel surfaces is the so-called "lamella technique". This is a typically adhesive process—the surface before immersion is prepared to obtain the greatest possible degree of development. In addition to the methods mentioned above, sherardization (where the coating has a structure similar to hot-dip galvanizing) and electro-galvanizing are also applied [4–6].

The most commonly used method for protecting industrial steel parts is hot-dip galvanizing [7–9]. Because hot-dip zinc galvanizing ensures high quality and long-term protection against corrosion, examples of its application can be found in every environment (marine, rural, industrial) and different kinds of industry (shipbuilding, machine, agriculture)—in most environments corrosion protection lasts on average 25–75 years, depending on the thickness of the coating correlated with the environment of exposure to corrosive agents [7].

Fasteners are protected primarily by hot-dip zinc galvanizing [5]—the thickness of the deposited coatings is usually in the range of 45–65 μm [4], and the coating microstructure has a typical diffusive character and consists of several intermetallic phases of the Fe–Zn system (η, ζ, δ, Γ) [10]. In addition to hot-dip galvanizing, electro-galvanizing is also used for anticorrosion protection for different kinds of bolts, especially those with a smaller diameter. However, the coating obtained by this method is usually much thinner, which usually results in lower corrosion resistance. This process also entails the risk of reducing the mechanical properties of the base material due to hydrogen embrittlement [11] and has a negative impact on the environment [12]. To achieve a thinner zinc coating on the surface of bolts characterized by comparable corrosion resistance to the standard hot-dip zinc galvanizing, the high-temperature galvanized method is applied. As a result of this process, the zinc coating is also composed of Fe–Zn intermetallic phases. Since galvanizing takes place at a temperature about 100 °C higher than during standard hot-dip galvanizing, the process requires the use of ceramic or ceramic lining baths (instead of the steel ones used in most galvanizing plants), which results in higher costs of energy and equipment [13]. In addition, in the case of galvanizing bolts made in the higher strength classes, the high-temperature method carries the risk of deterioration of the bolts' strength properties as a result of tempering [4]. A coating with a similar structure, as in the case of high-temperature galvanizing (composed of intermetallic phases), is also deposited using thermo-diffusion galvanizing [14]. As a result, parts showing very good anti-corrosion protection, without hydrogen embrittlement [15], are obtained. However, in comparison to hot-dip galvanizing, this method has limitations related to the size of the parts which can be subjected to this process (it is not possible to galvanize parts as large as those in the case of hot-dip zinc coating in galvanizing baths) and zinc plating takes more time (several, or up to even a dozen, hours).

One of the weaknesses of the hot-dip zinc coating is the relatively low hardness. The hardness measured on the zinc surface of the hot-dip coating, i.e., the hardness of the η layer, according to various sources oscillates around the value of 50 HV (50–57 HV [16]; 52 HV [17]; 41 HV [18]). One of the methods of improving the properties of commonly used coatings is heat treatment. In the literature on materials subjected to elevated temperatures, the attention is focused mainly on oxidation mechanisms [19–22], influence of different elements on the oxidation resistance [23,24] and the mechanical properties of oxidized materials [25]. In addition to the above, the subject of the coatings' controlled heat treatment and its impact on the properties of coatings is also increasingly taken up by researchers testing various types of coatings (Zn-Ni-P-W [26], Ni-W [27]; diamond-like [28]; bronze [29]; Ni-B [30]). In the presented article, the HT impact on the structure of coatings and their key properties related to exploitation conditions are analyzed.

Azadeh et al. [31] analyzed the effect of heat treatment on the deformability of hot-dip galvanized DC01 steel. Samples were subjected to temperature exposures of: 500, 510, 520, 530 and 540 °C. The treatment time was from 10 to 180 s. After the heat treatment, the samples were cooled in water. The presented results of the research (SEM, EDS, XRD, FLD) showed that the applied parameters (temperature and time) of the treatment result in structure changes in the coating. It was found that the use of higher temperatures in combination with shorter processing times allows for improving the plasticity of the zinc coating.

In the research of Szabadi et al. [32], the subject of which was the abrasive wear of the hot-dip zinc coating applied to S235JRG2 steel, the results of microhardness measurements using the Vickers method as well as tests of the condition and qualitative composition of the surface are presented (SEM-EDS). The authors compared data obtained for the samples galvanized in the crude state and the ones galvanized after being subjected to heat treatment. Measurements of abrasive wear were conducted in a specially designed abrasion tester. The researchers claim that the heat treatment caused a change in the structure of the zinc coating, which in turn increased abrasion resistance. The suggested reason for increased abrasion resistance is the intensification of iron diffusion. The zinc coating on the base sample had an average hardness value of about 47 HV. The heat treatment resulted in an increase in the hardness of the zinc coating to about 106 HV. In the work, however, there is no detailed data on the basic parameters of the heat treatment (time, temperature).

In the work of P. P. Chung et al. [33], the authors note the phenomenon of the beneficial effect of heat treatment on zinc coatings applied to fasteners. The research concerned zinc coating deposited onto 1022 steel bolts and flat samples made of CA3SN-G steel. The researchers heat-treated the galvanized parts using a temperature of 340 °C and a time of 10, 15 and 30 min. Heat treatment was carried out in ambient air. Corrosion resistance was tested by the potentiodynamic method in a 5% NaCl solution. Microscopic examinations (optical LOM, scanning SEM with EDS analysis) and structure analysis (XRD) were also carried out. The authors claim that the heat treatment caused changes in the structure of intermetallic phases, which contributed to the improvement of corrosion resistance because they could become a barrier to progressive corrosion processes. The researchers also observed an increase in the thickness of the treated coatings in relation to samples that were not heated at 340 °C.

So, considering the above analysis and the opinion of authors of the article [34], we can state that corrosion resistance is the most important property of zinc coating. How zinc protects iron alloys (steel, cast iron) against corrosion depends on its specific physico-chemical properties [35]: in relation to iron, zinc becomes the anode that is first attacked by corrosion; zinc can create a corrosion resistance barrier and shows a high passivation ability.

The anti-corrosion properties of hot-dip zinc coating depend on many factors, including microstructure [36,37] and thickness [37–40]. There is no absolute uniqueness among data regarding the corrosion resistance of individual layers of zinc coating. There is also a statement in the literature [33,41] confirming that corrosion rate in alloyed zone is usually lower and thicker coatings exhibit higher corrosion resistance due to the presence of a larger fraction of Fe–Zn phases. However, most publications confirm the lower corrosion resistance of alloyed layers in comparison to pure zinc η layer in different environment—both in fresh, hardened and chloride-contaminated concrete [42] and in wet–dry cyclic conditions, in chloride-containing environments [43]. In article [44], authors claim that the low-temperature annealing (225 °C) causes the formation of intermetallic Fe/Zn compounds, a transformation of amorphous oxide inclusions to the crystalline form and a decrease in the Zn lattice parameter for Zn–Co and Zn–Fe alloys, which also results in a decrease in corrosion resistance.

As shown in articles [45,46], an increase in HT temperature results in an increase in the thickness of the iron-rich layers (ζ, δ, Γ) in the zinc coating, and after the HT at 460 °C there is practically no pure η phase [45].

Considering the data from the literature and results of our own research presented in the previous article [47], where both disk samples and bolts were tested and the effect of heat treatment on the change of zinc coating microhardness, microstructure and coefficient of friction was studied, it seems crucial to assess the impact of the proposed heat treatment on the change in corrosion resistance of the tested coatings.

The aim of the presented research was to verify the possibility of increasing the hardness of zinc hot-dip coating by HT (up to 130–150 HV), with an acceptable reduction in corrosion resistance not exceeding 10%. During the experiment the impact of the heat treatment on the zinc coating deposited on typical steel bolts (M12x60) was assessed. Changes in microhardness (HV 0.02) values, microstructure character and corrosion resistance (NSS test) allowed to choose the optimal treatment conditions. The research was conducted in accordance with the adopted fractional plan, generated in the DOE module of the Statistica software. The optimal heat treatment parameters (ensuring hardness at the level of 140 HV.02 and corrosion resistance reduced by no more than 10%) were determined by the conjugate gradient method. The conducted studies confirmed that the controlled heat treatment allows a significant increase in the hardness of the zinc coating deposited on the tested bolts without a serious deterioration in the key property of the coating, i.e., corrosion resistance.

2. Materials and Methods

Steel bolts M12x60—8.8U (23MnB4 steel) were selected for the tests. On the one hand, the selected bolt diameter facilitated hot-dip zinc galvanizing (reducing the risk of zinc flashes formed on the thread). On the other hand, it met the requirements of the method used in the previous article for the friction coefficient determination (Schatz Analyze M12 testing machine system—where bolts with a maximum diameter of 12 mm can be tested). Samples were hot-dip galvanized according to EN ISO 10684 [48]—etching in 12% HCl, fluxing, dipping in Zn bath with Al, Bi and Ni at temp. of 460 °C within the time of 1.5 min and cooling in water.

Heat treatment was carried out in temperature range t = 270–430 °C in an electric chamber furnace. The bolts were held at the treatment temperature in time τ = 7–11 min. The parameters of heat treatment were selected based on the preliminary tests performed at work [47] with the use of disc-shaped samples measuring 25 mm in diameter and 4 mm in thickness and compared with the kinetics of heating disc samples and bolts (M12x60) using a thermal imaging camera. Table 1 presents values of the heat treatment parameters applied in the tests together with the variant numbers (1–9), i.e., the trivalent fractional plan (three values of variable parameters) for two input quantities (temperature and time of heat treatment). The plan was generated in the DOE (Design of Experiments) module of the Statistica software.

Table 1. Values of the heat treatment parameters and numbers of fractional plan variant.

Parameter of Heat Treatment		No. of Fractional Plan Variant
Temperature t, °C	**Time τ, min.**	
270	7	1
	9	2
	11	3
350	7	4
	9	5
	11	6
430	7	7
	9	8
	11	9

After the heat treatment samples were taken out of the furnace chamber and were air-cooled to the ambient temperature. The following parameters were analyzed during investigations: the heating process (chamber furnace FCF 75HM, Flir E95—thermal imaging camera), the microstructure of zinc coating structure and steel—with the use of a scanning microscope with EDS analysis (scanning electron microscope EVO 25 MA Zeiss with an EDS attachment), the microhardness changes measured on the bolts head (Vicker's HV 0.02, Mitutoyo Micro-Vickers HM-210A device 810–401 D), corrosion resistance (salt chamber—Liebisch Labortechnik, type S 1000 M-TR, with capacity 1 000 l). To analyze the microstructure, chemical composition, phase identification in thin coatings, advanced devices (X-ray diffraction (XRD) and scanning electronic microscope (SEM) with EDS) were usually used [49,50]. In this study, to identify individual phases occurring in the zinc coating, only a scanning microscope with an EDS attachment was used, which seems to be quite sufficient due to the very characteristic and at the same time diverse morphology of the analyzed phases (η, ζ, δ, Γ).

Considering the planned effects of the heat treatment (coating hardness increase to the value in the range: 130–150 HV and acceptable reduction in corrosion resistance less than 10%), the optimal heat treatment parameters were determined using the conjugate gradient method [51].

3. Results and Discussion

3.1. Coating Thickness and Microhardness Measurements

Measurements of the thickness of the coatings deposited on galvanized bolts M12 × 60—8.8U were made by the magnetic method, in accordance with PN-EN ISO 2178:2016-06 [52]. The measurements were carried out in places defined by standard on the bolts heads before and after their heat treatment, using an electronic coating thickness gauge. On each bolt, 10 measurements were made, the results of which were averaged. For every bolt, the average coating thickness was in the range of 59.76 ± 1.27 μm, which corresponds to the requirements of the fasteners [4]. It follows from the achieved results that the applied heat treatment does not affect the change in the thickness of the zinc coating.

Microhardness measurements were carried out on the bolt's head (outer layer) using the Vickers HV 0.02 method. Five measurements were made on the surface of each of the bolts. The obtained averaged results are shown in Figure 1.

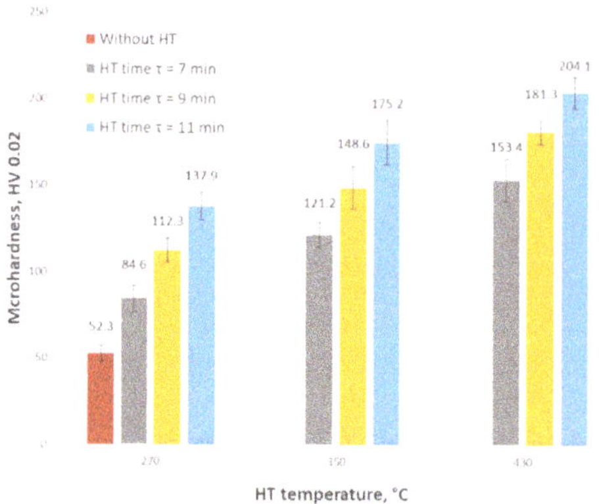

Figure 1. Comparison of the microhardness of zinc coatings measured on bolts surface before and after heat treatment.

Heat treatment conducted at fixed values (t = 270 °C, τ = 9 min) caused almost a twofold increase in the microhardness of the outer zinc coating layer. The highest—almost fourfold increase in microhardness (in comparison to the microhardness of coatings deposited on base bolts, without HT)—was obtained for coatings applied to bolts subjected to the treatment at the highest temperature and the longest time (t = 430 °C, τ = 11 min). Comparing the hardness measured after heat treatment at 430 °C and time τ = 11 min) (variant 9—Table 1) to the literature data (Table 2), it can be concluded that a phase ζ already appeared in the outer layer of the coating. Classical hot-dip zinc coating is composed of four phases according to the Fe–Zn diagram [17,53,54]: —Fe_3Zn_{10}, δ—$FeZn_7$, ζ—$FeZn_{13}$ and an iron solid solution in zinc—η (Figure 2a), which is settled on the surface during pulling out of the bath. After heat treatment carried out using parameters corresponding to the other variants (2–8), the outer layer of the coating was composed of a mixture of the phases η and ζ. The above statement was additionally verified on the basis of the microscopic observations and EDS analysis—Figure 2, the results of which are been related to the literature data—Table 3.

Table 2. Hardness of individual intermetallic phases of zinc coating deposited by hot-dip zinc galvanizing according to: Evans [8] (**a**) and Pokorný et al. [55] (**b**).

Phase Kind		Phase Hardness	
		(a)	(b)
	, Fe_3Zn_{10}	326 HB	326
	$_1$, Fe_5Zn_{21}	-	505 HV
δ	δ, $FeZn_{10}$	-	358 HV
	δ, $FeZn_7$	270 HB	-
ζ	ζ, $FeZn_{13}$	220 HB	208 HV
η	η, Zn(Fe)	70 HB	52 HV

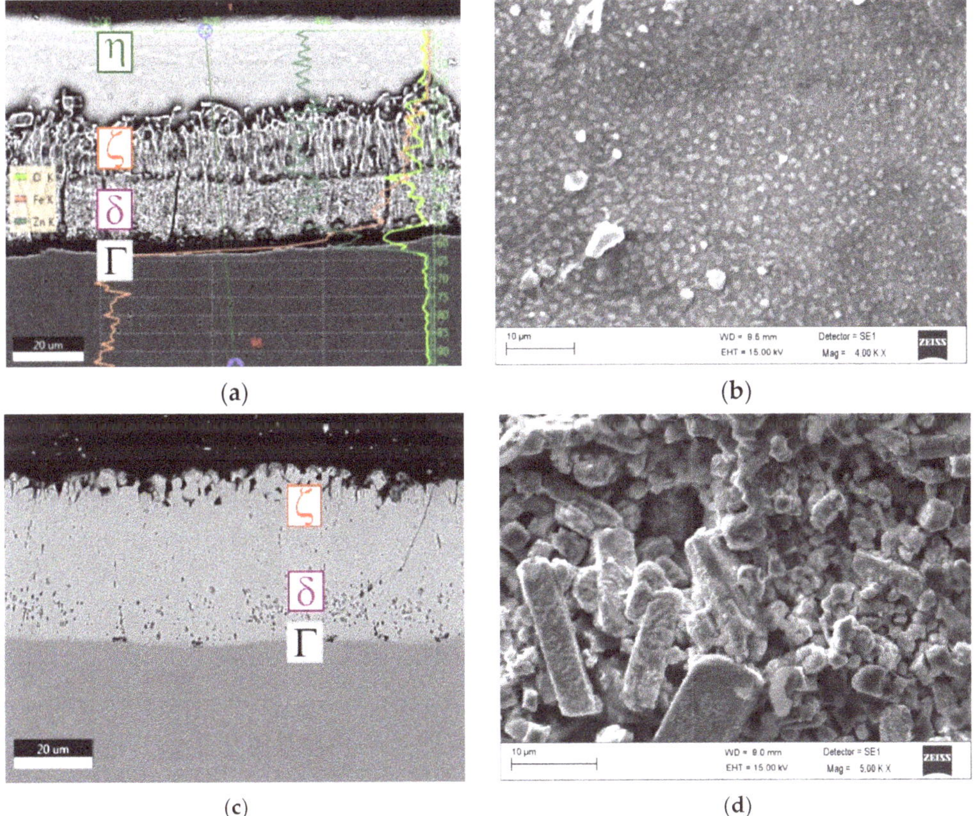

Figure 2. The microstructure observed in the cross section (**a**,**c**) and on the outer surface of zinc coating deposited on heads bolts (**b**,**d**) ((**a**,**b**)—without HT; (**c**,**d**)—t = 430 °C, τ = 11 min).

Table 3. The content of iron in individual intermetallic phases of the zinc coating, deposited by ho-dip zinc galvanizing, according to: J. D. Culcasi et al. [54] (**a**), D. Kopyciński et al. [56] (**b**) and H. Kania et al. [57] (**c**).

Phase Kind	Content Fe,%	Phase Kind	Content Fe, %	Phase Kind	Content Fe, %
(a)		(b)		(c)	
, Fe_3Zn_{10} **$_1$, Fe_5Zn_{21}**	18–31 19–24	$_1$, Fe_3Zn_{10}	21–28	, Fe_3Zn_{10}	23.7–31.5
δ, $FeZn_{10}$δ, $FeZn_7$	8–137–10 [58]	δ, $FeZn_{10}$δ, $FeZn_7$ [59]	7–11.510.87	$δ_1$, $FeZn_{10}$	8.1–13.4
ζ, $FeZn_{13}$	6–7	ζ, $FeZn_{13}$	5–6	ζ, $FeZn_{13}$	6.5–7.5
η, Zn(Fe)	0.04	η, Zn(Fe)	0	η, Zn(Fe)	0.03

Figure 2b,d shows a view of the outer surface of the zinc coatings obtained under extreme conditions: without heat treatment (Figure 2b and after HT), using the longest time and the highest temperature (τ = 11 min., t = 430 °C; Figure 2d). Iron content was determined by EDS analysis in five randomly selected places on the outer surface of the zinc coating and the obtained results were averaged. The error of a single measurement did not exceed 5%. An increase in HT temperature results in an increase in the thickness of the iron-rich layers (ζ, δ)—the measured Fe content in areas on the outer surface (Figure 2b,d) was equal accordingly to 0.17 (standard deviation σ = 0.02) and 7.31% (σ = 0.27). Additionally, the linear EDS analysis confirmed the extension of the ζ, δ, Γ phases range in comparison to zinc coating without HT. The outer surface of the coating without heat treatment is composed of solid solution η (smooth and compact), while the surface after heat treatment is composed of a column-crystalline phase ζ (more diverse, heterogeneous, less compact). The conditions for zinc coating structure recomposition during heat treatment are completely different than for hot-dip zinc coating deposition during galvanizing—the Zn amount in the coating is limited. It is quite possible that during the HT, the liquid solutions diffuse along the channels between the cells of ζ and δ phases (like in the model presented by Wołczyński [60]), but a more probable is the occurrence of bulk diffusion that results in the thickening of ζ and δ phases, which in the extreme cases leads to the occurrence of one phase ζ on the outer surface of the coating (Figure 2c).

The analysis of the obtained results using the DOE module of the Statistica software made it possible to present the impact of heat treatment parameters on the change in the hardness of the applied coatings in the entire range of used parameters both in the form of a graph (Figure 3) and an equation (Equation (1)). The coefficient of determination (fitting equation to experimental data) for microhardness results is R^2 = 0.99.

$$\hat{y}_{HV} = -192.39 + 0.716t - 0.000416t^2 + 18.3\tau - 0.2916\tau^2 \quad (1)$$

3.2. Corrosion Resistance Measurements

Both the bolts without HT and those that have been treated in accordance with the fractional experiment plan were placed inside the salt chamber. Samples were placed on a support made of chemically inert plastic in relation to the tested objects, allowing them to be positioned at an angle of ~20° from the vertical (Figure 4a) in accordance with the requirements presented in standard ISO 9227:2017 [61]. Five bolts treated at each of the selected temperatures and times were placed in the chamber. The appearance of red corrosion signs was controlled at intervals of 24 h.

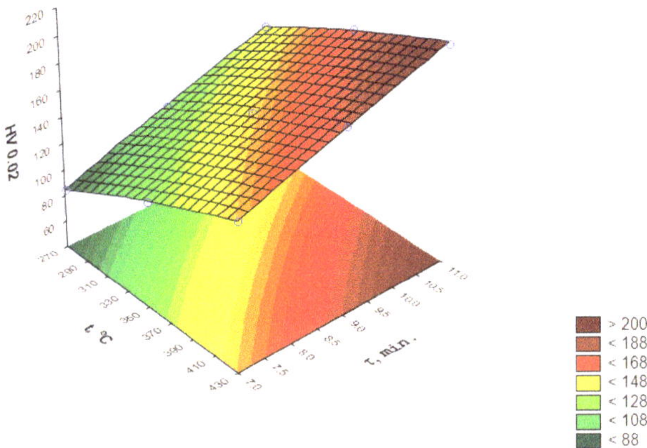

Figure 3. Influence of the heat treatment parameters on the microhardness of the zinc coating deposited on the head of M12 × 60 bolts—determined with "Statistica" software.

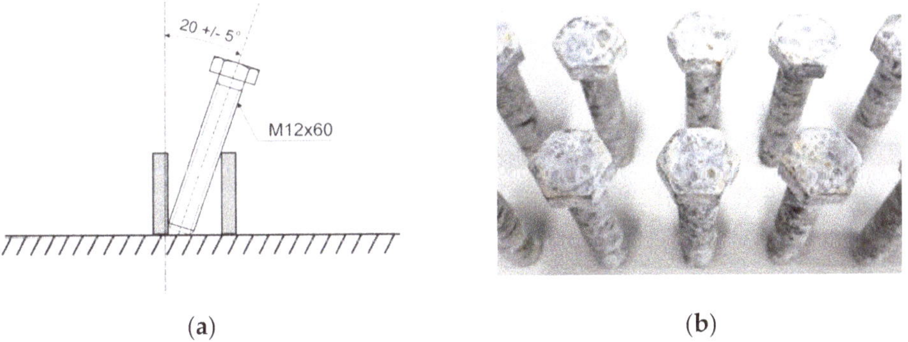

(**a**) (**b**)

Figure 4. Bolt position during the NSS test (**a**) and their appearance after the test (**b**).

Bolts were subjected to an accelerated corrosion resistance test in a neutral salt spray (NSS). The parameters of accelerated corrosion tests in the salt chamber were as follows: 5% NaCl, condensate pH 6.8–7.0, temperature 35 ± 1 °C, salt spray fall rate 1.5 mL/h. Figure 4b shows an example of the appearance of bolts after the corrosion resistance test. The achieved results are presented in Table 4. The bolts without the heat treatment were characterized by corrosion resistance of 360 h. This result was a reference point during the calculation of the reduction of corrosion resistance as a result of the conducted heat treatment (100% corresponds to the corrosion resistance of the bolts without HT).

Table 4. Values of the heat treatment parameters and corrosion resistance of zinc coating after HT.

Parameter of Heat Treatment		Time to Red Corrosion Appearance, Hours/Days
Temperature t, °C	Time τ, min.	
0	0	360/15
270	7	360/15
	9	360/15
	11	336/14
350	7	360/15
	9	336/14
	11	312/13
430	7	336/147
	9	312/13
	11	288/12

Changes in the corrosion resistance expressed as a percentage reduction of the starting value are presented in Figure 5. The same resistance as the one determined for bolts without heat treatment was shown by the bolts treated in the following conditions: t = 270 °C, τ = 7 min; t = 270 °C, τ = 9 min; and t = 350 °C, τ = 7 min. In the case of bolts treated under the conditions as follows: t = 270 °C, τ = 11 min; t = 350 °C, τ = 9 min; t = 350 °C, τ = 11 min; and t = 430 °C, τ = 7 min, corrosion resistance decreased by 24 h (i.e., 6.6%). A reduction in corrosion resistance by 48 h, corresponding to 13.3%, was observed for bolts subjected to heat treatment under the following temperature and time conditions: t = 350 °C, τ = 11 min and t = 430 °C, τ = 9 min. In turn, the shortest time to the appearance of red corrosion was measured for bolts heat-treated under conditions: t = 430 °C, τ = 11 min, i.e., there was a reduction in resistance time under corrosive conditions by 72 h in comparison to bolts that were not heat treated (so the resistance decreased by 20%). Therefore, it can be concluded that the heat treatment carried out at the highest temperature in combination with the longest holding time results in the greatest loss of corrosion resistance. The reason for the decrease in corrosion resistance of the coating after heat treatment may be explained by the expansion of the range of occurrence of iron-enriched phases (which, according to some researchers, are characterized by lower corrosion resistance [33,41]), and in the extreme case, the appearance on the surface of only the phase ζ, which is less homogeneous and less compact than the phase η. As a result, surface development and surface roughness also increase. On the other hand, heat treatment can also lead to the appearance of microcracks visible in the cross-section of the coating (reducing the compactness). The obtained results are similar to those achieved in the previous tests with application of disk samples [47]. This also proves that the conducted analysis of heating kinetics was correct, and the heat treatment parameters selected on its basis made it possible to obtain comparable results on both flat samples and M12x60 bolts.

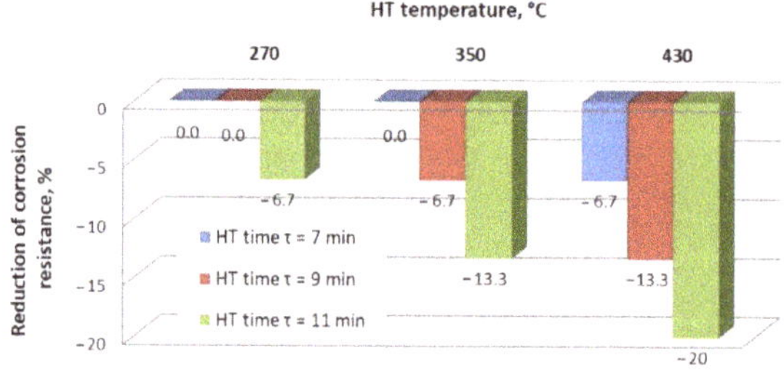

Figure 5. Changes in the corrosion resistance as a result of heat treatment expressed as a percentage reduction of the starting value.

The analysis of the obtained results using the DOE module of the Statistica software made it possible to present the impact of heat treatment parameters on the change in the corrosion resistance of the applied coatings in the entire range of used parameters both in the form of a graph (Figure 6) and an equation (Equation (2)).

$$\hat{y}_{NSS} = 358.60 + 0.18749\text{T} - 0.0006249\text{T}^2 + 7.9\tau - 0.9\tau^2 \quad (2)$$

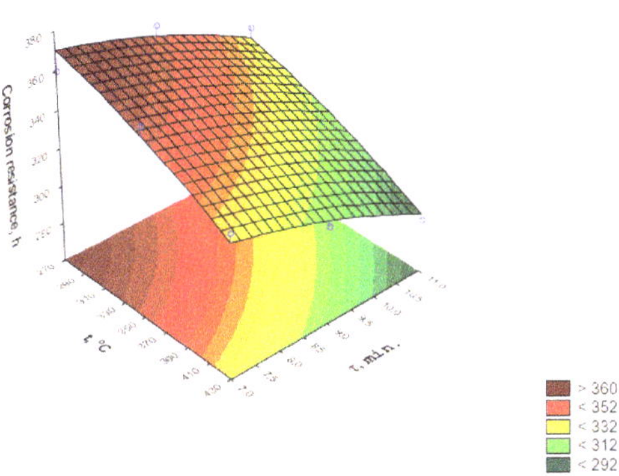

Figure 6. Influence of the heat treatment parameters on the corrosion resistance of the zinc coating deposited on the f M12 x 60 bolts—determined in "Statistica" software.

The coefficient of determination (fitting equation to experimental data) for corrosion resistance results is $R^2 = 0.95$. Because the heat treatment has the opposite influence on the changes in hardness and corrosion resistance of the tested zinc coatings, the analyzed issue is an excellent area for the application of optimization solutions.

3.3. Solving the Optimization Problem

Regression equations determined on the basis of data obtained as a result of experimental studies (Figures 3 and 6, Equations (1) and (2)) were used to formulate the optimization problem. In the conducted experiments, the influence of two parameters of the heat treatment was investigated (decision variables), i.e., temperature t and time τ, on microhardness, expressed in HV scale (criterion q_1), and the corrosion resistance, expressed in hours of duration of the corrosion resistance test (criterion q_2). The optimization problem was therefore formulated as a two-criteria task with component criteria q_1 and q_2.

To solve the above task, the concept of target programming was used, according to which the original two-criteria task is replaced by a single-criteria task, in which the cumulative criterion is the distance from the ideal point in the criteria space $\widetilde{Q}$ ($\widetilde{q}_1$, $\widetilde{q}_2$) in the sense of the square of the Euclid metric [62]. The following values were accepted as the coordinates of the ideal point: $\widetilde{q}_1$ = 140 HV i $\widetilde{q}_2$ = 324 h. In the first step, in the space of decision variables, a set of acceptable solutions Φ was defined, which is determined by the permissible intervals of variation of heat treatment parameters (t = 270–430 °C, τ = 7–11 min) fixed on the basis of the analysis of the results of preliminary tests [47]. The permissible set took the shape of a rectangle shown in Figure 7. Having regression Equations (1) and (2), the task was transformed into a criteria space mapping the permissible set Φ in a set of permissible values of criteria Ψ (Figure 8).

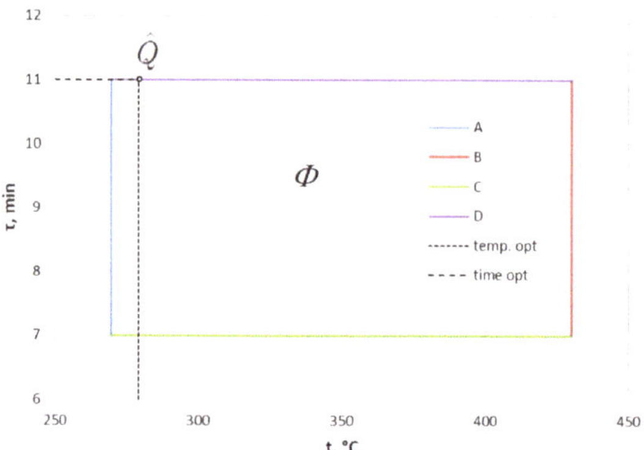

Figure 7. Set of acceptable HT parameters.

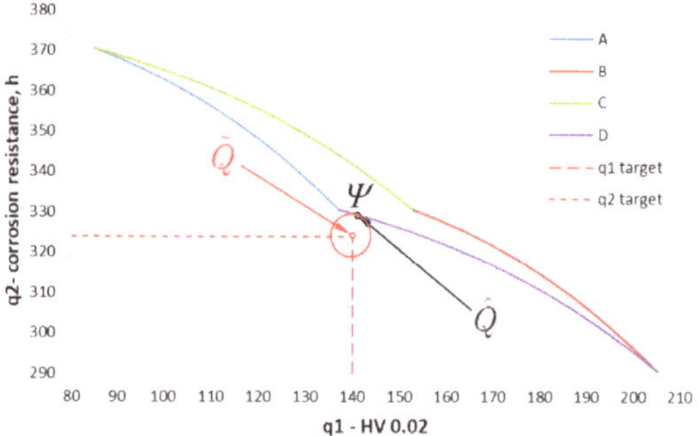

Figure 8. Set of acceptable criteria values.

Each point in the set Φ was assigned a point $q = [q_1(x),\ q_2(x)]^T$, adopting microhardness as $q_1 = q_1$ (t, τ) and corrosion resistance as $q_2 = q_2$ (t, τ). Then, due to the formulated assumptions regarding the postulated value of microhardness and the postulated reduction of corrosion resistance, the ideal point was determined, $\widetilde{Q} = [\widetilde{q}_1,\ \widetilde{q}_2]^T$. Taking into account the thesis put forward in the work, the postulated value of microhardness ($\widetilde{q}_1$) should be in the range of 130–150 HV, and the ideal value of this criterion was the average value $\widetilde{q}_1$ = 140 HV. It was further assumed that the postulated permissible reduction of corrosion resistance should not exceed 10% of the total corrosion resistance of the coatings applied to bolts without heat treatment (360 h). This value was accepted as a reference point and gave the ideal value of the criterion of $\widetilde{q}_2$ = 324 h. Assuming that both component criteria are equally valid, the aggregate criterion expressing the distance from the ideal point takes the form as follows:

$$Q = (\widetilde{q}_1 - q_1)^2 + (\widetilde{q}_2 - q_2)^2 \to min \tag{3}$$

The solution to the task of bi-criteria optimization in the goal space is the point $\hat{Q}$ belonging to a set of permissible values of criteria Ψ located the closest to the ideal point $\widetilde{Q}$. This point has coordinates q_1 = 142 HV and q_2 = 329 h (Figure 8). The exact coordinates of the optimal solution were achieved using conjugate gradient method [51,63].

After the inverse transformation of the task from the criteria space to the decision variable space using regression equations (Equations (1) and (2)), optimal point $\hat{Q}$ has the coordinates: temperature 280 °C and time 11 min (Figure 7), which corresponds to optimal hardness, q_1 = 142 HV, and optimal corrosion resistance, q_2 = 329 h (the duration of the test until the appearance of red corrosion).

4. Conclusions

(1) Microhardness measured on the heads of bolts subjected to the HT in the tested temperature range showed max. 4 times increase in the HV in the outer surface layer (85–204 HV 0.02) in relation to the reference samples (52 HV).

(2) The observed changes in the hardness and corrosion resistance of the zinc coating are a consequence of changes in its structure. As the treatment time and temperature increase, the range of occurrence of the harder phase ζ increases too, at the expense of the η phase range (the total thickness of the coating does not change).

(3) Only after a heat treatment conducted at 270 °C and a duration of 7 and 9 min and a heat treatment at 350 °C and a duration of 7 min, the corrosion resistance of the tested coatings did not decrease concerning coatings without heat treatment.

(4) The obtained results confirm that it is possible to increase the hardness of the tested zinc coatings deposited on the bolts to a value of approx. 120 HV without reducing their corrosion resistance.

(5) The obtained solution of the two-criteria optimization task allows for determining the most favorable/optimal parameters of heat treatment of hot-dip zinc coating (temperature and time) with reference to the postulated values of microhardness and corrosion resistance.

(6) The determined optimal parameters of for the heat treatment of bolts (fulfilled the assumptions formulated in the work, i.e., HV = 140; corrosion resistance reduction ≤10%) in the tested range of variability of parameters are as follows: temperature 280 °C and time 11 min. The presented values of heat treatment parameters allow obtaining a coating hardness of 141 HV and coating corrosion resistance of 329 h.

Author Contributions: Conceptualization, D.J.; methodology, D.J.; validation, D.J. and E.S.; formal analysis, D.J.; investigation, D.J. and E.S.; resources, D.J.; data curation, D.J. and E.S.; writing—original draft preparation, D.J.; writing—review and editing, D.J. and E.S; visualization, D.J. and E.S.; supervision, D.J.; project administration, D.J.; funding acquisition, D.J. and E.S. All authors have read and agreed to the published version of the manuscript.

Funding: This research received no external funding.

Institutional Review Board Statement: Not applicable.

Informed Consent Statement: Not applicable.

Data Availability Statement: Data is contained within the article.

Conflicts of Interest: The authors declare no conflict of interest. The funders had no role in the design of the study; in the collection, analyses, or interpretation of data; in the writing of the manuscript; or in the decision to publish the results.

References

1. Metals Price List—London Metal Exchange. Available online: www.lme.com (accessed on 27 July 2022).
2. Maaß, P.; Peißker, P. *Handbook of Hot-Dip Galvanization*; Wiley-VCH Verlag GmbH & Co. KGaA: Weinheim, Germany, 2011.
3. Report of the International Lead and Zinc Study Group. Available online: www.ilzsg.org/static/statistics.aspx (accessed on 26 August 2021).
4. Kania, H.; Sipa, J. Microstructure Characterization and Corrosion Resistance of Zinc Coating Obtained on High-Strength Grade 10.9 Bolts Using a New Thermal Diffusion Process. *Materials* **2019**, *12*, 1400. [CrossRef]
5. Votava, J. Anticorrosion protection of strength bolts. *Acta Univ. Agric. et Silvic. Mendel. Brun.* **2012**, *60*, 181–188. [CrossRef]
6. Chung, P.P.; Wang, J.; Durandet, Y. Deposition processes and properties of coatings on steel fasteners—A review. *Friction* **2019**, *7*, 389–416. [CrossRef]
7. Zhmurkin, D. *Corrosion Resistance of Bolt Coatings*; Tyco Electronics Corporation: Kawasaki, Japan, 2009; pp. 1–7.
8. Evans, D.W. Next Generation Technology for Corrosion Protection in Ground Support. In Proceedings of the 14th Coal Operators' Conference, Wollongong, Australia, 12–14 February 2014; The University of Wollongong: Wollongong, Australia, 2014; pp. 177–185.
9. Meikle, T.; Tadolini, S.C.; Sainsbury, B.-A.; Bolton, J. Laboratory and field testing of bolting systems subjected to highly corrosive environments. *Int. J. Min. Sci. Technol.* **2017**, *27*, 101–106. [CrossRef]
10. Królikowska, A.; Zubielewicz, M. Różnorodność cynkowych powłok zanurzeniowych i jej wpływ na właściwości antykorozyjne. *Ochr. Przed Korozją* **2013**, *56*, 430–435.

11. Georges, C.; Colinet, B.; Sturel, T.; Allely, C.; Lhoist, V.; Cornette, D. Development of electro galvanized AHSS with tensile strength ranging from 1200 to 1500 MPa for automotive application with no risk of delayed fracture. In Proceedings of the Conference Galvatech 2017, Tokyo, Japan, 11 December 2017.
12. Kwiatkowski, L. Cynk i ochrona przed korozją. *Ochr. Przed Korozją* **2004**, *47*, 256–257.
13. Kania, H. Cynkowanie wysokotemperaturowe stali z zakresu Sandelina. *Ochr. Przed Korozją* **2011**, *54*, 587–590.
14. Jiang, J.H.; Bin Ma, A.; Du Fan, X.; Gong, M.Z.; Zhang, L.Y. Sherardizing and Characteristic of Zinc Protective Coating on High-Strength Steel Bridge Cable Wires. *Adv. Mater. Res.* **2010**, *97–101*, 1368–1372. [CrossRef]
15. Natrup, F.; Graf, W. Sherardizing: Corrosion protection of steels by zinc diffusion coatings. *Thermochem. Surf. Eng. Steels* **2015**, *62*, 737–750. [CrossRef]
16. Shibli, S.; Manu, R.; Beegum, S. Studies on the influence of metal oxides on the galvanic characteristics of hot-dip zinc coating. *Surf. Coatings Technol.* **2008**, *202*, 1733–1737. [CrossRef]
17. Marder, A. The metallurgy of zinc-coated steel. *Prog. Mater. Sci.* **2000**, *45*, 191–271. [CrossRef]
18. Jordan, C.E.; Marder, A.R. Fe-Zn phase formation in interstitial-free steels hot-dip galvanized at 450°C: Part I 0.00 wt% Al-Zn baths. *J. Mater. Sci.* **1997**, *32*, 5593–5602. [CrossRef]
19. Pu, D.; Pan, Y. First-principles investigation of oxidation mechanism of Al-doped Mo5Si3 silicide. *Ceram. Int.* **2022**, *48*, 11518–11526. [CrossRef]
20. Pan, Y.; Wang, S. Insight into the oxidation mechanism of MoSi2: Ab-initio calculations. *Ceram. Int.* **2018**, *44*, 19583–19589. [CrossRef]
21. Pan, Y.; Pu, D. First-principles investigation of oxidation behavior of Mo5SiB2. *Ceram. Int.* **2020**, *46*, 6698–6702. [CrossRef]
22. Pan, Y.; Wen, M. Insight into the oxidation mechanism of Nb3Si(111) surface: First-principles calculations. *Mater. Res. Bull.* **2018**, *107*, 484–491. [CrossRef]
23. Pan, Y.; Pu, D.; Li, Y.; Zheng, Q. Origin of the antioxidation mechanism of RuAl(1 1 0) surface from first-principles calculations. *Mater. Sci. Eng. B* **2020**, *259*, 114580. [CrossRef]
24. Wang, S.; Pan, Y.; Lin, Y. First-principles study of the effect of Cr and Al on the oxidation resistance of WSi2. *Chem. Phys. Lett.* **2018**, *698*, 211–217. [CrossRef]
25. Pan, Y.; Wang, J.; Wang, D.; Deng, H. Influence of vacancy on the elastic properties, ductility and electronic properties of hexagonal C40 MoSi2 from first-principles calculations. *Vacuum* **2020**, *179*, 109438. [CrossRef]
26. Popczyk, M.; Łosiewicz, B. Wpływ obróbki cieplnej na odporność korozyjną powłok stopowych Zn-Ni-P-W. *Ochr. Przed Korozją* **2012**, *55*, 132–134.
27. Łosiewicz, B.; Popczyk, M. Wpływ obróbki cieplnej na odporność korozyjną elektrolitycznych powłok Ni-W. *Ochr. Przed Korozją* **2012**, *55*, 128–131.
28. Staszewska-Samson, K.; Scendo, M. Wpływ podwyższonej i wysokiej temperatury na właściwości antykorozyjne powłoki diamentopodobnej na stali S355. *Ochr. Przed Korozją* **2019**, *62*, 197–203.
29. Li, W.; Li, C.-J.; Liao, H.; Coddet, C. Effect of heat treatment on the microstructure and microhardness of cold-sprayed tin bronze coating. *Appl. Surf. Sci.* **2007**, *253*, 5967–5971. [CrossRef]
30. Mukhopadhyay, A.; Barman, T.K.; Sahoo, P. Effects of heat treatment on tribological behavior of electroless ni–b coating at elevated temperatures. *Surf. Rev. Lett.* **2017**, *24*, 1850014. [CrossRef]
31. Azadeh, M.; Toroghinejad, M.R. Effect of Heat Treatment on Formability of Hot-dip Galvanized Low Carbon Steel Sheet. *ISIJ Int.* **2009**, *49*, 1945–1951. [CrossRef]
32. Szabadi, L.; Kalácska, G.; Pék, L.; Pálinkás, I. Abrasive wear of different hot-dip galvanized multilayers. *Sustain. Constr. Des.* **2011**, *2*, 82–91. [CrossRef]
33. Chung, P.; Esfahani, M.; Wang, J.; Cook, P.; Durandet, Y. Effects of heat treatment on microstructure evolution and corrosion performance of mechanically plated zinc coatings. *Surf. Coatings Technol.* **2019**, *377*, 124916. [CrossRef]
34. Tatarek, P.; Liberski, H.; Kania, H.; Formanek, B.; Podolski, P. Porównawcze badania odporności korozyjnej powłok cynkowych otrzymanych różnymi metodami. *Ochr. Przed Korozją* **2008**, *4–5*, 131–134.
35. Kuklík, V.; Kudláček, J. *Hot-Dip Galvanizing of Steel Structures;* Elsevier Science: Oxford, UK, 2016. [CrossRef]
36. Kania, H. Wpływ dodatku Pb do kąpieli cynkowej na odporność korozyjną powłok. *Ochr. Przed Korozją* **2016**, *10*, 368–371. [CrossRef]
37. Kania, H.; Liberski, P.; Kwiatkowski, L. Wpływ dodatku niklu w kąpieli do cynkowania na odporność korozyjną powłok zanurzeniowych. *Ochr. Przed Korozją* **2010**, *10*, 471–474.
38. Kania, H.; Liberski, P. Wpływ dodatku Sn na strukturę i odporność korozyjną powłok otrzymywanych na stalach z zakresu Sandelina. *Ochr. Przed Korozją* **2009**, *10*, 388–391.
39. Liberski, P.; Kania, H. Wpływ dodatków stopowych w kąpieli cynkowej na jakość otrzymywanych powłok. *Inżynieria Mater.* **2009**, *30*, 182–188.
40. Pistofidis, N.; Vourlias, G.; Konidaris, S.; Pavlidou, E.; Stergiou, A.; Stergioudis, G. The effect of bismuth on the structure of zinc hot-dip galvanized coatings. *Mater. Lett.* **2007**, *61*, 994–997. [CrossRef]
41. Liberski, P.; Podolski, P.; Kama, H.; Gierek, A.; Mendala, J. Corrosion Resistance of Zinc Coatings Obtained in High-Temperature Baths. *Mater. Sci.* **2003**, *39*, 652–657. [CrossRef]
42. Pokorný, P. Influence of Fe-Zn intermetallic layer on corrosion behaviour of galvanized concrete reinforcement. *Koroze a Ochr. Mater.* **2016**, *60*, 91–100. [CrossRef]

43. Yadav, A.; Katayama, H.; Noda, K.; Masuda, H.; Nishikata, A.; Tsuru, T. Effect of Fe–Zn alloy layer on the corrosion resistance of galvanized steel in chloride containing environments. *Corros. Sci.* **2007**, *49*, 3716–3731. [CrossRef]
44. Ramanauskas, R.; Juškėnas, R.; Kaliničenko, A.; Garfias-Mesias, L.F. Microstructure and corrosion resistance of electrodeposited zinc alloy coatings. *J. Solid State Electrochem.* **2004**, *8*, 416–421. [CrossRef]
45. Jędrzejczyk, D.; Szatkowska, E. Changes in Properties of Hot-Dip Zinc Coating Resulting from Heat Treatment. *Arch. Metall. Mater.* **2019**, *64*, 201–206. [CrossRef]
46. Jędrzejczyk, D.; Skotnicki, W. Comparison of the Tribological Properties of the Thermal Diffusion Zinc Coating to the Classic and Heat Treated Hot-Dip Zinc Coatings. *Materials* **2021**, *14*, 1655. [CrossRef]
47. Jędrzejczyk, D.; Szatkowska, E. The Impact of Heat Treatment on the Behavior of a Hot-Dip Zinc Coating Applied to Steel During Dry Friction. *Materials* **2021**, *14*, 660. [CrossRef]
48. *PN-EN ISO 10684:2006*; Fasteners, Zinc Coatings Applied by Hot-Dip Method. Polish Committee for Standardization: Warsaw, Poland, 2006.
49. Abbas, M.M.; Rasheed, M. Solid State Reaction Synthesis and Characterization of Cu doped TiO_2 Nanomaterials. *J. Physics: Conf. Ser.* **2021**, *1795*, 012059. [CrossRef]
50. Aukštuolis, A.; Girtan, M.; Mousdis, G.A.; Mallet, R.; Socol, M.; Rasheed, M.; Stanculescu, A. Measurement of charge carrier mobility in perovskite nanowire films by photo-celiv method. *Rom. Acad. Ser. A—Math. Phys. Tech. Sci. Inf. Sci.* **2017**, *18*, 34–41.
51. Hestenes, M.R.; Stiefel, E. Methods of Conjugate Gradients for Solving Linear Systems. *J. Res. Natl. Bur. Stand.* **1952**, *49*, 409–436. [CrossRef]
52. *PN-EN ISO 2178:2016-06*; Non-Magnetic Coatings on Magnetic Substrate. Coating Thickness Measurement. Magnetic Method. Polish Committee for Standardization: Warsaw, Poland, 2016.
53. Massalski, T.B.; Okamoto, H.; Subramanian, P.R.; Kacprzak, L. *Binary Alloy Phase Diagrams*; ASM International: Novelty, OH, USA, 1990.
54. Culcasi, J.; Seré, P.; Elsner, C.; Di Sarli, A. Control of the growth of zinc–iron phases in the hot-dip galvanizing process. *Surf. Coatings Technol.* **1999**, *122*, 21–23. [CrossRef]
55. Pokorný, P.; Cinert, J.; Pala, Z. Fe-Zn intermetallic phases prepared by diffusion annealing and spark-plasma sintering. *Mater. Teh.* **2016**, *50*, 253–256. [CrossRef]
56. Kopyciński, D.; Siekaniec, D.; Szczęsny, A.; Guzik, E. Oddziaływanie warstwy wierzchniej odlewu z żeliwa sferoidalnego na wzrost powłoki cynkowej podczas metalizacji zanurzeniowej. *Inżynieria Mater.* **2015**, *2*, 65–68. [CrossRef]
57. Kania, H.; Mendala, J.; Kozuba, J.; Saternus, M. Development of Bath Chemical Composition for Batch Hot-Dip Galvanizing—A Review. *Materials* **2020**, *13*, 4168. [CrossRef]
58. Chatterjee, B. Sherardizing. *Met. Finish.* **2004**, *102*, 40–46. [CrossRef]
59. Pokorny, P.; Kolisko, J.; Balik, L.; Novak, P. Reaction kinetics of the formation of intermetallic Fe-Zn during hot-dip galvanizing of steel. *Metalurgija* **2016**, *55*, 111–114.
60. Wołczyński, W.; Kucharska, B.; Garzeł, G.; Sypień, A.; Pogoda, Z.; Okane, T. Part III. Kinetics of the (Zn)—Coating Deposition During Stable and Meta-Stable Solidifications. *Arch. Met. Mater.* **2015**, *60*, 199–207. [CrossRef]
61. *PN-EN ISO 9227:2017-06*; Corrosion Tests in Artificial Atmospheres—Salt Spray Tests. Polish Committee for Standardization: Warsaw, Poland, 2017.
62. Stadnicki, J. *Teoria I Praktyka Rozwiązywania Zadań Optymalizacj*; Wydawnictwo Naukowo-Techniczne: Warszawa, Poland, 2006.
63. Lisowski, J. *Metoda Gradientu Sprzężonego Hestenesa-Stiefela*; Akademia Morska w Gdyni: Gdynia, Poland, 2017.

 *materials* 

Article

Tribocorrosion and Abrasive Wear Test of 22MnCrB5 Hot-Formed Steel

Dariusz Ulbrich [1,*], Arkadiusz Stachowiak [1], Jakub Kowalczyk [1], Daniel Wieczorek [1] and Waldemar Matysiak [2]

[1] Faculty of Civil and Transport Engineering, Poznan University of Technology, Marii Sklodowskiej-Curie 5 sq., 60-965 Poznan, Poland; arkadiusz.stachowiak@put.poznan.pl (A.S.); jakub.kowalczyk@put.poznan.pl (J.K.); daniel.wieczorek@put.poznan.pl (D.W.)

[2] Faculty of Mechanical Engineering, Poznan University of Technology, Marii Sklodowskiej-Curie 5 sq., 60-965 Poznan, Poland; waldemar.matysiak@put.poznan.pl

* Correspondence: dariusz.ulbrich@put.poznan.pl; Tel.: +48-61-665-2248

Abstract: The article presents the results of research on abrasive and tribocorrosion wear of boron steel. This type of steel is used in the automotive and agricultural industries for the production of tools working in soil. The main goal of the article is the evaluation of tribocorrosion and abrasive wear for hot-formed 22MnCrB5 steel and a comparison of the obtained results with test results for steel in a cold-formed state. The spinning bowl method to determine the wear of samples working in the abrasive mass was used. Furthermore, a stand developed based on the ball-on-plate system allows to determine the wear during the interaction of friction and corrosion. After the hot-forming process, 22MnCrB5 steel was three times more resistant for the abrasive wear than steel without this treatment. The average wear intensity for 22MnCrB5 untreated steel was 0.00046 g per km, while for 22MnCrB5 hot-formed steel it was 0.00014 g per km. The tribocorrosion tests show that the wear trace of hot-formed 22MnCrB5 steel was about 7.03 μm, and for cold-formed 22MnCrB5 steel a 12.11 μm trace was noticed. The hot-forming method allows to obtain the desired shape of the machine element and improves the anti-wear and anti-corrosion properties for boron steel.

Keywords: tribocorrosion; abrasive wear; hot-forming process; boron steel

Citation: Ulbrich, D.; Stachowiak, A.; Kowalczyk, J.; Wieczorek, D.; Matysiak, W. Tribocorrosion and Abrasive Wear Test of 22MnCrB5 Hot-Formed Steel. *Materials* **2022**, *15*, 3892. https://doi.org/10.3390/ma15113892

Academic Editors: Yong Sun and Andrea Di Schino

Received: 20 April 2022
Accepted: 29 May 2022
Published: 30 May 2022

1. Introduction

Manufacturers of machines and vehicles seek to extend their failure-free operation through scheduled maintenance, which reduces the wear of individual elements. Excessive wear of cooperating parts results in disturbances in proper operation and the necessity to perform repairs. This situation generates machine downtime and additional costs. The wear of machine elements cannot be stopped; therefore, it is aimed at limiting it with various methods. For many years, both manufacturers of machines as well as scientists have conducted research on reducing the parts' wear [1–3]. As well as the wear of machine elements during their operation, corrosive processes take place which have a destructive effect on the condition of the elements [4,5].

An example of a complex wear process is tribocorrosion [6,7]. In this process, the material loss of friction nodes in machines as a result of simultaneous friction and corrosion effects occurs. Generally, tribocorrosion is the synergy of friction and corrosion that determines the loss of material. Most often, such a mechanism of the synergy effect is observed for materials covered with a layer of passive oxides in a corrosive environment [8–11]. This protective layer is removed due to the frictional effects. Intensive electrochemical processes are initiated on the exposed surface of the material. On the other hand, in the case of materials that do not show the ability of passivation, the factors causing greater corrosion wear in the friction area may be the movement of the electrolyte or a change in the stress state in the surface layers [12,13].

The increase in resistance to only mechanical wear (abrasion) is usually achieved by increasing the hardness of the material. Unfortunately, most of the classic technological treatments that produce such an effect (heat treatment, plastic working) reduce the corrosion resistance [14]. The selection of the optimal material solution to minimize tribocorrosive wear is difficult [15–17].

One of the methods of improving the durability and strength of machine elements is to use a different material with better mechanical properties [18]. Nevertheless, this type of change is not always possible, especially in the case of the influence on the steel elements of several factors. An example of this problem is the wear of elements working in the ground (soil), where both abrasive and corrosive factors act together. In this case, additional modification of the material or its surface layer can be applied, which increase the resistance to wear and corrosion [19,20].

The wear of steel elements of agricultural machinery in soil mass is defined as a phenomenon of shearing, grooving, scratching, or cutting processes, causing the destruction of the surface layer of the material [21]. In addition, in abrasive wear, the properties of the abrasive mass, such as grain size composition, moisture, and compactness of the soil, should be considered [22–24]. Napiórkowski [25] presented a comprehensive study to understand the relationship between the physicochemical properties of soil, agrotechnical extortion, or properties of materials, and the wear process of the element of an agricultural machine. The obtained dependencies allowed for changes in the process of the design and selection of steel for the ploughshare, depending on various soil conditions. Similar studies were carried out by Natsis et al. [26], in which a strong correlation between soil type and share wear for different values of share hardness was established. Other factors influencing abrasive wear include the working time and the depth of the element's immersion in the abrasive mass, as well as the hardness of the material [27–29].

22MnB5 steel has been used for several years in the construction of car bodies, especially for elements that require high strength. The main advantages of this material solution (steel + forming technology) are:

- High abrasion resistance of boron steel.
- Large possibilities of forming complex shapes (greater than in the case of cold-forming).

Therefore, as part of various tests, the properties of this steel are verified, especially after heat treatment processes [30]. Merklein [31] found that the temperature increase during the hot-stamping process for 22MnB5 steel results in a significant decrease of the flow stress values and the slope of the initial strain hardening. Escosa et al. [32] conducted a wear test of 22MnB5 steel using the pin-on-disc test. The research result showed that AlSi reduces the friction coefficient during the test. The electrochemical behavior of 22MnB5 steel coated with hot-dip AlSi before and after the hot-stamping process was investigated by Couto et al. [32]. These studies confirmed that the hot-stamping process changes the coating morphology, and consequently, the electrochemical behavior. A significant change in the anodic/cathodic coupling of the coating/steel due to iron enrichment in the coating layer was observed. Other studies assess the effect of the applied coating on steel corrosion [33,34]. Allely et al. [35] tested anticorrosion mechanisms of aluminized steel for hot stamping. The main conclusion from the test is that the AlSi coating can be applied for the 22MnB6 substrate to reduce the corrosion potential. Additionally, Park et al. [36] showed that hot-forming can provide better corrosion resistance to steel than cold-forming.

There are various methods of improving abrasive resistance, such as pad welding [37,38], hardening [39], laser treatment [40], or application of coatings with the required technological properties [41,42]. Other methods of increasing the resistance to abrasive wear include the use of cemented carbides [43] and oxide ceramics [44]. Considering the above technological possibilities, the authors proposed the hot-forming process as a method that allows for the modification of the structure and the surface layer, which may increase the corrosion resistance and wear resistance in relation to the basic material.

This article compares hot- and cold-formed 22MnCrB5 steel in terms of wear resistance under the following conditions:

- Intense abrasion in the abrasive mass—the influence of the shaping technology on the material properties (especially hardness) determining the abrasion resistance was analyzed.
- Tribocorrosion—combined action of friction and corrosive environment, where the influence of the shaping technology on the complex properties (hardness, corrosion resistance) was analyzed.

The tests were performed for samples cut from ready-made factory products. The subject of the analysis was the wear resistance under the conditions mentioned above, but not the material properties.

Boron steels are used in the construction of elements of agricultural and heavy machinery working in the ground (soil). Therefore, it is important to evaluate the influence of the hot-forming process on the resistance of tribocorrosion and abrasive wear of steel. The hot-forming process is used not only to change the shape of the elements, but also to change the internal structure, which can have a positive effect on wear and corrosion resistance. The tests were carried out in laboratory conditions on prepared stands that allowed to obtain the same conditions (humidity, pH, grain size) during the research.

The research was divided in two main stages. One of them included the tribocorrosion test, and the second one focused only on the abrasive wear test. The main goal of the article was achieved by performing the following tasks:

- Determination of material loss for tested samples of 22MnCrB5 (after the hot-forming process) under tribocorrosion conditions on a laboratory stand.
- Determination of material loss for tested samples of 22MnCrB5 (after the hot-forming process) under the abrasive wear test in a spinning bowl unit.
- Comparison of sample wear in the tribocorrosion test and abrasive wear test without the influence of corrosive factors, with results for 22MnCrB5 cold-formed steel.

The above tests allow to determine how the hot-forming process affects the wear resistance and corrosion resistance of boron steels. It is important from the point of view of the application of these steels, especially in the agricultural industry. Limiting the process of steel wear as well as corrosion through the hot-forming operation will allow for longer periods of use of agricultural tools, which is extremely important during field work.

2. Materials

The resistance to tribocorrosion and abrasive wear was assessed for hot-formed 22MnCrB5 steel and compared to untreated boron steel. Materials used in the research contain boron. The mechanical properties of the material are presented in Table 1. Table 2 includes the chemical composition of the tested steel. The microstructure of the steel is presented in Figure 1, which includes a view of the 22MnCrB5 steel before the hot-forming process (cold-formed state). There is a hard pearlite phase embedded in ferrite. This is one of the boron-alloyed quenched and tempered steels. Steel is characterized by good formability in the hot-rolled state and high strength after the heat treatment. Low boron additions to the composition of the steel ensure high hardenability, better mechanical properties, and weight savings of up to 50% compared to conventional steel used in the automotive and agriculture industry. Boron is also added to the steel to increase the hardenability as it retards the heterogeneous nucleation of ferrite at austenite grain boundaries. In addition, in the quenching process, shrinkage occurs. Therefore, parts are resistant to the aggressive conditions, including adhesive wear, abrasion, thermal stresses, and fatigue.

Table 1. Mechanical properties of the tested materials.

Properties	22MnCrB5
Hardness	165–235 HV, after hot-forming, 400–520 HV
Tensile strength (MPa)	400–570, after hot-forming, 1300–1650
Yield strength (MPa)	200
Elongation A_{80} (%)	15

Table 2. Chemical composition.

Material	C	Si	P	S	Mn	Al	Cr	Ti	B
22MnCrB5	0.19–0.25	0.4	0.025	0.015	1.10–1.40	0.02–0.06	0.15–0.25	0.02–0.05	0.0008–0.005

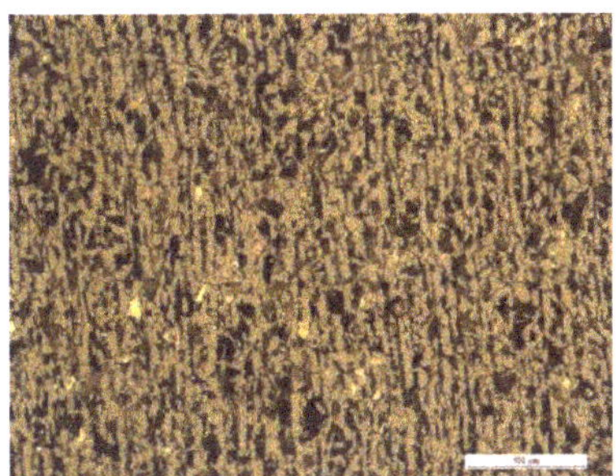

Figure 1. Microstructure of the tested 22MnCrB5 steel in the cold-formed state.

The hot-forming method is gaining importance due to the possibility of changing properties inside the material, in accordance with technical requirements in terms of strength [45–47]. It allows to obtain a shape similar to the final product, as well as to adjust the microstructural properties in one production step [48]. To obtain specific properties of the final part in the form of a burr, it is necessary to use parameters such as time and temperature in a controlled manner. Thanks to this, it is possible to maintain appropriate structural changes in the material during the heating, annealing, and cooling process. This approach allows gradation of hardness and strength along the thickness or length of the part. The scope of the transition zone and the achievement of precisely located hard and soft areas with different zone properties was possible mainly due to the ability to control the exact range of temperature during the heating of the blank. The main advantages of the hot-forming process are the excellent shape accuracy of the burr, as well as the ability to produce ultra-high-strength parts without spring-back. It is due to the conversion of austenite to martensite during the pressing operation. In the literature, there are test results that present the hot-forming technology of new materials [49,50] as well as those that improve the entire process [51].

The hot-forming process consists of various stages: austenitization, blank transfer, hot-forming, and cutting. The first step is to heat the blank to 900–950 °C for a few minutes. At this temperature, the steels are very ductile and can easily be formed into complex shapes. The heating time depends on the thickness of the blanks (for our process and thickness of the steel, the heating time was around 25 s). Then, the operation of transferring the hot blank is performed. It is important to control the temperature of the blank so that it does not fall below 780 °C, as bainite and/or ferrite begin to form. During the stamping process, the hot blank is placed by the robotic arm in a tool—a die—which is cooled with water at ambient temperature. The tools are cooled down for approximately 15 s on the hydraulic press. The part is then taken from the tool at a temperature of around 80 °C to guarantee a good geometry after the last air-quench. The last stage is cutting the product to its final dimensions. The development of laser cutters, especially 3D, resulted in the use of these machines for cutting moldings made with the hot-forming technology.

Six identical samples of the tested steel, in accordance with the dimensions shown in Figure 2 (for the abrasive wear test), were prepared. The samples of 22MnCrB5 steel were cut from the finished machine element, which was manufactured in the hot-forming process. For untreated steel (cold-formed state), samples from a sheet of metal were cut. To evaluate the tribocorrosion of the tested steel, samples with a diameter of 10 mm and a thickness of 2 mm were prepared.

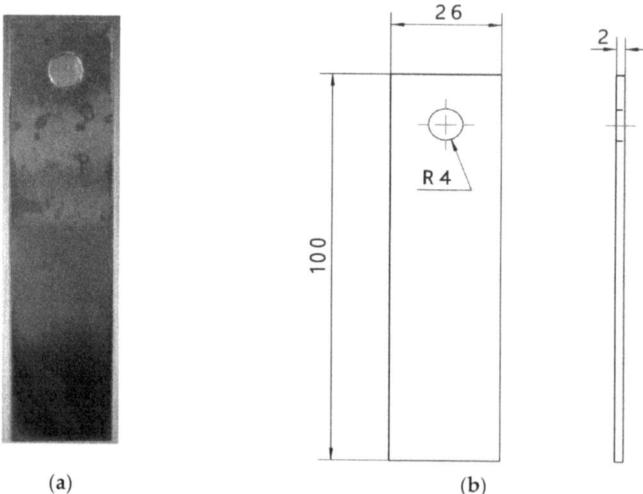

Figure 2. View (**a**) and dimensions (in mm) (**b**) of the samples used during the abrasive wear test.

3. Research Methods

3.1. Tribocorrosion Test

Tribocorrosion tests were performed for a ball-on-plate model node using the stand (Figure 3) presented in the previous publication [52]. A sample made of the tested material had the shape of a cylinder with a diameter of 10 mm. Using a special holder, the sample was mounted inside a chamber filled with a corrosive environment (3.5% NaCl was used in the tests). The ball was moving on the surface of the sample in a reciprocating motion. The stroke length (distance between the extreme positions of the ball) was about 6 mm. In the tests, Al_2O_3 balls with a diameter of 7.0 mm and a load of about 9.6 N were used. Spheres with Al_2O_3 (as counter-samples) were used in the tests. The hardness of the balls was approximately 1800 HV. The balls were made in the G25 accuracy class. The contact force and ball diameter were selected according to the hardness of the tested materials [53].

Three research tests (each one lasted about 1 h) were performed for each of the tested materials. During this test, the ball made about 14,400 displacements (movement frequency 2 Hz). The duration of the test and the frequency of ball movement were selected to achieve measurable sample wear (above 2 μm). The wear of the sample was determined after the end of the tests based on profilometric measurements of the wear trace. The maximum depth of the wear trace (determined in the middle of the friction path) was assumed as the measure of wear. Measurements were performed using a Carl Zeiss ME-10 (Carl Zeiss AG, Jena, Germany) profilometer. The device made it possible to estimate the depth with an accuracy of about 0.01 μm.

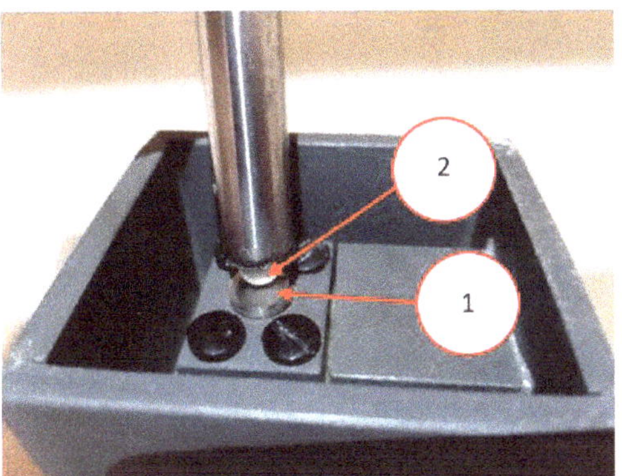

Figure 3. View of the sample during the tribocorrosion test: (**1**)—sample, (**2**)—Al_2O_3 ball.

Electrochemical measurements (determination of polarization curves) in a three-electrode system were carried out. The ATLAS 9833 potentiostat was used for this purpose. The reference electrode was the calomel electrode (SCE). The role of the auxiliary electrode was played by a platinum mesh. The area of the sample, except for the wear place, was covered with a tape made of non-conductive material. The essence of the tribocorrosion process is the occurrence of the synergy of friction and corrosion (ΔZ) according to the relationship:

$$Z_T = Z_M + Z_K + \Delta Z \quad (1)$$

where Z_T is material loss under tribocorrosion conditions, Z_M is material loss caused only by mechanical effects, and Z_K is material loss caused only by corrosive effects.

To estimate the value of the component ΔZ, tests under the conditions of only mechanical impact were carried out. The cooperation of the ball–sample friction association took place in a corrosive environment, but with cathodic polarization eliminating the corrosive processes. The material loss determined in these conditions corresponds to the component Z_M. The component Z_K was analytically calculated using the Faraday equation.

3.2. Abrasive Wear Test

The abrasive wear test on the laboratory stand, by the "spinning bowl unit" method, was performed. The view of the test stand is shown in Figure 4. The tests were carried out in sand with a grain diameter of 0.2–0.8 mm. The humidity level of the abrasive mass was also controlled and equaled around 1.5%. The rotational speed of the bowl unit allowed to obtain the working conditions in the field, which equaled an average speed of 6–7 km/h. The total distance covered by one sample during the test was 300 km. The weight of the samples was measured before starting the tests and after every 100 km of the distance test in the abrasive environment. After each measurement, the samples were replaced on the stand. It was caused by the distribution of samples on the holder (Figure 5), and samples covered different distances (radius R_1, R_2, and R_3).

(**a**)

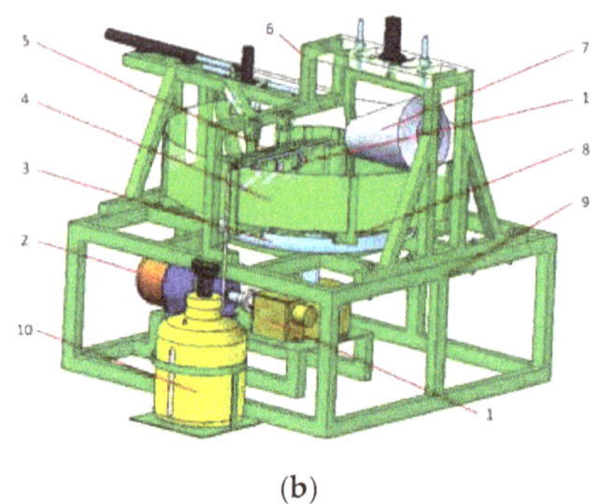

(**b**)

Figure 4. Spinning bowl unit: (**a**) view of the stand, (**b**) model of the stand: 1—transmission shaft, 2—engine, 3—travel rail, 4—bowl, 5—sample holder, 6—compacting roller frame, 7—compacting roller, 8—frame, 9—main frame, 10—water tank.

(**a**)

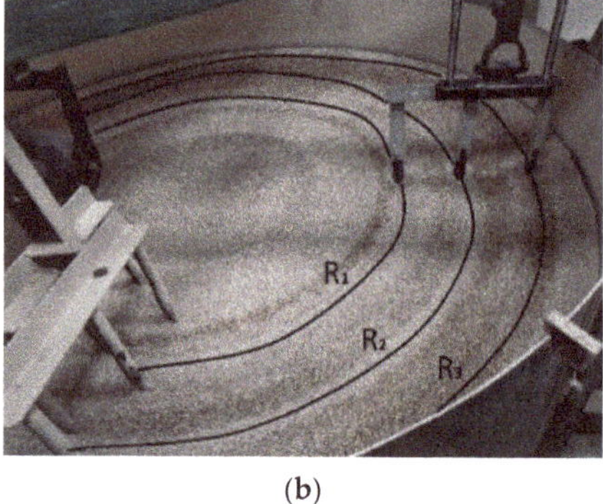

(**b**)

Figure 5. Distribution of samples on the stand: (**a**) view, (**b**) radius R_1, R_2, and R_3, along which the sample moves on the stand.

Samples' weight measurements were each time preceded by cleaning in an ultrasonic bath to remove impurities. Then, the samples were dried at the temperature of 80 °C for 20 min. The samples prepared in this way were weighed on the device with the accuracy of ±0.001 g. The samples were placed in the same direction each time on the scale. Based on the test results, the weight loss of the samples in accordance with Equation (2) was determined:

$$Z_{pw} = m_w - m_i \ [\text{g}] \tag{2}$$

where m_w is the initial sample mass before the test, in g, and m_i is the sample mass after the friction test, in g.

Based on the above equation and data, the intensity of mass wear was calculated (3):

$$I_{pw} = \frac{Z_{pw}}{S} \left[\frac{\text{g}}{\text{km}}\right] \tag{3}$$

where S is the friction distance, in km.

Additionally, the topography (surface roughness profile) for the samples tested in the abrasive wear process was measured. The T8000 contact profilograph and HommelMap Expert software by Hommelwerke were used to study the topography of the surface. The measuring tip TKL100/17 was used with a cone-shaped diamond needle with an angle of 90° and a tip radius of 2 µm. The probe travel speed was 0.15 mm/s. The dimensions of the recorded area were 1.5 × 1.5 mm. It consisted of 301 tracks separated by 0.005 mm. After registration, the surface was leveled, and then roughness was filtered using a cut-off wavelength of 0.25 mm. As a result of filtration, the outer surface fragments of 0.125 mm on each side were cut-off (the analyzed area was reduced to 1.25 × 1.25 mm).

4. Results of Research

4.1. Tribocorrosion Test Result

Figure 6 shows the martensitic structure, which was created in the hot-forming process. Additionally, it shows the AlSi layer with around 50 μm thickness that was applied to the steel surface to prevent oxidation during the hot-forming process.

Figure 6. Microstructure of the tested 22MnCrB5 steel after the hot-forming process with the AlSi layer.

Before the tribocorrosion tests, the polarization curves of the tested materials were determined. The obtained results are shown in Figure 7. Three tests were performed for each material. To quantify the resistance of the tested steels to the corrosive action of 3.5% NaCl, the corrosion potential and the corrosion current density were determined. The Tafel method and the AtlasLab software were used for this purpose. The results are presented in Table 3. The obtained results have shown that the hot-forming process improves the material's resistance to corrosion with 3.5% NaCl. The corrosion current density for 22MnCrB5 steel is several times lower for the same type of steel but in the cold-formed state.

Table 3. Corrosion potential (Ecorr) and corrosion current density (icorr).

	22MnCrB5—Cold-Formed State	22MnCrB5 after Hot-Forming
Ecorr (mV) (SCE)	−710 ± 20	−620 ± 20
icorr ($\mu A/cm^2$)	13 ± 1	2.3 ± 0.3

Table 4 presents the material loss of the tested steels for the tribocorrosion process (Z_T) and only mechanical wear (Z_M). Only mechanical wear was determined under the conditions of cathodic polarization, using the polarization potential of 200 mV lower than the corrosion potential. For the comparative analysis, it was assumed that the main measure of material loss was the depth of the wear pattern. Figure 8 shows exemplary wear patterns. Table 5 contains the wear volume in the tribocorrosion test for a 6 mm track length.

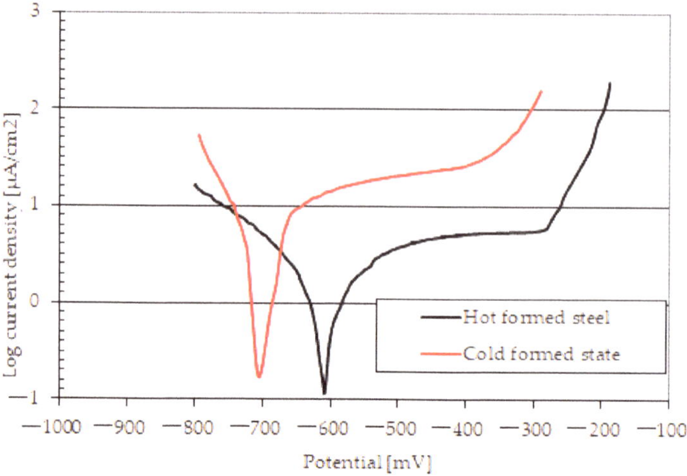

Figure 7. Polarization curves of tested materials in 3.5% NaCl (5 mV/s).

Table 4. Experimental research results.

Material	Material Loss in the Tribocorrosion (Z_T)	Mechanical Wear (Z_M)	($\Delta Z = Z_T - Z_M$)	$\Delta Z/Z_T$
	(μm)	(μm)	(μm)	(%)
22MnCrB5 in cold-formed state	12.1 ± 0.2	7.9 ± 0.2	4.20	35
22MnCrB5 after hot-forming	7.0 ± 0.2	5.0 ± 0.2	2.02	28

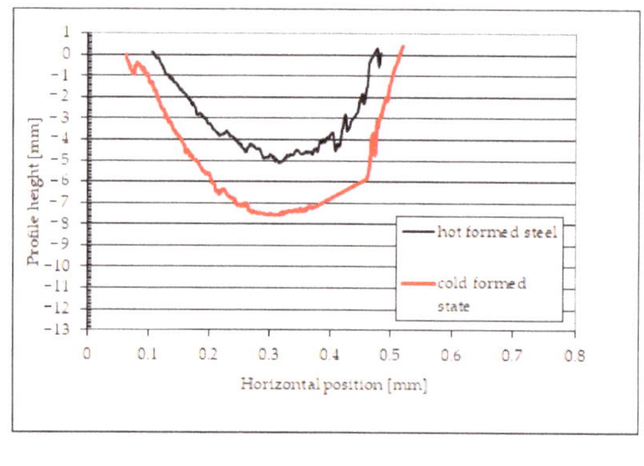

(**a**)

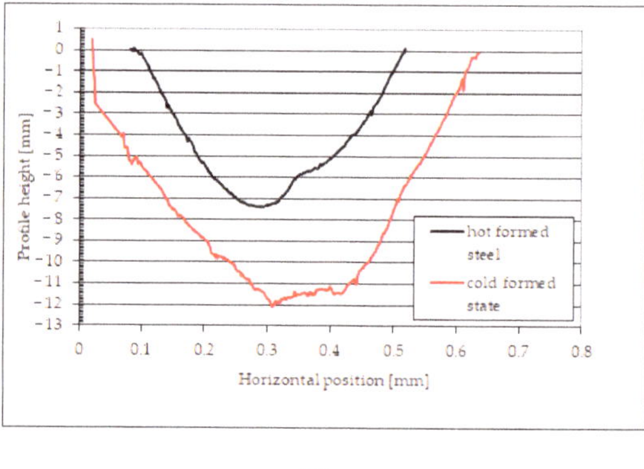

(**b**)

Figure 8. Examples of wear patterns: (**a**) only mechanical wear, (**b**) tribocorrosion.

Table 5. Calculation of wear volume.

Material	Material Loss in the Tribocorrosion (Z_T)	Mechanical Wear (Z_M)
	(mm^3)	(mm^3)
22MnCrB5 in cold-formed state	0.028 ± 0.001	0.015 ± 0.001
22MnCrB5 after hot-forming	0.013 ± 0.001	0.008 ± 0.001

During tribocorrosion research, the wear of 22MnCrB5 steel after the hot-forming process was about 7.03 μm, and for steel in the cold-formed state, about 12.11 μm. Under the conditions of only mechanical wear, the wear trace depth for the hot-formed 22MnCrB5 steel was approximately 5.01 μm, while for the untreated steel it was approximately 7.91 μm.

Based on the test results, the synergy effect of friction and corrosion (ΔZ) was also calculated. According to the information provided in Section 3.1, the material loss caused only by the corrosive effects (Z_K) was estimated using the Faraday equation. The calculations were performed for the corrosion current density (Table 3) and the test duration (1 h). The depth of the removed material layer was approximately 0.003 μm for 22MnCrB5 hot-formed steel and 0.016 μm for steel in the cold-formed state. Due to the very small values of the component (Z_K), the synergy effect of friction and corrosion was estimated as the difference ($\Delta Z = Z_T - Z_M$).

4.2. Abrasive Wear Test Result

The test results in the form of weight loss (wear) in grams for individual samples are shown in Figures 9 and 10. Figure 11 presents a comparison of the average wear of the samples depending on the distance affected by the abrasive mass. The average wear intensity for cold-formed steel was 0.00046 g/km, while for the 22MnCrB5 steel sheet after the hot-forming process it was 0.00014 g/km. This means there was a three times greater resistance to the abrasive wear of steel after the hot-forming process. The results of the wear of samples are presented in grams because the sand pressure occurred from the front of the sample on the surface, regardless of the material from which it was made.

In addition to the weight loss test, the surface of the sample was verified to determine the wear model and the dominant damage mechanism of the surface layer. A selected view of the samples after the test in the abrasive mass is shown in Figure 12.

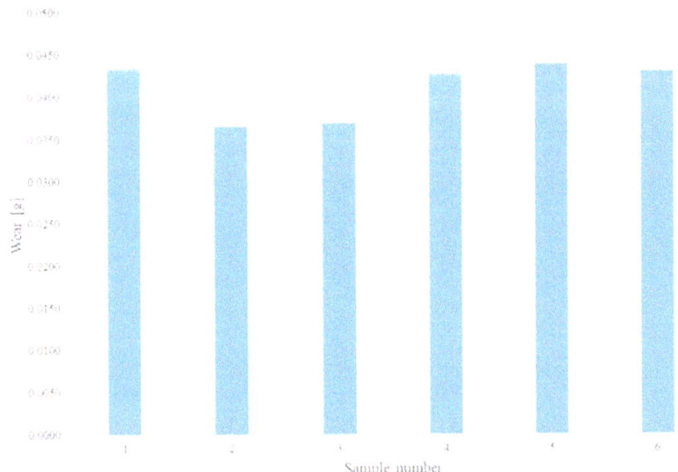

Figure 9. Wear of all samples, in g, for 22MnCrB5 steel after the hot-forming process.

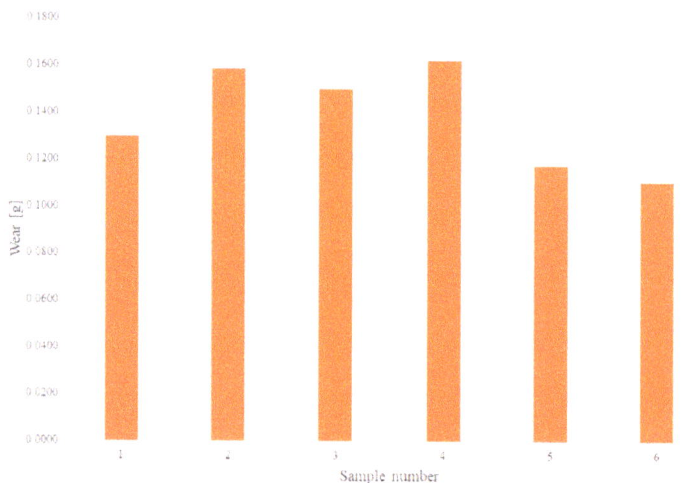

Figure 10. Wear of all samples, in g, for cold-formed 22MnCrB5 steel.

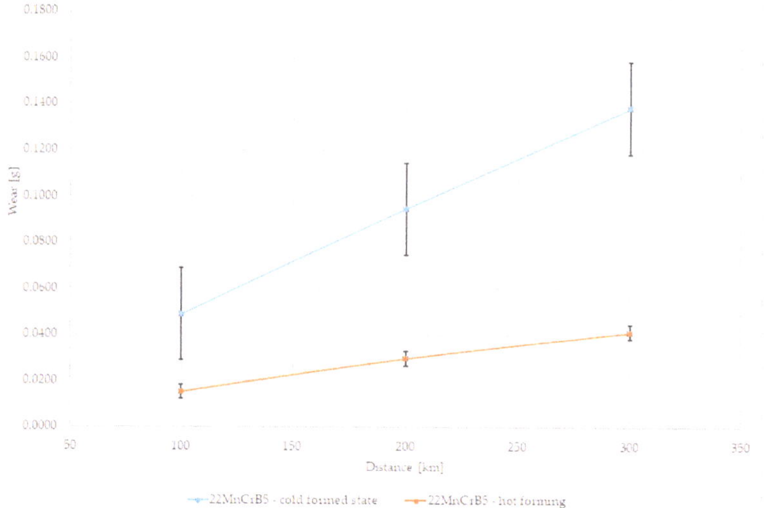

Figure 11. Comparison of the results of the average wear of samples depending on the distance in the abrasive mass on the spinning bowl unit.

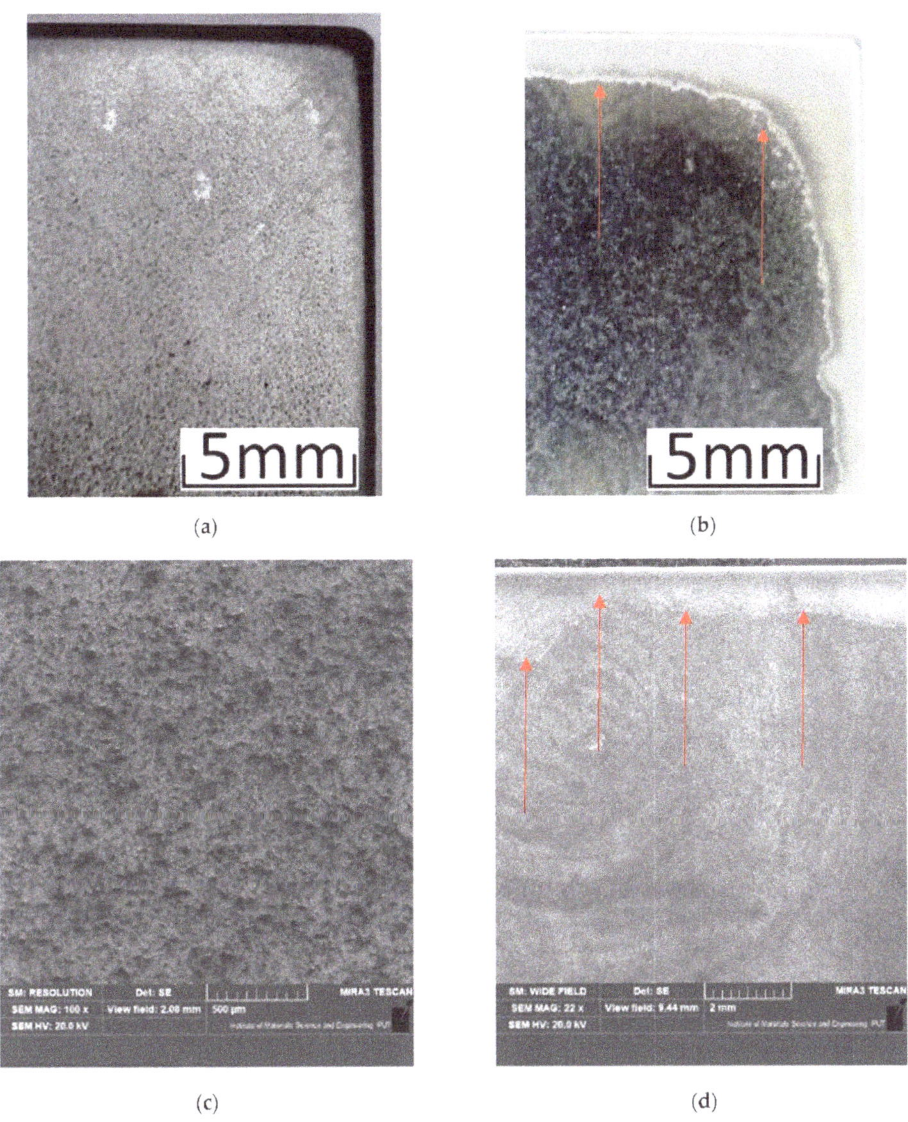

Figure 12. View of sample surface wear: (**a**,**c**) 22MnCrB5 in the cold-formed state, (**b**,**d**) 22MnCrB5 after the hot-forming process.

5. Discussion

The results of wear tests indicate that 22MnCrB5 hot-formed steel is characterized by a greater resistance to tribocorrosion (Tables 4 and 5). In the case of this steel, a significantly smaller material loss was found under the conditions of simultaneous frictional and corrosive effects. The main reason for this is a significant increase in the hardness of the material obtained as a result of the hot-forming process. The higher hardness of hot-formed steel resulted in shallower wear marks (than for 22MnCrB5 steel in the cold-formed state) after tests under only mechanical impacts.

In the case of both tested steels, a clear synergy effect of friction and corrosion was found in the tribocorrosion process. Most likely, this effect was caused by the influence of the frictional interactions in the ball-on-plate sliding association on the corrosion processes. The synergy effect (ΔZ) was smaller for the 22MnCrB5 hot-formed steel, around 28%. The lower effects of the interaction of friction and corrosion resulted from the greater resistance of the hot-formed 22MnCrB5 steel to the corrosive action of 3.5% NaCl (Figure 6). It is worth emphasizing that in the analyzed case, the hot-forming process improved the hardness and corrosion resistance of the material. Consequently, it may be suitable for steels used in the construction of friction junctions operated in tribocorrosion conditions.

Based on the microscopic observations of the surface of the wear trace, an attempt was made to identify the elementary mechanisms causing removal of the material. Sample views of the surface after tests are presented in Figure 13. In the case of only mechanical interactions, micro-cutting dominates for both samples. This is evidenced by parallel grooves in the direction of the ball sliding (counter samples). These traces were clearer and deeper for 22MnCrB5 steel in the cold-formed state (lower hardness of the sample). After tribocorrosion tests, traces of micro-cutting were less visible. This may be the result of the removal of the surface layers of the material by corrosive interactions. In the case of 22MnCrB5 hot-formed steel, numerous cracks are visible. This material is characterized by high hardness and may be prone to local cracking under conditions of high unit pressures. Such conditions prevail in a ball-on-plate association. The presence of the electrolyte and its corrosive effect favor the propagation of microcracks in the subsurface layers of the material.

An additional layer of AlSi, which was applied to the surface of the sheet before the hot-forming process, had a significant influence on corrosion. This is a standard operation that primarily prevents oxidation of the sheet during the hot-stamping process. Therefore, Figure 14 shows the chemical composition of the surface layer before and after the test, depending on the treatment (cold-formed state and hot-formed state). In the case of the cold-formed state sample, the minimum Al values were observed. For samples after the hot-forming process, the content of this element was many times higher due to the additional coating. The influence of the tribocorrosion phenomenon causes the partial abrasion of this layer and the decrease of the Al content by about half in relation to the original value.

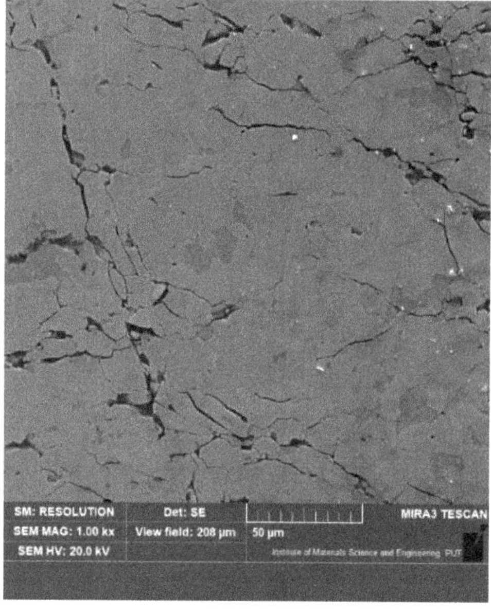

(a)

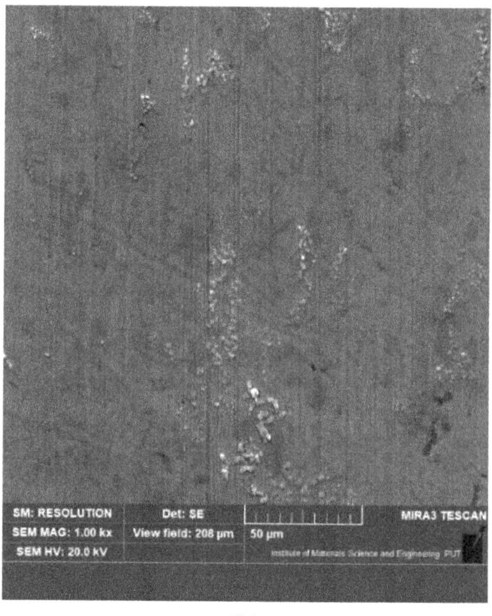

(b)

Figure 13. *Cont.*

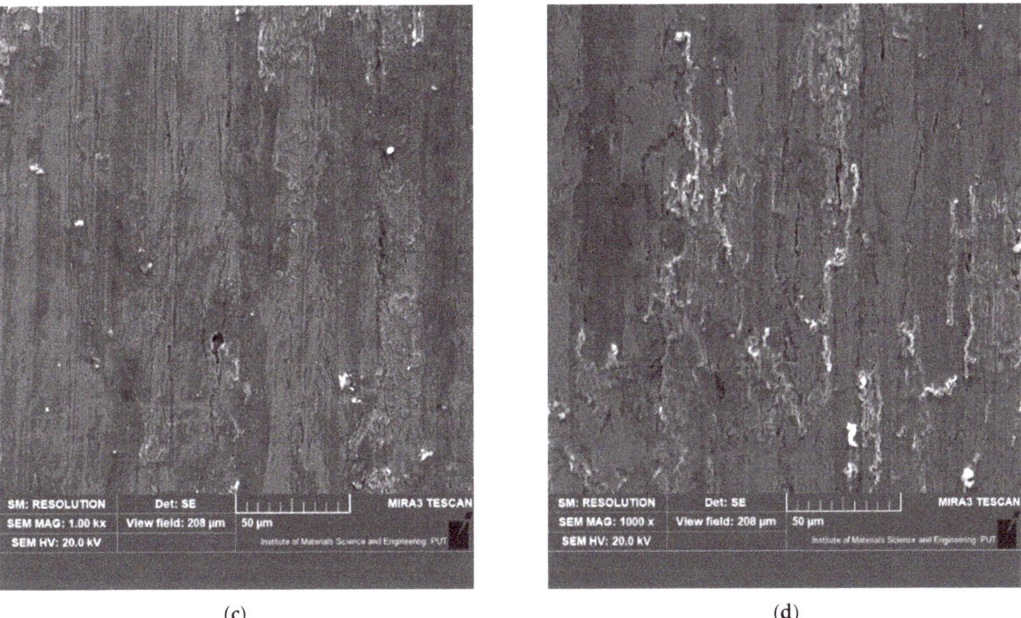

(c) (d)

Figure 13. View of the surface after the wear tests: (**a**) 22MnCrB5 hot-formed steel—tribocorrosion test, (**b**) 22MnCrB5 hot-formed steel—mechanical wear test, (**c**) 22MnCrB5 cold-formed state—tribocorrosion test, (**d**) 22MnCrB5 cold-formed state—mechanical wear test.

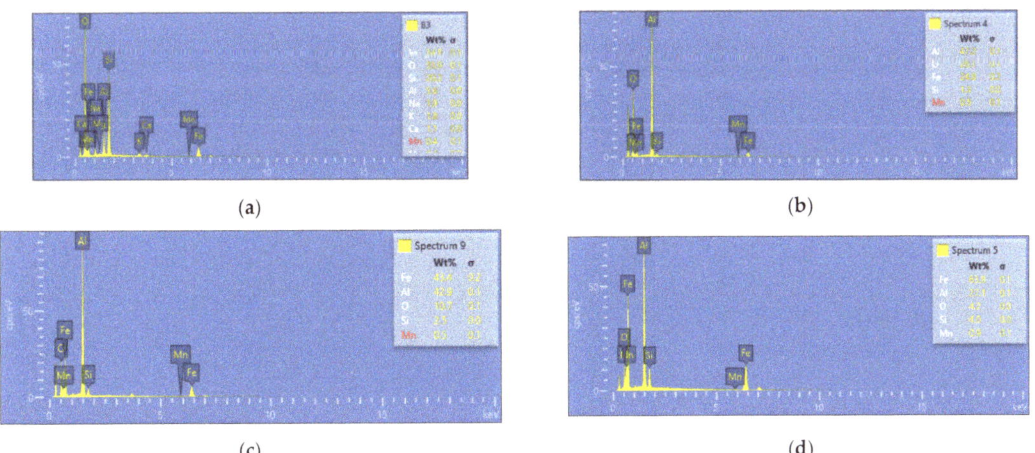

(a) (b)

(c) (d)

Figure 14. View of the chemical composition of the surface for 22MnCrB5 steel: (**a**) for cold-formed steel, and (**b**–**d**) for hot-formed steel. (**a**,**b**) The base material, (**c**) after the corrosion condition, and (**d**) after the tribocorrosion condition.

Based on the wear tests of the samples in the abrasive mass on the spinning bowl unit, it was found that for the 22MnCrB5 steel (after the hot-forming process), over three times lower wear was determined than for the 22MnCrB5 cold-formed state steel. This fact can be due to the presence of martensite in the material structure and the additional AlSi layer. It was noticed that the wear intensity for 22MnCrB5 untreated steel was constant, while for

22MnCrB5 hot-formed steel, a slight decrease was observed. Tests were carried out over a distance of 300 km. The wear measurements were conducted every 100 km of the test. For 22MnCrB5 steel in the cold-formed state, it was found that the average wear of the samples was 0.00049 g/km for the first 100 km, and for the next 100 and 200 km, 0.00046 and 0.00044 g/km, respectively. For 22MnCrB5 steel (after the hot-forming process), the wear was 0.00015, 0.00014, and 0.00011 g/km, for each 100 km of the test in the spinning bowl unit. The average wear intensity for 22MnCrB5 untreated steel was 0.00046 g per km, while for 22MnCrB5 hot-formed steel it was 0.00014 g per km. Based on the view of the surface structure after the wear tests (Figure 15) in the abrasive mass, it was found that the main mechanism destroying the structure of the material was the crushing of particles from the surface layer. In the case of samples after the hot-forming process, damages of the AlSi layer were additionally noticed (Figure 11b,d and Figure 13b—the area of damage to the layer is marked with arrows). Moreover, for 22MnCrB5 cold-formed state steel, cracks in the surface layer of the material were visible in several places. A smaller number of cracks were also noticed for the 22MnCrB5 hot-formed steel, which occurred in the area where the AlSi coating was removed. However, in the case of the two tested samples, there were no observed particles in the surface layer, which was shown in [54] for boron steel.

In addition, in the case of testing the wear resistance of materials in the abrasive mass, the surface roughness profile (surface topography) was verified both before and after the test using the spinning bowl method (Figure 16). For the roughness profile study, the surface was analyzed in the most worn area before the fault, and near the hole in the least worn area. Based on the obtained results, it was found that there were local increases in the roughness profile. This may be caused by impacts of the abrasive particles, which locally damaged the surface and crushed the material (this was also confirmed by cracks on the surface of the samples). This was especially visible for 22MnCrB5 cold-formed steel. In the case of steel after the hot-forming process, a greater smoothing of the surface of the samples was visible. Moreover, detachment of the AlSi coating was found. The wear of this coating is additionally illustrated in Figure 17. The additional coating allowed for the reduction of material wear.

Additional treatment in the form of the hot-forming process allowed to improve the anti-wear and anti-corrosion properties. Similar results of increasing the wear resistance were obtained by the authors of [55], who reduced the wear of boron steel (30MnCrB4) used for rotavator blades via cryogenic treatment. The main reason why the hot-formed samples have better corrosion and tribocorrosion resistance than cold-formed samples is the change in the microstructure of the steel. The pearlitic structure, which is the basic structure of 22MnCrB5 steel, was changed into a martensitic structure during the hot-forming process. Accordingly, the hardness of the samples changed. Additionally, to avoid oxidation during the hot-forming process, an AlSi layer was applied. As the test results show, this coating can additionally protect the material against corrosive agents.

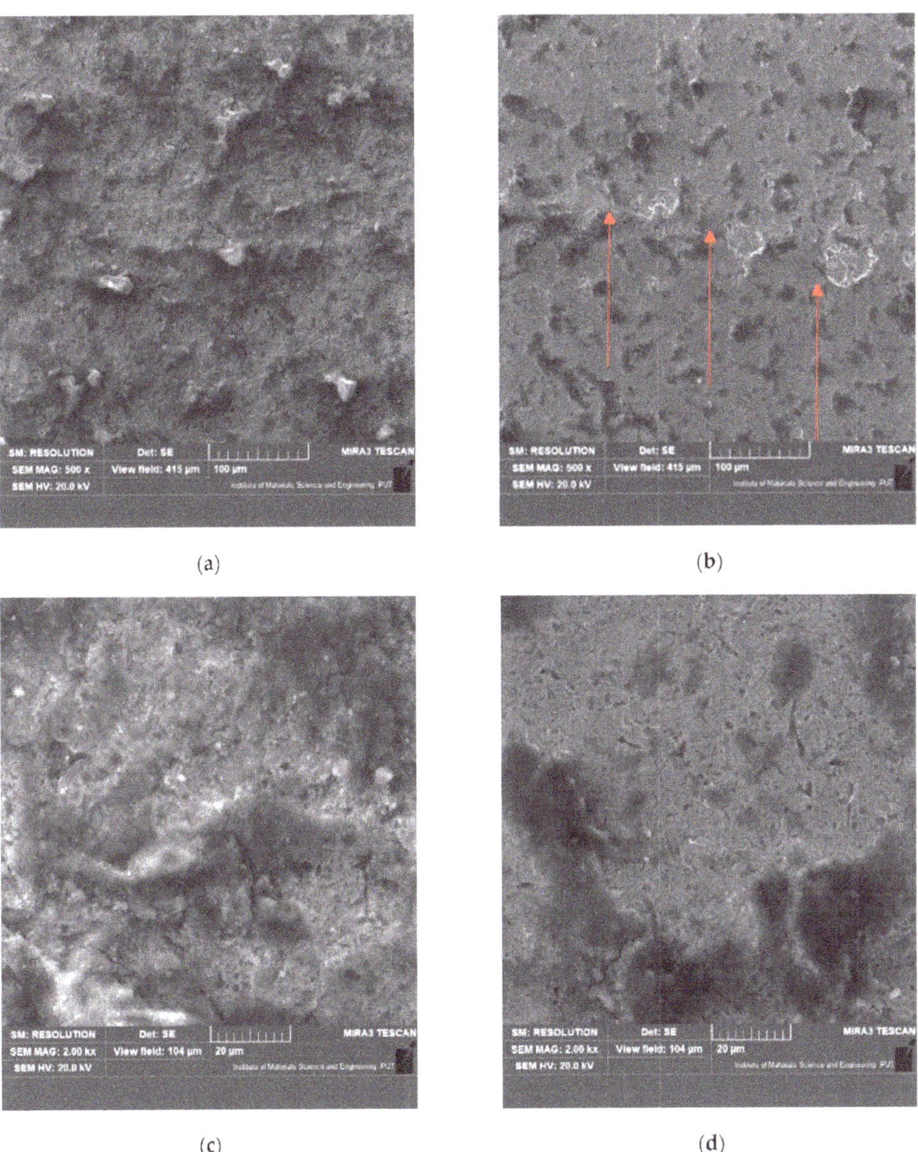

Figure 15. View of the surface after the wear tests: (**a**,**c**) 22MnCrB5 steel in cold-formed state, (**b**,**d**) 22MnCrB5 hot-formed steel.

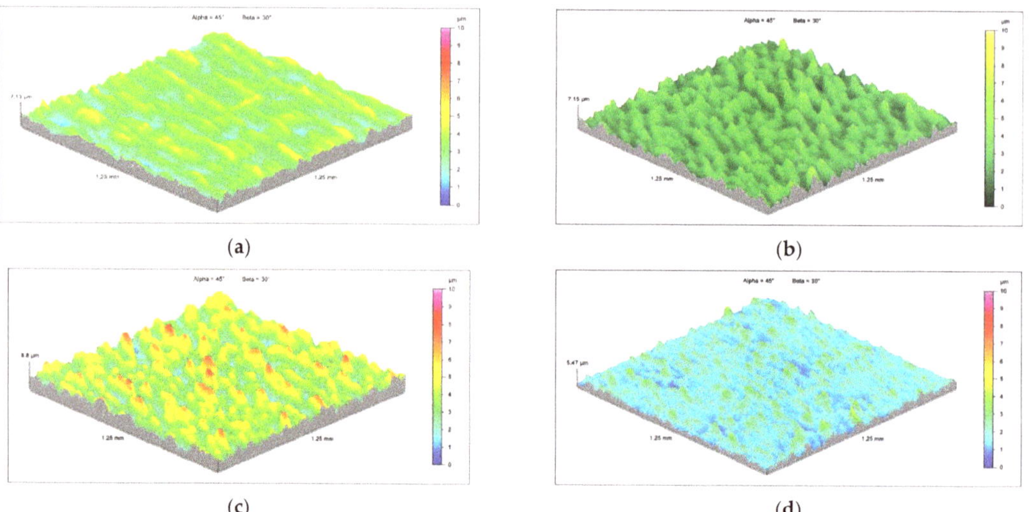

Figure 16. Roughness profile topography: (**a**) 22MnCrB5 cold-formed steel before the test, (**b**) 22MnCrB5 cold-formed steel after the test, (**c**) 22MnCrB5 hot-formed steel before the test, (**d**) 22MnCrB5 hot-formed steel after the test.

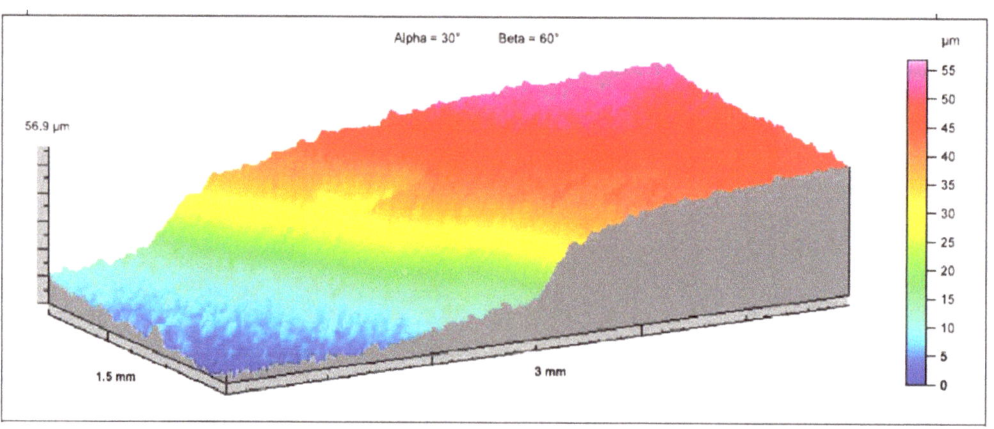

Figure 17. Roughness profile topography for the 22MnCrB5 hot-formed sample after the test with the decrease of the surface profile.

6. Conclusions

Based on the performed research, the following conclusions can be made:

- In the case of both tested materials (in the cold-formed state and the hot-forming process), a clear synergy effect of friction and corrosion was identified in the tribocorrosion process. This effect was most likely caused by the influence of frictional interactions on the course of electrochemical phenomena on the material surface.
- The test results indicated that 22MnCrB5 hot-formed steel obtained significantly greater resistance to tribocorrosion. For this material, a smaller material loss was found after tribocorrosion tests, as well as a smaller share of the friction–corrosion synergy effect in total wear.

- The hot-forming process significantly improved the anti-wear properties of boron steel due to the change in the internal structure (mechanical properties) and an additional AlSi coating applied on the surface of the sample.
- The performed laboratory tests showed that the use of the hot-forming technology for boron steel significantly reduced the abrasive wear and limited the corrosion steel of the process. This may result in an increase in the operation time of agricultural and machinery components working in soil.

A disadvantage (limitation) related to the use of the technology of shaping machine elements by plastic deformation may be the differentiation of the structure, and consequently, the properties of the material on the surface. Unfortunately, such a state favors the development of corrosion (the presence of corrosive micro-cells). Therefore, based on the performed tests and the analysis of their results, within a certain range of conditions affecting the tested steels, the analyzed material solution and the hot-forming process can be used.

The directions of further research should include the comparison of wear and tribocorrosion of steel subjected to the hot-formed process with steel that has undergone various heat treatments. This will allow to optimize the production of selected machine parts, especially in terms of wear and corrosion resistance, which may have a significant impact on the lifetime of a selected machine element.

Author Contributions: Conceptualization, D.U. and A.S.; methodology, D.U. and A.S.; software, J.K. and W.M.; validation, D.U., D.W. and J.K.; formal analysis, J.K. and D.W.; investigation, D.U., A.S. and D.W.; resources, A.S. and W.M.; data curation, D.U. and J.K.; writing—original draft preparation, D.U., A.S. and W.M.; writing—review and editing, D.U.; visualization, J.K. and A.S.; supervision, A.S.; project administration, J.K., D.W. and W.M.; funding acquisition, A.S. All authors have read and agreed to the published version of the manuscript.

Funding: This work has been financed by the Statutory Activities Fund of the Institute of Machines and Motor Vehicles, Poznan University of Technology (PL), 0414/SBAD/3612.

Institutional Review Board Statement: Not applicable.

Informed Consent Statement: Not applicable.

Data Availability Statement: Not applicable.

Conflicts of Interest: The authors declare no conflict of interest.

References

1. Chintha, A.R. Metallurgical aspects of steels designed to resist abrasion, and impact-abrasion wear. *Mater. Sci. Technol.* **2019**, *35*, 1133–1148. [CrossRef]
2. Krawiec, P.; Waluś, K.; Warguła, Ł.; Adamiec, J. Wear evaluation of elements of V-belt transmission with the application of optical microscope. *MATEC Web Conf.* **2018**, *157*, 01009. [CrossRef]
3. Trembach, B.; Grin, A.; Subbotina, V.; Vynar, V.; Knyazev, S.; Zakiev, V.; Trembach, I.; Kabatskyi, O. Effect of Exothermic Addition (CuO-Al) on the Structure, Mechanical Properties and Abrasive Wear Resistance of the Deposited Metal During Self-Shielded Flux-Cored Arc Welding. *Tribol. Ind.* **2021**, *43*, 452–464. [CrossRef]
4. Ulbrich, D.; Kowalczyk, J.; Stachowiak, A.; Sawczuk, W.; Selech, J. The Influence of Surface Preparation of the Steel during the Renovation of the Car Body on Its Corrosion Resistance. *Coatings* **2021**, *4*, 384. [CrossRef]
5. Devimeenakhi, S.; Anandhi, M.; Velkannan, V.; Balaji, G. Effect of green tea in artificial saliva on the corrosion resistance behaviour of stainless steel. *Mater. Today Proc.* 2021; *in press*. [CrossRef]
6. Munoz, A.I.; Espallargas, N.; Mischler, S. *Tribocorrosion*; Springer: Berlin/Heidelberg, Germany, 2020.
7. Kajdas, C.z.; Hiratsuka, K. (Eds.) *Tribocatalysis, Tribochemistry, and Tribocorrosion*; Pan Stanford Publishing: Singapore, 2018.
8. Landolt, D.; Mischler, S. (Eds.) *Tribocorrosion of Passive Metals and Coatings*; Woodhead Publishing in Materials: Cambridge, UK, 2011.
9. Jemmely, P.; Mischler, S.; Landolt, D. Electrochemical modeling of passivation phenomena in tribocorrosion. *Wear* **2000**, *237*, 63–76. [CrossRef]
10. Olsson, C.-O.A.; Landolt, D. Passive films on stainless steels—Chemistry, structure and growth. *Electrochim. Acta* **2003**, *48*, 1093–1104. [CrossRef]

11. Mischler, S. Triboelectrochemical techniques and interpretation methods in tribocorrosion: A comparative evaluation. *Tribol. Int.* **2008**, *41*, 573–583. [CrossRef]
12. Papageorgiou, N.; Mischler, S. Electrochemical simulation of the current and potential response in sliding tribocorrosion. *Tribol. Lett.* **2012**, *48*, 281–283. [CrossRef]
13. Songbo, Y.; Li, D.Y. A new phenomenpn observed in determining the wear-corrosion synergy during a corrosive sliding wear test. *Tribol. Lett.* **2008**, *29*, 45–52.
14. Landolt, D.; Mischler, S.; Stemp, M. Electrochemical methods in tribocorrosion: A critical appraisal. *Electrochim. Acta* **2001**, *46*, 3913–3929. [CrossRef]
15. Stack, M.M. Mapping tribo-corrosion processes in dry and in aqueous conditions: Some new directions for the new millennium. *Tribol. Int.* **2022**, *35*, 681–689. [CrossRef]
16. Mindivan, F.; Yildirim, M.P.; Bayindir, F.; Mindivan, H. Corrosion and Tribocorrosion Behavior of Cast and Machine Milled Co-Cr Alloys for Biomedical Applications. *Acta Phys. Pol. A* **2016**, *129*, 701–704. [CrossRef]
17. Stachowiak, A.; Tyczewski, P.; Zwierzycki, W. The application of wear maps for analyzing the results of research into tribocorrosion. *Wear* **2016**, *352–353*, 146–154. [CrossRef]
18. Han, X.; Zhang, Z.; Barber, G.; Thrush, S.; Li, X. Wear Resistance of Medium Carbon Steel with Different Microstructures. *Materials* **2021**, *14*, 2015. [CrossRef] [PubMed]
19. Trembach, B.; Grin, A.; Turchanin, M.; Makarenko, N.; Markov, O.; Trembach, I. Application of Taguchi method and ANOVA analysis for optimization of process parameters and exothermic addition (CuO-Al) introduction in the core filler during self-shielded flux-cored arc welding. *Int. J. Adv. Manuf. Technol.* **2021**, *114*, 1099–1118. [CrossRef]
20. Dutta Majumdar, J.; Ramesh Chandra, B.; Manna, I. Laser composite surfacing of AISI 304 stainless steel with titanium boride for improved wear resistance. *Tribol. Int.* **2007**, *40*, 146–152. [CrossRef]
21. Devaraju, A. A critical review on different types of wear of materials. *Int. J. Mech. Eng. Technol.* **2015**, *6*, 77–83.
22. Barzegari, G.; Uromeihy, A.; Zhao, J. Parametric study of soil abrasivity for predicting wear issue in TBM tunneling projects. *Tunn. Undergr. Space Technol.* **2015**, *48*, 43–57. [CrossRef]
23. Mosleh, M.; Gharahbagh, E.A.; Rostami, J. Effects of relative hardness and moisture on tool wear in soil excavation operations. *Wear* **2013**, *1*, 1555–1559. [CrossRef]
24. Napiórkowski, J. Analiza właściwości glebowej masy ściernej w aspekcie oddziaływania zużyciowego. *Tribologia* **2010**, *5*, 53–62.
25. Napiórkowski, J. Zużyciowe oddziaływanie gleby na elementy robocze narzędzi rolniczych. *Inżynieria Rol.* **2005**, *9*, 3–171.
26. Natsis, A.; Petropoulos, G.; Pandazaras, C. Influence of local soil conditions on mouldboard ploughshare abrasive wear. *Tribol. Int.* **2008**, *41*, 151–157. [CrossRef]
27. Yu, H.-J.; Bhole, S.D. Development of a prototype abrasive wear tester for tillage tool materials. *Tribol. Int.* **1990**, *23*, 309–316. [CrossRef]
28. Capanidis, D. The influence of hardness of polyurethane on its abrasive wear resistance. *Tribologia* **2016**, *4*, 29–40. [CrossRef]
29. Yazici, A. Investigation of the reduction of mouldboard ploughshare wear through hot stamping and hardfacing processes. *Turk. J. Agric. For.* **2011**, *35*, 461–468.
30. Venturato, G.; Novella, M.; Bruschi, S.; Ghiotti, A.; Shivpuri, R. Effects of phase transformation in hot stamping of 22MnB5 high strength steel. *Proceedia Eng.* **2017**, *183*, 316–321. [CrossRef]
31. Merklein, M.; Lechler, J. Investigation of the thermo-mechanical properties of hot stamping steels. *J. Mater. Processing Technol.* **2006**, *177*, 452–455. [CrossRef]
32. Gracia-Escosa, E.; Garcia, I.; de Damborenea, J.J.; Conde, A. Friction and wear behaviour of tool steels sliding against 22MnB5 steel. *J. Mater. Res. Technol.* **2017**, *6*, 241–250. [CrossRef]
33. Cho, L.; Golem, L.; Jung Seo, E.; Bhattacharya, D.; Speer, J.G.; Findley, K.O. Microstructural characteristics and mechanical properties of the AleSi coating on press hardened 22MnB5 steel. *J. Alloy. Compd.* **2020**, *846*, 156349. [CrossRef]
34. Dosdat, L.; Petitjean, J.; Vietoris, T.; Clauzeau, O. Corrosion Resistance of Different Metallic Coatings on Press-Hardened Steels for Automotive. *Steel Res. Int.* **2011**, *82*, 726–733. [CrossRef]
35. Allely, C.; Dosdat, L.; Clauzeau, O.; Ogle, K.; Volovitch, P. Anticorrosion mechanisms of aluminized steel for hot stamping. *Surf. Coat. Technol.* **2014**, *238*, 188–196. [CrossRef]
36. Park, A.; Kim, J.G.; He, Y.S.; Shin, K.S.; Yoon, J.B. Comparative Study on the Corrosion Behavior of the Cold Rolled and Hot Rolled LowAlloy Steels Containing Copper and Antimony in Flue Gas Desulfurization Environment. *Phys. Met. Metallogr.* **2014**, *115*, 1285–1294. [CrossRef]
37. Romek, D.; Selech, J.; Ulbrich, D.; Felusiak, A.; Kieruj, P.; Janeba-Bartosiewicz, E.; Pieniak, D. The impact of padding weld shape of agricultural machinery tools on their abrasive wear. *Tribologia* **2020**, *290*, 55–62. [CrossRef]
38. Hrabe, P.; Muller, M. Research of overlays influence on ploughshare lifetime. *Res. Agric. Eng.* **2013**, *59*, 147–152. [CrossRef]
39. Konstencki, P.; Stawicki, T.; Królicka, A.; Sedłak, P. Wear of cultivator coulters reinforced with cemented-carbideplates and hardfacing. *Wear* **2019**, *438*, 203063. [CrossRef]
40. Paczkowska, M.; Selech, J.; Piasecki, A. Effect of surface treatment on abrasive wear resistance of seeder coulerflap. *Surf. Rev. Lett.* **2016**, *23*, 1650007. [CrossRef]
41. Aramide, B.; Pityana, S.; Sadiku, R.; Jamiru, T.; Popoola, P. Improving the durability of tillage tools through surface modification—A review. *Int. J. Adv. Manuf. Technol.* **2021**, *116*, 83–98. [CrossRef]

42. Dilay, J.; Guney, B.; Ozkan, A.; Oz, A. Microstructure and wear properties of WC-10Co-4Cr coating to cultivator blades by DJ-HVOF. *Emerg. Mater. Res.* **2021**, *10*, 278–288. [CrossRef]
43. Kostencki, P.; Stawicki, T.; Białobrzeska, B. Durability and wear geometry of subsoiler shanks provided with sintered carbide plates. *Tribol. Int.* **2016**, *104*, 19–35. [CrossRef]
44. Foley, A.G.; Chisholm, C.J.; Mclees, V.A. Wear of ceramic-protected agricultural subsoilers. *Tribol. Int.* **1988**, *21*, 97–103. [CrossRef]
45. Steinhoff, K.; Weidig, U.; Scholtes, B.; Zinn, W. Innovative flexible metal forming processes based on hybrid thermo-mechanical interaction. *Steel Res. Int.* **2005**, *76*, 154–159. [CrossRef]
46. Karbasian, H.; Tekkaya, A.E. A review on hot stamping. *J. Mater. Process. Technol.* **2010**, *210*, 2103–2118. [CrossRef]
47. Behrens, B.A.; Brunotte, K.; Weste, H.; Kock, C. Experimental investigations on the interactions between the process parameters of hot forming and the resulting residual stresses in the component. *Procedia Manuf.* **2020**, *50*, 706–712. [CrossRef]
48. Yusoff, A.R.; Lim, S.K.; Ramadan, M. Microstructure and Mechanical Properties of Boron Sheet Metal Steels in Hot Press Forming Process With Nanofluid as a Coolant. *Encycl. Smart Mater.* **2022**, *3*, 266–280.
49. Rong, H.; Hu, P.; Ying, L.; Hou, W.; Dai, M. Modeling the anisotropic plasticity and damage of AA7075 alloy in hot forming. *Int. J. Mech. Sci.* **2022**, *215*, 106951. [CrossRef]
50. Wu, Y.; Fan, R.; Qin, Z.; Chen, M. Shape controlling and property optimization of TA32 titanium alloy thin-walled part prepared by hot forming. *Trans. Nonferrous Met. Soc. China* **2021**, *31*, 2336–2357. [CrossRef]
51. Bong, H.J.; Yoo, D.H.; Kim, D.; Kwon, Y.N.; Lee, J. Correlative Study on Plastic Response and Formability of Ti-6Al-4V Sheets under Hot Forming Conditions. *J. Manuf. Processes* **2020**, *58*, 775–786. [CrossRef]
52. Stachowiak, A.; Zwierzycki, W. Analysis of the tribocorrosion mechanisms in a pin-on-plate combination on the example of AISI304 steel. *Wear* **2012**, *294–295*, 277–285. [CrossRef]
53. Ghanbarzadeh, A.; Salehi, F.M.; Bryant, M.; Neville, A. A New Asperity-Scale Mechanistic Model of Tribocorrosive Wear: Synergistic Effects of Mechanical Wear and Corrosion. *J. Tribol.* **2019**, *141*, 021601. [CrossRef]
54. Hernandez, S.; Hardell, J.; Winkelmann, H.; Rodriguez Ripoll, M.; Prakash, B. Influence of temperature on abrasive wear of boron steel and hot forming tool steels. *Wear* **2015**, *338–339*, 27–35. [CrossRef]
55. Singh, T.P.; Singla, A.K.; Singh, J.; Singh, K.; Gupta, M.K.; Ji, H.; Song, Q.; Liu, Z.; Pruncu, C.I. Abrasive Wear Behavior of Cryogenically Treated Boron Steel (30MnCrB4) Used for Rotavator Blades. *Materials* **2020**, *13*, 436. [CrossRef] [PubMed]

Article

Hydrothermal Corrosion of Double Layer Glass/Ceramic Coatings Obtained from Preceramic Polymers

Ivana Parchovianská [1,*], Milan Parchovianský [1], Hana Kaňková [1], Aleksandra Nowicka [1] and Dušan Galusek [1,2]

[1] Centre for Functional and Surface Functionalised Glass, Alexander Dubček University of Trenčín, Študentská 2, 911 50 Trencin, Slovakia; milan.parchoviansky@tnuni.sk (M.P.); hana.kankova@tnuni.sk (H.K.); aleksandra.nowicka@tnuni.sk (A.N.); dusan.galusek@tnuni.sk (D.G.)

[2] Joint Glass Centre of the Institute of Inorganic Chemistry Slovak Academy of Sciences, Alexander Dubček University of Trenčín and Faculty of Chemical and Food Technology Slovak University of Technology, Študentská 2, 911 50 Trencin, Slovakia

* Correspondence: ivana.parchovianska@tnuni.sk

Abstract: Polysilazane-based double layer composite coatings consisting of a polymer-derived ceramic (PDC) bond-coat and a PDC top-coat that contains ceramic passive and glass fillers were developed. To investigate the environmental protection ability of the prepared coatings, quasi-dynamic corrosion tests under hydrothermal conditions were conducted at 200 °C for 48–192 h. The tested PDC coatings exhibited significant mass loss of up to 2.25 mg/cm^2 after 192 h of corrosion tests, which was attributed to the leaching of elements from the PDC coatings to the corrosion medium. Analysis of corrosion solutions by inductively coupled plasma optical emission spectrometry (ICP-OES) confirmed the presence of Ba, Al, Si, Y, Zr, and Cr, the main component of the steel substrate, in the corrosion medium. Scanning electron microscopy (SEM) of the corroded surfaces revealed randomly distributed globular crystallites approximately 3.5 μm in diameter. Energy-dispersive X-ray spectroscopy (EDXS) of the precipitates showed the presence of Ba, Al, Si, and O. The predominant phases detected after corrosion tests by X-ray powder diffraction analysis (XRD) were monoclinic and cubic ZrO_2, originating from the used passive fillers. In addition, the crystalline phase of $BaAl_2Si_2O_8$ was also identified, which is in accordance with the results of EDXS analysis of the precipitates formed on the coating surface.

Keywords: polymer derived ceramics; coatings; fillers; stainless steel; hydrothermal corrosion

Citation: Parchovianská, I.; Parchovianský, M.; Kaňková, H.; Nowicka, A.; Galusek, D. Hydrothermal Corrosion of Double Layer Glass/Ceramic Coatings Obtained from Preceramic Polymers. *Materials* **2021**, *14*, 7777. https://doi.org/10.3390/ma14247777

Academic Editor: Yong Sun

Received: 22 November 2021
Accepted: 13 December 2021
Published: 16 December 2021

1. Introduction

Ferritic stainless steels are frequently used as components in high temperature industrial processes, such as solid oxide fuel cells, exhaust gas elements, waste incineration plants or in the chemical industry, because of their excellent corrosion and oxidation resistance as well as their relatively low cost. However, even stainless steels are susceptible to corrosion under extreme conditions and interact with the environment in which they are used, leading to a deterioration of their physical and mechanical properties which adversely affects their further applicability. Thus, corrosion resistance is one of the most important factors for determining the overall material performance [1]. With the help of protective coatings, the surfaces of metals can be adapted for a specific application and their service life can be drastically extended. Advanced silicon-based ceramics are suitable for operation in aggressive chemical environments, as they resist molten metals, molten salts, hot gases, hot solutions of acids, bases and salts, or supercritical water [2]. For example, their corrosion resistance was shown to be significantly higher than that of metals, even at ultrahigh temperatures [3].

In recent years, the polymer-derived ceramic (PDC) synthesis route has gained attention as an economical and easy approach for producing ceramic coatings by the pyrolysis of suitable organoelemental precursors (preceramic polymers) [4–7]. The chemical composition of the ceramic after pyrolysis is based on the chemistry of the starting precursors, the preparation conditions (temperature, pressure, pyrolysis atmosphere) and the possible incorporation of other additives or modifiers (fillers, metal alkoxides). Polysiloxanes [8,9], polycarbosilanes [10,11] and polysilazanes [12], characterized by Si–O, Si–N and Si–C chains, allow the preparation of amorphous $Si_xO_yC_z$, $Si_xC_yN_z$ and Si_xC_y ceramics after pyrolysis in an inert atmosphere [4,13]. The pyrolysis mechanism involved in the ceramization process is quite complex and involves structural rearrangement and radical reactions that result in the cleavage of Si–H, Si–C and C–H chemical bonds, the release of organic functional groups (CH_4, C_6H_6, CH_3NH_2) and the formation of an inorganic network. However, the greatest disadvantage of this approach is high shrinkage of the organosilicon polymer (up to 50 vol. %, depending on the precursor) during pyrolysis [14]. The shrinkage of the precursor can be compensated by adding active ($ZrSi_2$ [15] or $TiSi_2$ [16]) and/or passive fillers (BN [17], Si_3N_4 [18], ZrO_2 [19], NbC [20] or Al_2O_3 [21]). The use of fillers not only reduces the shrinkage of the polymer during pyrolysis, but also allows the preparation of composites with special properties, e.g., high mechanical strength, thermal and electrical conductivity, and improved corrosion and oxidation resistance. When preparing PDCs with the addition of passive fillers, the coefficient of thermal expansion (CTE) of the passive filler must be taken into account, as it affects the total CTE of the final composite. The ability to adapt the CTE is particularly useful in the preparation of coatings on metals, because a large mismatch of the CTE between the coating and a metallic substrate can cause the formation of cracks. This can adversely affect the final mechanical properties and eventually even cause the spallation of the coating from the substrate. The application of polymer precursors represents one of the greatest advantages of PDCs, since all common methods used for application and shaping of conventional polymers can, in principle, also be applied to polymer precursors. Several authors have prepared PDC coatings by spraying [22], dip coating [23,24], the doctor blade method [25], tape casting [26], or spin coating [27]. These features of the PDC can be successfully utilized in the protection of metals, especially steels, used in the construction of heat exchangers in thermal power plants and municipal waste incinerators, which are exposed to the long-term corrosive action of combustion gases at high temperatures. A variety of PDC coatings, such as SiOC [15,16], SiCN [19,28], and SiON [28,29], have been investigated in order to improve the corrosion and oxidation resistance of metallic substrates. However, there is a lack of information about their corrosion behavior in hot, aqueous solutions under subcritical conditions.

The objective of the present study is to demonstrate the possibility of fabrication of environmental barrier coatings from preceramic polymers as well as to asses and discuss their corrosion behavior under hydrothermal conditions at 200 °C. Thick (~40 μm) PDC coatings were prepared, as the increased coating thickness is expected to act as a more robust protective barrier and provide better separation between corrosion medium and the steel substrate. The extent of corrosion in the coatings was monitored by the weight change of the corroded samples, followed by detailed study of the microstructure, phase and chemical composition before and after corrosion tests, and the formation of corrosion products. Analysis of specific elements released from the tested specimens into the corrosion medium was also performed. Our previous work identified this type of PDC glass/ceramic coatings as a suitable coating system for the protection of stainless steel from oxidation in synthetic air and water vapor atmospheres up to 950 °C [30]. The results presented in this paper introduce new information on the corrosion behavior of thick PDC coatings suitable for applications in the field of chemical industry to protect metals against corrosion.

2. Materials and Methods

2.1. Coatings Processing

The low carbon ferritic stainless steel AISI 441 was selected as a substrate. This material is widely used as a structural material in systems containing water at elevated temperatures and pressures, with the maximum recommended temperature of use 950 °C in dry air.

The 1 mm thick steel substrates were dimensioned as 10×15 mm^2 sheets and ultrasonically cleaned in acetone, ethanol, and deionized water (10 min for each) in order to remove the impurities and degrease the surface. Two types of organosilicon precursors were selected: perhydropolysilazane Durazane2250 (Merck KGaA, Darmstadt, Germany) was used to prepare the ceramic interlayer (bond-coat), and Durazane1800 (Merck KGaA, Darmstadt, Germany), liquid organosilazane, was used to prepare the ceramic top-coat. The bond-coat was obtained by the conventional dip-coating technique (dip-coater RDC 15, Relamatic, Glattbrugg, Switzerland): a deposition time of 20 s and a hoisting speed of 0.3 m/min were employed. After deposition, the bond-coat was heat treated in air at 450 °C for 1 h using a heating rate of 5 °C/min (Nabertherm® N41/H, Nabertherm, Lilienthal, Germany). Two prepared preceramic suspensions (denoted as C2c and D4), both containing ZrO_2 stabilized with 8 mol. % Y_2O_3 (8YSZ, Inframat® Advanced Materials TM, Manchester, CT, USA) and a commercial barium aluminosilicate glass (G018-281, Schott AG, Mainz, Germany) as passive fillers, were used to deposit the top-coat onto the pyrolyzed bond-coat by spray-coating (Nordson EFD 781S-SS). The composite glass/ceramic coatings were cured in air at 850 °C at a heating rate of 3 °C/min (Classic 0718E, Prague, Czech Republic) with a 1 h isothermal dwell. In the case of the D4 suspension, a polycrystalline Al2O3-Y2O3-ZrO2 (AYZ) precursor powder prepared by the modified Pechini sol–gel method was used as an additional filler. The filler material was included in the mixture to modify the CTE of the top-coat to closely match the CTE of the steel substrate (11×10^{-6}/K) and to minimize volume changes during the polymer to ceramic transformation. The compositions of the prepared coatings are listed in Table 1. The detailed preparation of glass/ceramic coatings as well as AYZ precursor powder and its basic characterization was described in our previous work [30,31]. The basic properties of the filler materials are listed in Table 2.

Table 1. Compositions of prepared composite coatings (vol. %) and their coefficients of thermal expansion (CTE).

Compositions	Durazane1800	YSZ	Glass G018-281	AYZ	CTE (10^{-6}/K)
C2c	30	35	35	-	10.1
D4	30	17.5	35	17.5	9.4

Table 2. Basic properties of filler materials.

Passive Fillers	d50 (μm)	ϱ (g/cm^3)	CTE (10^{-6}/K)
8YSZ	0.5	6.1	11.5
AYZ	1–10	4.6	8.6
Glass G018-281	0.5–5	2.7	12.1

2.2. Hydrothermal Corrosion Tests

The hydrothermal corrosion resistance of the prepared glass/ceramic coatings was monitored under quasi-dynamic conditions. The corrosion tests were carried out in stainless steel pressure reactors with an inner Teflon lining that prevented direct contact of the corrosion medium with the steel reactor; 18 ml of deionized water was used as the corrosion medium. Coated samples and uncoated steel substrates were placed in the steel reactors in special holders to ensure maximum contact area of the specimens with the corrosion medium. Quasi-dynamic tests were carried out in a laboratory oven at 200 °C for 192 h. The temperature was selected to fill the gap in knowledge on the corrosion resistance of PDC coatings under subcritical hydrothermal conditions, and to provide new information on the corrosion behavior of the PDC coatings that may be used in e.g., the chemical industry. At defined time intervals (every 48 h), corroded samples and corrosion medium were removed from the reactors. The coated and uncoated samples were rinsed with deionized water, dried in an oven at 100 °C for 2 h and weighed. The pH value of the corrosion medium after each time interval was determined using a pH meter (Mettler ToledoSevenEasy, Columbus, OH, USA). Next, the corrosion medium was stabilized for chemical analysis using nitric acid to pH < 2 at room temperature. Finally, the tested samples were returned to the steel reactors and the corrosion medium was replaced with fresh deionized water. The tests were carried out in three reactors in parallel, i.e., three samples of the same composition were tested under identical conditions in order to ensure the reproducibility of the corrosion results. Moreover, tests without any sample were performed for the given time intervals, providing a blank solution. The concentrations of the elements identified in the blank were subtracted from the concentrations determined in the corrosion medium obtained after the corrosion tests with the coated and uncoated sample.

2.3. Characterization Methods

The weight changes of the tested samples as a function of corrosion time were measured using an electronic balance with an accuracy of ±0.0001 g (Axis AD 1000, Gdansk, Poland). The content of metallic elements released from the coated and uncoated samples into the corrosion medium were determined by ICP-OES (Agilent 5100 SVDV, Santa Clara, CA, USA). In this work, the amount of Y, Zr, Si, Ba, Al, and Cr, the main component of the steel substrate, leached from the tested samples was considered for each parallel experiment. The amount of elements released into the solution was calculated according to the Formula (1):

$$Q_i^t = \frac{(c_i - c_{blank}).V}{S} + Q_i^{t-\Delta t} \quad (1)$$

where Q_i is the total amount of leached element i in time t [mg/m^2], c_i is the concentration of leached element i in time t [mg/dm^3], c_{blank} is the concentration of element i in the blank [mg/dm^3], V is the volume of the corrosion medium [dm^3], and S is the sample surface in contact with the corrosion medium [m^2]. The microstructure of both uncorroded and hydrothermally corroded samples was examined in detail by SEM (JEOL JSM 7600 F, JEOL, Tokyo, Japan). The chemical composition of the investigated samples and corrosion products formed at the corroded surfaces was determined by EDXS (Oxford instruments, Abingdon, UK). XRD (PANalytical Empyrean DY1098 (Panalytical, BV, Almelo, The Netherlands)) with a Cu anode and an X ray wavelength of λ = 1.5405 Å over 2θ angles of 10–80° was used to determine the phase composition of the stainless steel and prepared coatings before and after corrosion tests. Diffraction records were evaluated using HighScore Plus (v. 3.0.4, Panalytical, Almelo, The Netherlands) with the use of the Crystallographic Open Database COD2019. Raman spectra of the uncoated steel substrates were recorded in the range of the Raman shift (200–800) cm^{-1} by a RENISHAW inVia Reflex Raman spectrometer (RENISHAW, Wotton-under-Edge, UK).

3. Results and Discussion

3.1. Weight Changes Measurement

Corrosion may be defined as the physical and chemical alteration of a material due to its interaction with the surrounding environment, which leads to the loss of functional properties of the material of interest. Corrosion may be broadly divided into two modes: active and passive. The active form is characterized by the loss of material in contact with the environment, which is accompanied by the decrease in size and weight of the specimen. The losses can be in the form of gaseous or dissolved species. The passive form involves processes where the material reacts with the environment to form a new condensed phase covering the surface (layer or scale) and is associated with a weight gain of the component [32]. Hence, the weight gain or loss is an important parameter indicating the lifetime prediction and corrosion mechanism involved.

Figure 1 displays the time dependencies of the cumulative weight changes normalized with respect to the corroded surface area for the coated and uncoated samples tested in deionized water at 200 °C. Both tested coatings, C2c and D4, exhibited significant mass loss during corrosion tests in deionized water compared to the uncoated steel substrates. This indicates an active corrosion mode, which is a typical behavior of Si-based ceramics upon exposure to a hydrothermal environment [33,34]. The observed weight losses of the coated samples can be explained by the reaction of some of the glass-ceramic matrix components with the deionized water and the subsequent dissolution and release of elements from the coatings. In contrast, the uncoated steel substrates showed a relatively small mass gain after the tests due to the formation of corrosion products on their surface. The largest weight gain of uncoated steel exposed to deionized water was observed after 144 h of corrosion tests, reaching a value of only 0.121 mg/cm^{-2}. At the end of the experiment, i.e., after 192 h, the uncoated steel achieved a smaller weight gain of 0.051 mg/cm^2. In the first 48 h of corrosion tests, comparable values of weight losses were measured for both PDC coatings. After 48 h, the weight loss in the C2c coating was found to be higher than for the D4 coating for the rest of the experiment. The measured weight loss in the D4 coating after 192 h was 1.99 mg/cm^2. In the C2c coating the weight loss was higher at 2.25 mg/cm^{-2}. In the time interval up to 96 h, rapid mass loss in both coatings was observed, followed by a slower rate of mass loss until the end of the corrosion tests. This could indicate that a state of saturation was achieved after 96 h of exposure to deionized water. However, there are two phenomena that influence weight change measurements and the determination of corrosion mechanisms in the present case: the dissolution and release of elements from the coatings that cause a mass loss, and the almost immediate corrosion products formation causing mass gain. Other factors, especially inhomogeneous dissolution in some places (e.g., microcracks in the coatings) with simultaneous precipitation of reaction products and the formation of a passivation layer at other places, could influence the mass loss-time dependencies significantly. Therefore, in this case, we do not consider the mass change measurement to be a definitive parameter for evaluating corrosion mechanisms.

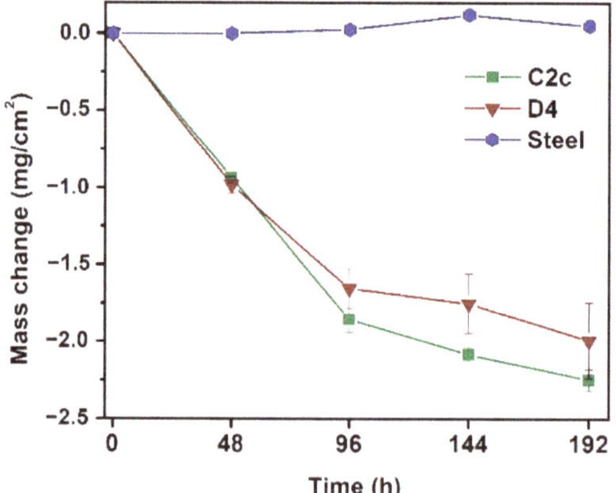

Figure 1. Mass changes measured for coated and uncoated samples tested in deionized water at 200 °C.

3.2. *Surface Morphologies of Corroded Samples*

SEM analysis was used for a detailed study of the surfaces of uncoated and coated samples after corrosion tests. Visual inspection of the uncoated steel substrates revealed that all tested steel samples exhibited a loss in brightness and the initial shiny silver surface turned into a red-yellowish color after the corrosion tests. Figure 2 shows the surface morphologies of stainless steel substrates without any coating after 96 h and 192 h of corrosion tests. As can be seen in Figure 2 (96 h), most of the surface exhibits thin oxide scales consisting of tiny rod-shaped crystals up to 1 µm in size rather than a continuous layer of corrosion products. After 192 h of corrosion testing, the bare substrate shows a homogeneous corrosion attack and growth of the rod-shaped corrosion products all over the surface. The EDXS analysis (not shown) of these crystals formed at the steel's surface showed the presence of Fe, Cr, and O, with a small amount of Mn, indicating

$(Mn, Cr, Fe)_3O_4$ spinel formation. However, due to the small size of crystals and the thickness of the layer of corrosion products, the EDXS analysis was probably affected by the underlying steel substrate and is, therefore, not considered as a suitable indication of the real composition of the crystallites. SEM examination also revealed a few randomly distributed crystallites with spherical morphology, approximately 2.5 µm in diameter, identified by EDXS analysis as a mixture of iron and chromium oxide (see Figure 2—192 h).

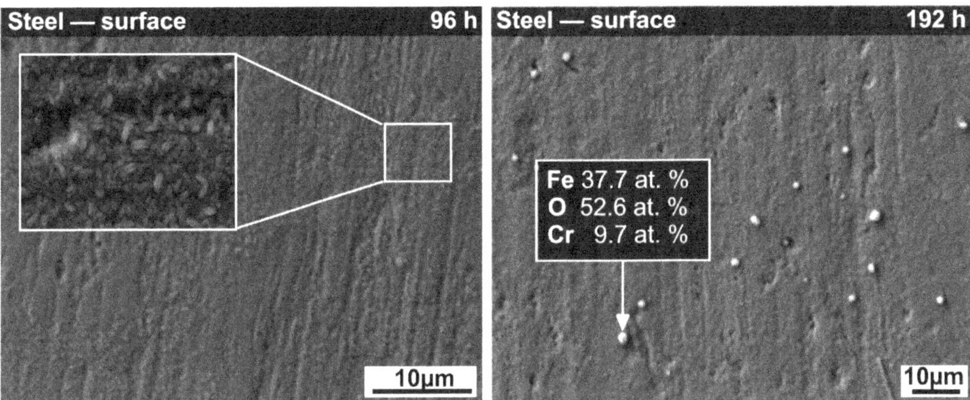

Figure 2. Scanning electron microscopy (SEM) micrographs of uncoated steel surfaces after corrosion tests in deionized water at 200 °C.

SEM micrographs of the C2c and D4 coated samples' surfaces, before and after corrosion tests, are shown in Figure 3. After pyrolysis in air at 850 °C for 1 h, homogeneous and almost dense protective coatings, with only small pores, were prepared. Both coatings, C2c and D4 (Figure 3), showed neither delamination nor significant cracks at the surface. Hence, they were expected to protect the steel substrate against corrosion in deionized water. According to the visual inspection of the coatings after corrosion tests, both compositions showed no signs of corrosion and the coatings adhered very well to the steel substrate. SEM examination of the corroded surfaces (Figure 3) revealed randomly distributed globular crystallites, approximately 3.5 µm in diameter. The precipitates were found to form after 96 h of testing, remaining spherically shaped during the whole experiment. In both coatings, the precipitates were morphologically similar, and from the point of view of their chemical composition, they were identical. The surface of the coating became less smooth with increasing time of exposure to the corrosion medium, which was attributed to crystal growth.

In Figure 4, the morphology of the D4 coating surface, including an EDXS elemental map, is displayed. EDXS analysis of the spherical crystals showed the presence of Ba, Al, Si, and O, indicating the formation of $BaAl_2Si_2O_8$ precipitates. The EDXS map of the surface of the D4 coating was similar to the C2c coating and is, therefore, the only one shown here (Figure 4).

3.3. *Analysis of Corrosion Solutions*

The amount of elements leached from the tested samples to the corrosion medium was determined by ICP-OES. Stainless steels contain Fe, C and at least 11 % of Cr, the element responsible for their corrosion resistance. In the case of uncoated steel (Figure 5), only the amounts of leached Cr and Si were considered because other elements contained in the steel were below the detection limit of the applied analytical method. Si was detected in the corrosion solutions, as Si, a common element in materials for elevated temperature applications, is also included in the composition of the studied ferritic AISI 441 stainless steel.

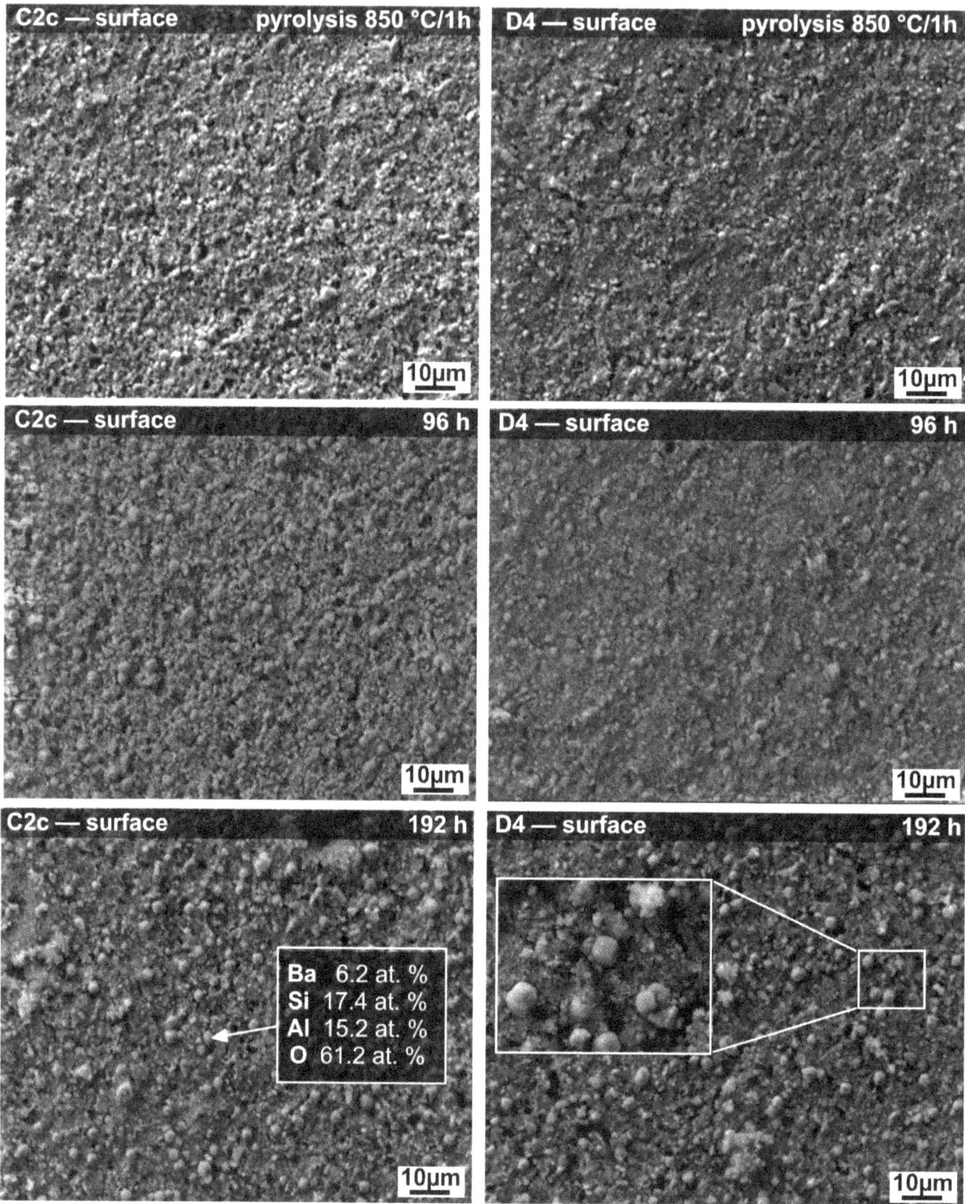

Figure 3. SEM micrographs of C2c and D4 coatings' surfaces before and after corrosion tests in deionized water at 200 °C.

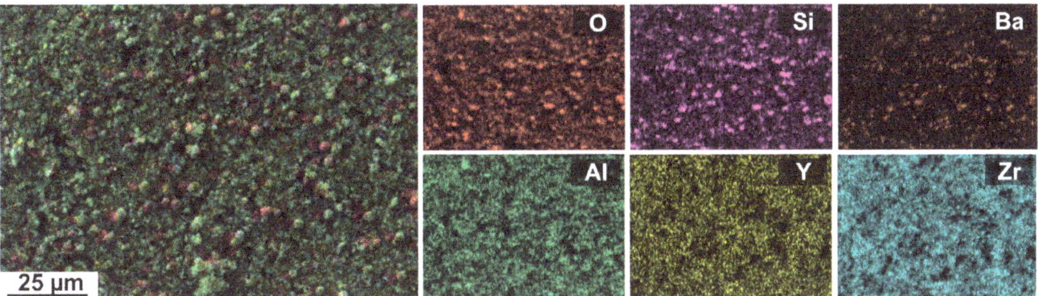

Figure 4. SEM/energy-dispersive X-ray spectroscopy (EDXS) analysis of the surface of the D4 coating after 192 h of corrosion tests in deionized water.

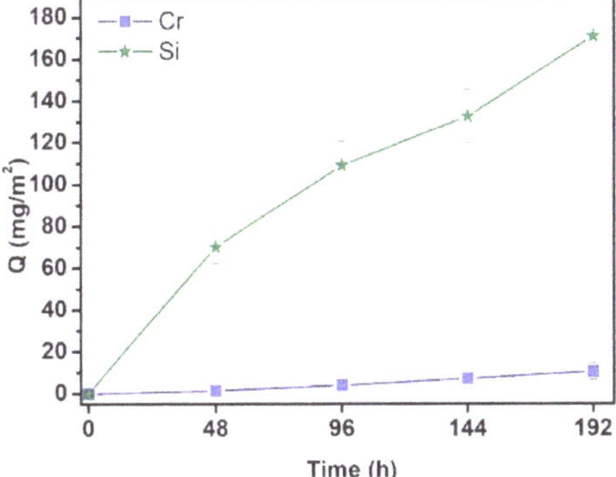

Figure 5. The amounts Q of elements (Si, Cr) leached from uncoated steel substrates in deionized water at 200 °C.

As for the coatings, analysis of the corrosion liquid by ICP-OES confirmed the presence of Ba, Al, Si, Zr, and Cr in the solution (Figure 6a–d). As with the uncoated steel substrates, the concentrations of Mn and Fe in deionized water were under the detection limit of ICP-OES. As can be seen in Figure 6c,d, only a negligible amount of Zr and Y in the corrosion solution was also observed. Very low concentrations of Zr detected in the corrosion solution suggest a high chemical durability of the YSZ used as a ceramic filler. Comparable amounts of Al were leached from both coatings to the solution at the applied quasi-dynamic conditions. Therefore, we suspect the AYZ filler was not the source of Al (it is contained only in the D4 coating) but the Al originates from barium aluminosilicate glass frits. These observations also suggest that AYZ filler is resistant to hydrothermal corrosion attack under the applied test conditions. As can be seen in Figure 6c,d, a small amount of Cr was identified in the corrosion solutions after 48 h and 96 h due to the outward diffusion of the Cr through the PDC coating. However, Cr content in both solutions dramatically increased after 144 h of corrosion test. This can be attributed to the faster outward diffusion of Cr ions probably due to higher occurrence of micropores and microcracks in the coatings that started to form after 144 h of exposure to the corrosive environment. Moreover, a higher content of Cr in the corrosion medium for both tested coatings was detected compared to the solution in which the uncoated steel was tested (Figure 5). The differences in Q values for Cr can be explained by the diffusion and release of Cr from steel and simultaneous precipitation of corrosion products containing Cr at the uncoated steel surface. In the case of the tested coatings, Cr remained dissolved in the corrosion medium since the coatings acted as a barrier for Cr migration back to the steel surface and prevented its precipitation.

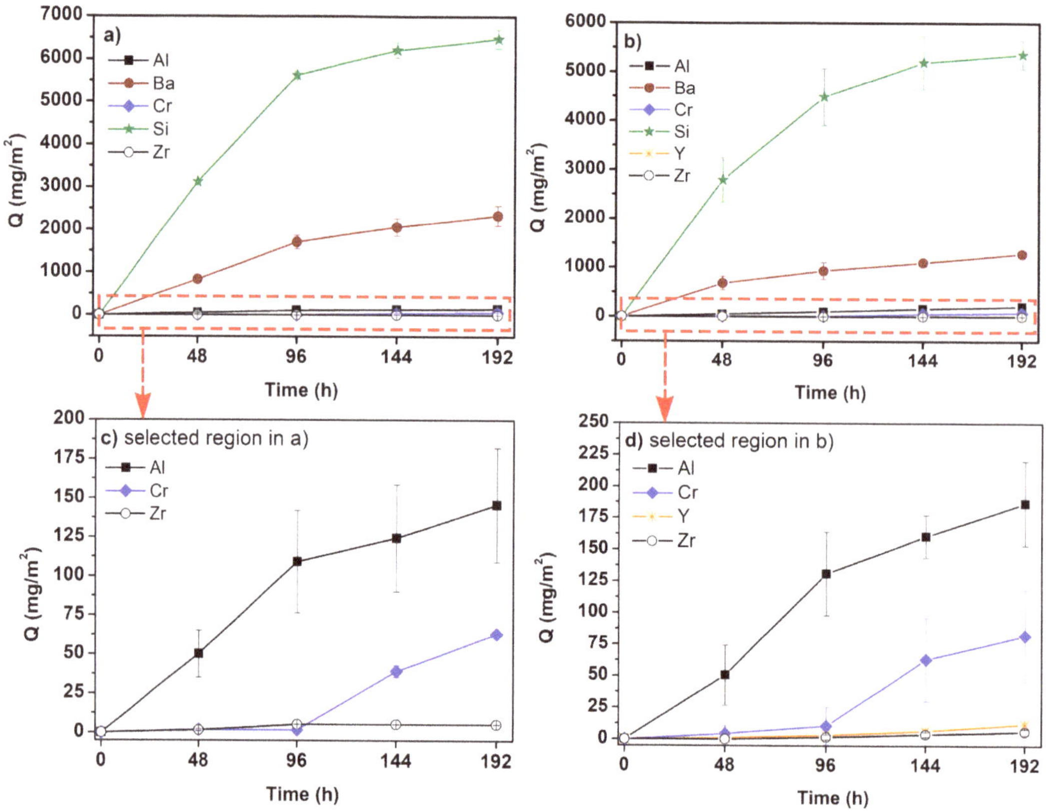

Figure 6. The amounts Q of elements (Si, Ba, Al, Zr, Y, Cr) leached from tested coatings in deionized water at 200 °C: (**a**) C2c coating; (**b**) D4 coating; (**c**) selected region in (**a**); (**d**) selected region in (**b**).

From Figure 6 it is evident that the Q values for Si are significantly higher than the values for Ba and other leached elements. Moreover, the amount of Si and Ba leached into the corrosion media was found to be higher for the C2c coating than for the D4 coating. The total amount of Si identified in the solution is probably the sum of the contributions of Si leached from the glass frits, steel substrate, bond-coat as well as from the PDC matrix. In both coatings, the content of Si leached into the solution grew quickly in the first 96 h of the corrosion test, then the dissolution reaction slowed down as the state of saturation was attained and the precipitates were formed at the coating surface. We suppose that the amount of released Si increased even further, since existing and newly forming micro-cracks or micro-pores acted as weak points through which the corrosive medium can pass, thus causing further dissolution from these places. As a result, equilibrium was achieved and accompanied by a precipitation of reaction products at the coating surface.

The pH values of corrosion solutions were determined before and after corrosion tests and are summarized in Table 3.

Table 3. Evolution of the pH values of the corrosion medium with increasing time of corrosion.

Sample	48 h	96 h	144 h	192 h
Steel	7.05	7.12	7.10	7.11
C2c	8.90	8.30	7.55	7.46
D4	8.64	8.24	7.51	7.43

The deionized water used in this study had a starting pH of 7.04 (19.5 °C) which increased upon the hydrothermal corrosion of the coated samples. In the time interval up to 96 h during the corrosion tests, the pH value of the corrosion solutions was found to be higher than 8 for both C2c and D4 coating tests. Considering the chemical composition of the two tested polysilazane-based (Si–C–N–O) coatings, this indicates that hydrolysis reactions in the PDC matrix occurred upon corrosion leading to the formation of silica and release of ammonia. Thus, we propose that the following process describes the hydrothermal corrosion of PDC-based coatings investigated in this work (2) (equation not balanced):

$$Si_xC_yN_zO + H_2O \rightarrow xSiO_2 + yCH_4 + zNH_3 \quad (2)$$

In the first step, Si–N and Si–C bonds are attacked by deionized water accompanied by the formation of silica and a release of methane and ammonia, which is a weak base highly soluble in water [35]. This is in agreement with the increase of the pH values of the corrosion solutions during the corrosion tests. In a second step, water reacts with Si–O bonds, and silica dissolution occurs according to Equation (3):

$$SiO_2 + H_2O \rightarrow Si(OH)_4 \quad (3)$$

Based on the results shown above, we assume that Si–N bonds in the investigated polysilazane-based coatings were attacked by hydrothermal corrosion while Si was released and present in the form of soluble $Si(OH)_4$ in the corrosion solution. Considerably high concentrations of Si released into the corrosion medium from both tested coatings (Figure 6) further support this observation. However, as mentioned earlier, Si detected in the solutions could originate not only from the cleavage of Si–N and Si–C bonds, but also from the dissolution of the glass frits used as fillers in the coatings. However, the presence of a glassy phase in the coating structure is important for the corrosion protection due to the additional protective barrier provided by the glasses. The amounts of Al and Ba, along with Si, released to the solution likely shift the equilibrium towards precipitation within a short time interval. Because of this, the solution becomes saturated with respect to some secondary phases, which in turn precipitate at the coating surface in the form of insoluble barium aluminosilicate crystals, as confirmed by EDXS analysis of the coating surface after 192 h of corrosion tests (Figure 4). This could lead to a decreased rate of active corrosion for the studied coatings. A comparison of the corrosion behavior of our samples with other PDC glass/ceramic coatings is difficult to perform because of the lack of standard procedures in the corrosion testing of this type of protective coatings under hydrothermal conditions. However, similar behavior was observed in the case of SiC, Si_3N_4 or SiOC-based material [33,34,36]. For instance, SiOC-based ceramic nanocomposites were investigated with respect to their hydrothermal corrosion behavior at 250 °C. The results show an active corrosion behavior, i.e., silica was leached out of the samples [34].

3.4. Identification of Corrosion Products

XRD patterns of stainless steel before and after corrosion tests at 200 °C in deionized water are shown in Figure 7. In the steel substrates before and after corrosion tests, only the Fe phase (PDF-96-901-3474) was identified from the XRD patterns. No diffraction peaks belonging to the newly formed corrosion products were detected. However, the corrosion systematically led to a decreased amount of the Fe phase with increasing time of corrosion. This explains the gradual formation of corrosion products covering the steel surface, thereby reducing the intensity of the diffused signal from Fe.

Literature data [37,38] suggests that the oxide scale formed at the stainless steel surface in hot aqueous solutions exhibit a duplex structure with a Cr-enriched inner layer while the outer layer is Fe-enriched. Chromia Cr_2O_3 inner layer acts as a diffusion barrier for other elements (e.g., Fe, Ni), which prevents the metal from further corrosion [39]. For a Cr_2O_3 layer to be protective, it must be dense and continuous and cover the entire metal surface. However, the chemical compositions of the steel, its microstructure, and the service environments are the major factors affecting the formation of the passive oxide film. It is well known that corrosion of stainless steels is accelerated in atmospheres containing H_2O.

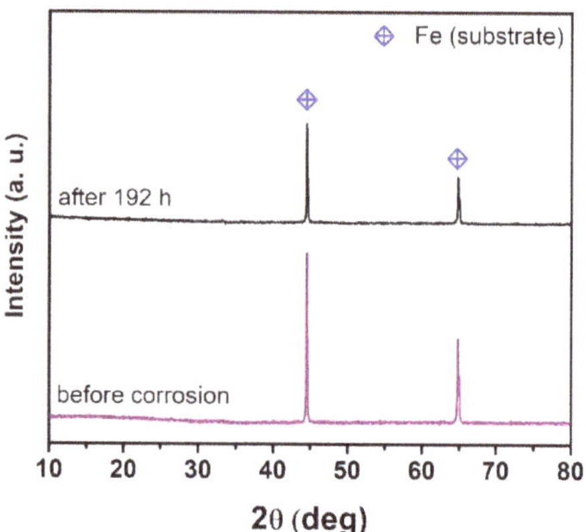

Figure 7. X-ray diffraction (XRD) patterns of uncoated steel before and after 192 h of corrosion tests.

In order to identify the corrosion products formed at the steel surface after the corrosion tests, the specimens were also analyzed by Raman spectroscopy, as shown in Figure 8. The most intense feature in the spectrum of the uncoated steel after 192 h of corrosion tests was observed at ~660 cm^{-1} and corresponds to (Mn, Cr, $Fe)_3O_4$ spinel [40]. The bands around ~295 cm^{-1} and ~405 cm^{-1} were attributed to Fe_2O_3 [41,42]. The Raman bands of Cr_2O_3, typically appearing at ~550 cm^{-1} [40,43], were not found in the measured results. This indicates that crystalline Cr_2O_3 was absent or only present at low concentrations. Therefore, two hypotheses of the formation of corrosion products on the steel surface were considered. First, the protective Cr_2O_3 scale did not form in the initial corrosion stage, and it could not provide effective protection of the steel from future corrosion. The second hypothesis is that the protective Cr_2O_3 layer was formed in the initial time interval according to the reaction (4). However, at some point it lost its protective behavior due to the reaction with the deionized water resulting in the formation of chromium hydroxides. Both hypotheses correlate well with the absence of Cr_2O_3 peak in the XRD as well as in the Raman spectrum of corroded steel substrate. Furthermore, we assume that the deionized water reacts with the diffusing metallic species (Cr, Mn and Fe) according to the reaction (5) [44]:

$$2Cr + 3H_2O \rightarrow Cr_2O_3 + 3H_2 \tag{4}$$

$$3(Mn, Cr, Fe) + 4H_2O \rightarrow (Mn, Cr, Fe)_3O_4 + 4H_2 \tag{5}$$

Based on the Raman results, we can conclude that a majority of the corrosion products are a mixture of Fe_2O_3 and Mn, Fe, Cr spinels, as reported by other authors for Fe–Cr ferritic steels tested in water-containing atmospheres [43,44].

In the coated samples, XRD was used to detect any secondary phases that could result from chemical reactions between the components of the steel substrate, coating, and corrosive agent. In both tested coatings, the dominant phases detected after pyrolysis by XRD are monoclinic (PDF- 96-901-6715) and cubic ZrO_2 (PDF-96-210-1235). Moreover, a peak located near the most intense diffraction peak of cubic ZrO_2, assigned as SiO_2 (quartz, PDF-96-901-2602), also appeared in the coatings, probably as a result of crystallization of the glass frit during pyrolysis. In the D4 coating, yttrium-aluminum garnet (YAG, PDF-96-431-2143) was also identified. This phase originates from the polycrystalline AYZ precursor powder used as passive filler. After 96 h of corrosion tests, new diffraction peaks are observed in both coatings, which are assigned to the crystalline phase $BaAl_2Si_2O_8$ (celsian, PDF-96-201-3138). This is in accordance with the results of EDXS analysis of the precipitates formed at the coating surface. It can be noted that after 96 h of corrosion testing, a new phase, namely $BaAl_2Si_2O_8$—hexacelsian (PDF-01-088-1049), a polymorph of celsian—is detected in the D4 coating. With increasing time, the XRD patterns show a decrease in the peak intensities of this phase (Figure 9b). The intensity of the diffraction peaks corresponding to cubic ZrO_2 increased with the time of corrosion tests, while the SiO_2 peak totally vanished after 48 h of exposure to the corrosion medium. This effect can be explained by the incorporation of SiO_2 into other crystalline/amorphous phase and/or to the dissolution of the silica to the corrosion medium according to Equation (3).

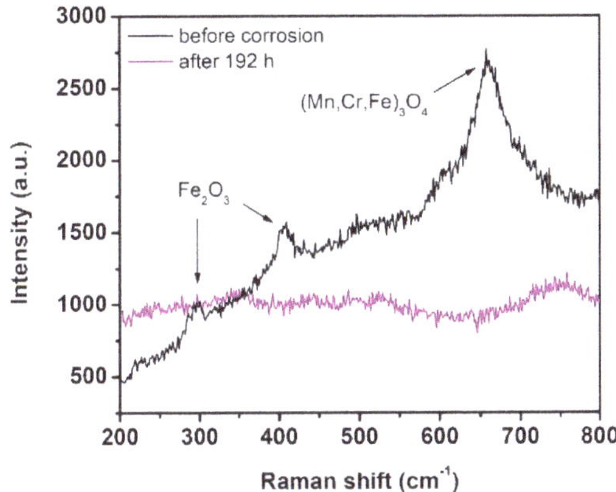

Figure 8. Raman spectra of uncoated steel substrates before and after 192 h of corrosion tests.

Considering the chemical composition of the fillers used and the tested PDC coatings, and the corrosion products found on the corroded surfaces, it was deduced that the chemical reactions (6)–(10) likely occurred during corrosion test in deionized water at 200 °C:

$$BaO + Al_2O_3 \rightarrow BaAl_2O_4 \quad (6)$$

$$2BaO + SiO_2 \rightarrow Ba_2SiO_4 \quad (7)$$

$$Ba_2SiO_4 + 3SiO_2 \rightarrow 2BaSi_2O_5 \quad (8)$$

$$BaAl_2O_4 + 2SiO_2 \rightarrow BaAl_2Si_2O_8 \quad (9)$$

$$BaSi_2O_5 + Al_2O_3 \rightarrow BaAl_2Si_2O_8 \quad (10)$$

Since the presence of these transient phases (BaO, $BaAl_2O_4$, Ba_2SiO_4) was not detected in the XRD patterns, it was hypothesized that they reacted with the glassy phase or at other places in the coating locally enriched in SiO_2 and Al_2O_3 as soon as they were formed to form the main corrosion product, $BaAl_2Si_2O_8$. A large part of the coating surface was then covered by celsian precipitates, with an additional very small amount of residual hexacelsian crystals on the D4 coating. However, the formation of barium aluminosilicates by reactions of other components contained in the coatings or in the corrosion solution cannot be completely ruled out.

Rietveld refinement of XRD data was used for the semi-quantitative analysis of the phase composition of the tested coatings. The time dependences of the phase composition of both C2c and D4 coatings are shown in Figure 10. The results indicate that the content of monoclinic ZrO_2 slightly decreased with increasing time. As shown in the XRD patterns (Figure 9), the SiO_2 phase disappeared in the early stage of corrosion testing. In contrast, an increasing content of celsian phase can be observed as a consequence of the chemical reactions between the individual components of the layers and/or elements leached to the corrosion medium.

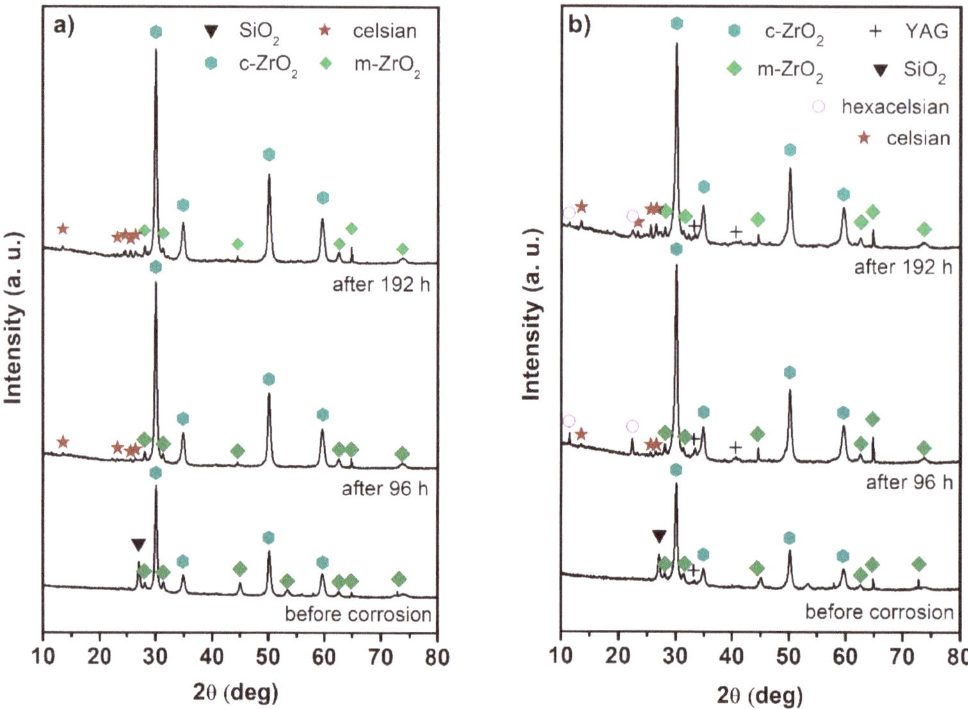

Figure 9. XRD patterns of the coatings before and after corrosion tests. (**a**) C2c coating; (**b**) D4 coating.

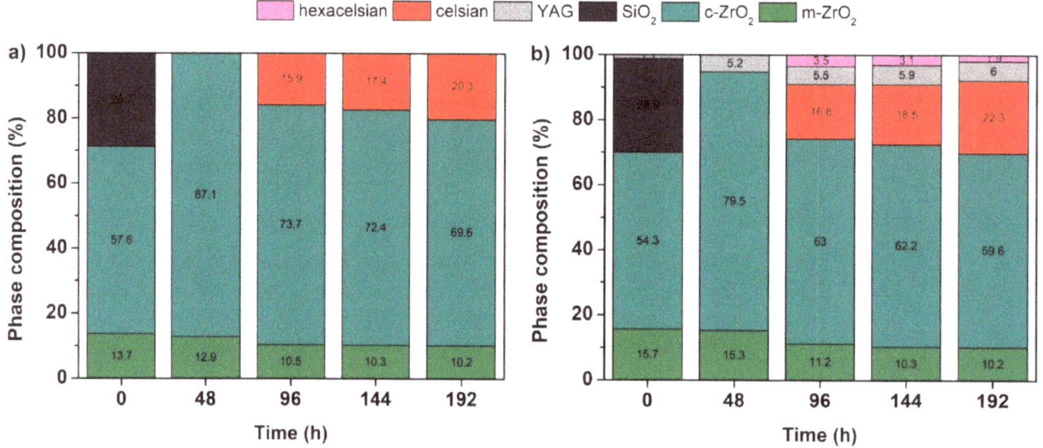

Figure 10. Time dependence of phase composition determined by the Rietveld refinement of XRD patterns acquired from: (**a**) C2c coating; (**b**) D4 coating.

3.5. Microstructures of Polymer-Derived Ceramic (PDC) Coatings

SEM/EDXS analysis was used for a detailed study of the cross-sections of coated samples before and after corrosion tests (Figure 11). Parchoviansky et al. [30,45] have already described the microstructure of the PDC coatings studied in this work after pyrolysis. Results of these investigations are briefly described here. After pyrolysis in air at 850 °C for 1 h, homogeneous and almost dense protective coatings, with only small pore sizes, were prepared. Both coatings, C2c and D4, showed good adhesion: no gaps or cracks propagating along the metal/coating interface were detected. On steel exposed to ambient environment, a natural oxide layer with chemically bonded water is always present. Because of the high reactivity of Durazane 2250 with hydroxyl groups, steel forms direct metal–O–Si chemical bonds with the precursor-based bond-coat, leading to strong adhesion [17]. As indicated by SEM and EDXS elemental mapping [30], the coating microstructure is composed of three main constituents: evenly distributed original filler particles, and residual porosity aggregated in the amorphous PDC phase. The coating C2c is characterized by higher residual porosity. However, such a microstructure with residual porosity is beneficial for the thermal stability of the coatings, as it contributes to the reduction of residual stresses during heating and cooling cycles [46].

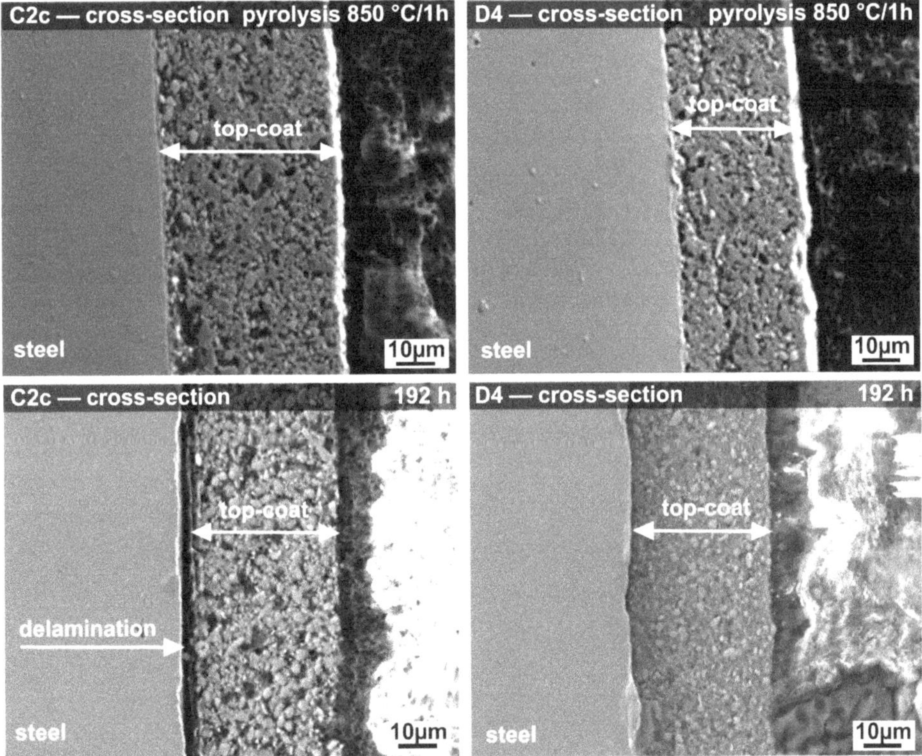

Figure 11. SEM cross-sections of C2c and D4 coatings before and after corrosion tests.

More promising results after corrosion tests were observed for the D4 coating, which contains the polycrystalline AYZ powder as passive filler. The D4 coating showed lower porosity than the C2c coating (Figure 11). SEM cross-sectional images also showed good coating adhesion, even after 192 h of corrosion tests. Overall, no significant visible corrosion damage was observed on D4 coating after the tests, confirming it acted as an efficient protection system. In the case of the C2c coating after corrosion tests, a significant increase in the porosity of the layer accompanied by the growth of pores was observed. Moreover, spalling of the C2c coating occurred after 192 h of exposure to deionized water.

There are several factors that could cause the better corrosion performance of the D4 coating. The first is the composition, the lower content of YSZ filler and the addition of polycrystalline AYZ powder in the D4 coating structure. The lower weight loss during the corrosion tests of the D4 coating could also be influenced by the use of AYZ filler, which obviously acts as reinforcing phase, resistant to the attack of deionized water and improving the hydrothermal stability of the PDC matrix. The better corrosion resistance of the D4 coating can be also attributed to the different microstructure of the coatings. The addition of the AYZ powder with irregular and angular particles helped create a solid and rigid structure which allowed outgassing of the preceramic polymer pyrolysis products from the system, thereby effectively reducing the size and amount of pores after pyrolysis. The absence of larger pores, and thus an increased density, led to a significantly more compact coating in comparison to the C2c composition pyrolyzed under the same conditions. The defects of a critical size in the C2c coating could induce macroscopic failure and thus reduce the adhesion strength resulting in the delamination of the ceramic layer during corrosion testing. The increased occurrence of pores and cracks in the C2c coating also allowed a faster penetration of deionized water through the coating to the metal substrate causing a subsequent spallation and disintegration of the layer. The contents of preceramic polymer (source of Si) and glass frits (source of Si and Ba) in both coatings are identical and cannot explain the difference of the leached amounts of these elements to the corrosion solutions. Therefore, the higher number of pores is likely to be responsible for higher dissolution rates of Si and Ba in deionized water, which could cause faster recession and failure of the C2c coating during the corrosion tests. Based on the observations shown above, it is proposed that the highly porous microstructure of the C2c coating is the main reason of failure for this coating.

EDXS mapping was performed on the cross-section of the D4 coating (Figure 12). The EDXS maps identified a homogeneous distribution of Zr, Y, Si, Ba, Al, and O in the top-coat. The presence of Fe and Cr is clearly seen in the stainless steel substrate. As confirmed by ICP-OES analysis of the corrosion solutions, no diffusion of Fe from the substrate through the coating was observed. This is also shown in the EDXS map where the presence of Fe ends exactly at the steel/coating interface. However, EDXS cross-sectional analysis revealed the presence of a small amount of Cr in the top-coat, which likely diffused out of the steel during the tests. This is also consistent with the presence of Cr in the corrosion solution detected by ICP-OES. No corrosion products were observed at the stainless steel/top-coat interfaces after corrosion tests.

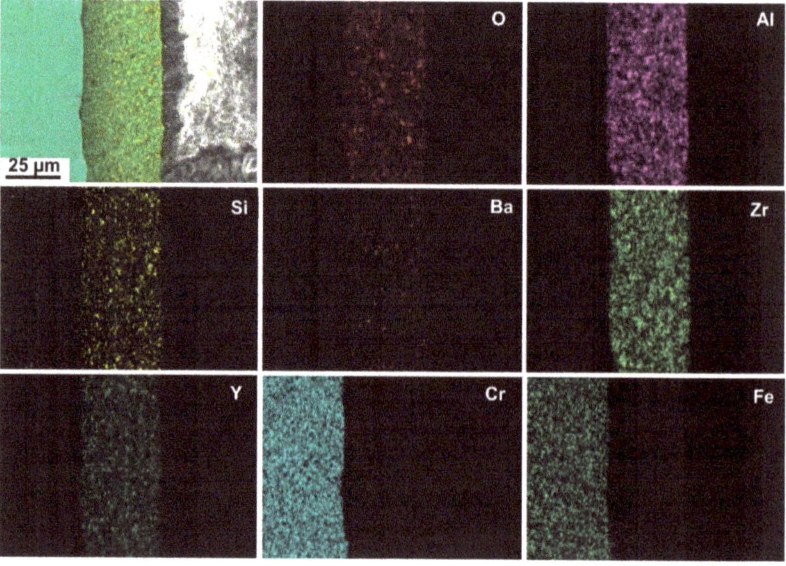

Figure 12. SEM/EDXS cross-sections of the D4 coating after 192 h of corrosion tests.

4. Conclusions

In this work, the hydrothermal corrosion resistance of ferritic stainless steel and two PDC-based coatings was studied systematically in deionized water. Two types of corrosion product morphologies were found on the exposed uncoated steel surface. Most of the surface was covered by a thin layer of rod-shaped crystallites identified by Raman spectroscopy as a mix of Fe_2O_3 and $(Mn, Cr, Fe)_3O_4$ spinels. Occasionally, Fe- and Cr-enriched globular crystallites were found on the corroded steel surface after the corrosion tests. In the case of the coated samples, the results of weight gain measurements together with SEM and ICP-OES indicate that a state of saturation was achieved in the early stage of the dissolution reactions. These reactions were followed by the precipitation of spherical-shaped $BaAl_2Si_2O_8$ crystals. M–ZrO_2, c–ZrO_2, YAG, and $BaAl_2Si_2O_8$ –hexacelsian, celsian phases were identified from XRD patterns of the corroded PDC coatings. The D4 coating exhibited better protective properties than the C2c coating in deionized water under quasi-dynamic conditions at 200 °C. No significant corrosion damage was observed for the D4 coating, but the C2c coating showed a highly porous structure, delamination of the protective layer and the loss of corrosion resistance. The results of this study indicate the high potential of the investigated PDC coating with AYZ passive filler for applications in harsh environments, such as hydrothermal corrosion conditions.

Author Contributions: I.P. prepared the coatings and precursor powder, performed corrosion tests and wrote the manuscript; M.P. performed the SEM and EDXS analysis and performed the XRD measurements; H.K. performed the analysis of corrosion solutions by ICP-OES and analyzed the data; A.N. performed Raman spectroscopy measurements and analyzed the data; D.G. conceived the study, supervised the project, gave research guidelines, and finalized the manuscript. All authors have read and agreed to the published version of the manuscript.

Funding: This paper was created in the frame of the project Centre for Functional and Surface Functionalised Glass (CEGLASS), ITMS code 313011R453, operational program Research and innovation, co-funded from European Regional Development Fund. This work is a part of dissemination activities of project FunGlass. This project has received funding from the European Union's Horizon 2020 research and innovation programme under grant agreement No 739566. Financial support of this work by the grants APVV 0014-15 and VEGA 1/0191/20 is gratefully acknowledged.

Institutional Review Board Statement: Not applicable.

Informed Consent Statement: Not applicable.

Data Availability Statement: Not applicable.

Acknowledgments: The authors would like to thank students Alexander Horcher and Mateus Lenz-Leite for their help to create the coating solutions. The authors would also like to thank Jacob Andrew Peterson for language revision of the manuscript.

Conflicts of Interest: The authors declare no conflict of interest.

References

1. Radhamani, A.; Lau, H.C.; Ramakrishna, S. Nanocomposite coatings on steel for enhancing the corrosion resistance: A review. *J. Compos. Mater.* **2020**, *54*, 681–701. [CrossRef]
2. Raj, R. Fundamental Research in Structural Ceramics for Service Near 2000 °C. *J. Am. Ceram. Soc.* **1993**, *76*, 2147–2174. [CrossRef]
3. Gogotsi, Y.G.; Lavrenko, V.A. *Corrosion of High-Performance Ceramics*; Springer: Berlin, Germany, 1992.
4. Colombo, P.; Mera, G.; Riedel, R.; Soraru, G.D. Polymer-derived ceramics: 40 Years of research and innovation in advanced ceramics. *J. Am. Ceram. Soc.* **2010**, *93*, 1805–1837. [CrossRef]
5. Ionescu, E. Polymer-Derived Ceramics. *Ceram. Sci. Technol.* **2013**, *3–4*, 457–500.
6. Barroso, G.; Li, Q.; Bordia, K. Polymeric and ceramic silicon-based coatings—A review. *J. Mater. Chem. A* **2018**, *7*, 1936–1963. [CrossRef]
7. Mera, G.; Gallei, M.; Bernard, S.; Ionescu, E. Ceramic nanocomposites from tailor-made preceramic polymers. *Nanomaterials* **2015**, *5*, 468–540. [CrossRef]
8. Torrey, J.D.; Bordia, R.K. Processing of polymer-derived ceramic composite coatings on steel. *J. Am. Ceram. Soc.* **2008**, *91*, 41–45. [CrossRef]

9. Torrey, J.D.; Bordia, R.K. Mechanical properties of polymer-derived ceramic composite coatings on steel. *J. Eur. Ceram. Soc.* **2008**, *28*, 253–257. [CrossRef]
10. Colombo, P.; Paulson, T.E.; Pantano, C.G. Synthesis of Silicon Carbide Thin Films with Polycarbosilane (PCS). *J. Am. Ceram. Soc.* **2005**, *40*, 2333–2340. [CrossRef]
11. Mucalo, M.R.; Milestone, N.B. Preparation of ceramic coatings from pre-ceramic precursors—Part II SiC on metal substrates. *J. Mater. Sci.* **1994**, *29*, 5934–5946. [CrossRef]
12. Blum, Y.D.; Schwartz, K.B.; Laine, R.M. Preceramic polymer pyrolysis. *J. Mater. Sci.* **1989**, *24*, 1707–1718. [CrossRef]
13. Greil, P. Polymer Derived Engineering Ceramics. *Adv. Eng. Mater.* **2000**, *2*, 339–348. [CrossRef]
14. Günthner, M.; Wang, K.; Bordia, R.K.; Motz, G. Conversion behaviour and resulting mechanical properties of polysilazane-based coatings. *J. Eur. Ceram. Soc.* **2012**, *32*, 1883–1892. [CrossRef]
15. Wang, K.; Günthner, M.; Motz, G.; Bordia, R.K. High performance environmental barrier coatings, Part II: Active filler loaded SiOC system for superalloys. *J. Eur. Ceram. Soc.* **2011**, *31*, 3011–3020. [CrossRef]
16. Wang, K.; Unger, J.; Torrey, J.D.; Flinn, B.D.; Bordia, R.K. Corrosion resistant polymer derived ceramic composite environmental barrier coatings. *J. Eur. Ceram. Soc.* **2014**, *34*, 3597–3606. [CrossRef]
17. Günthner, M.; Kraus, T.; Krenkel, W.; Motz, G.; Dierdorf, A.; Decker, D. Particle-Filled PHPS silazane-based coatings on steel. *Int. J. Appl. Ceram. Technol.* **2009**, *6*, 373–380. [CrossRef]
18. Bernardo, E.; Colombo, P.; Hampshire, S. SiAlON-based ceramics from filled preceramic polymers. *J. Am. Ceram. Soc.* **2006**, *89*, 3839–3842. [CrossRef]
19. Günthner, M.; Schütz, A.; Glatzel, U.; Wang, K.; Bordia, R.K.; Greißl, O.; Krenkel, W.; Motz, G. High performance environmental barrier coatings, Part I: Passive filler loaded SiCN system for steel. *J. Eur. Ceram. Soc.* **2011**, *31*, 3003–3010. [CrossRef]
20. Scheffler, M.; Dernovsek, O.; Schwarze, D.; Science, M.; Science, M. Polymer/filler derived NbC composite ceramics. *J. Mater. Sci.* **2003**, *8*, 4925–4931. [CrossRef]
21. Labrousse, M.; Nanot, M.; Boch, P.; Chassagneux, E. Ex-polymer SiC coatings with Al_2O_3 particulates as filler materials. *Ceram. Int.* **1993**, *19*, 259–267. [CrossRef]
22. Goerke, O.; Feike, E.; Heine, T.; Trampert, A.; Schubert, H. Ceramic coatings processed by spraying of siloxane precursors (polymer-spraying). *J. Eur. Ceram. Soc.* **2004**, *24*, 2141–2147. [CrossRef]
23. Konegger, T.; Tsai, C.; Bordia, R.K. Preparation of Polymer-Derived Ceramic Coatings by Dip-Coating Preparation of polymer-derived ceramic coatings by dip-coating. *Mater. Sci. Forum* **2015**, *825–826*, 645–652. [CrossRef]
24. Bill, J.; Heimann, D. Polymer-Derived Composites Ceramic Coatings on C/C-Sic. *J. Eur. Ceram. Soc.* **1996**, *2219*, 1115–1120. [CrossRef]
25. Barroso, G.S.; Krenkel, W.; Motz, G. Low thermal conductivity coating system for application up to 1000 °C by simple PDC processing with active and passive fillers. *J. Eur. Ceram. Soc.* **2015**, *35*, 3339–3348.
26. Hotza, D.; Geffroy, P.M. Tape casting of preceramic polymers toward advanced ceramics: A review. *Int. J. Ceram. Eng. Sci.* **2019**, *1*, 21–41. [CrossRef]
27. Morlier, A.; Cros, S.; Garandet, J.P.; Alberola, N. Thin gas-barrier silica layers from perhydropolysilazane obtained through low temperature curings: A comparative study. *Thin Solid Film.* **2012**, *524*, 62–66. [CrossRef]
28. Gunthner, M.; Albrecht, Y.; Motz, G. Polymeric and ceramic like coatings on the basis of SiN(C) precursors for the protection of metals against corrosion and oxidation. *Ceram. Eng. Sci. Proc.* **2007**, *27*, 277–284.
29. Riffard, F.; Joannet, E.; Buscail, H.; Rolland, R. Beneficial Effect of a Pre-ceramic Polymer Coating on the Protection at 900 °C of a Commercial AISI 304 Stainless Steel. *Oxiadation Met.* **2017**, *88*, 211–220. [CrossRef]
30. Parchovianský, M.; Parchovianská, I.; Švančárek, P.; Medved', D.; Lenz-Leite, M.; Motz, G.; Galusek, D. High-temperature oxidation resistance of pdc coatings in synthetic air and water vapor atmospheres. *Molecules* **2021**, *26*, 2388. [CrossRef] [PubMed]
31. Petríková, I.; Parchovianský, M.; Švančárek, P.; Lenz Leite, M.; Motz, G.; Galusek, D. Passive filler loaded polysilazane-derived glass/ceramic coating system applied to AISI 441 stainless steel, part 1: Processing and characterization. *Int. J. Appl. Ceram. Technol.* **2020**, *17*, 998–1009. [CrossRef]
32. Riedel, R. *Handbook of Ceramic Hard Materials*; Wiley-VCH: Weinheim, Germany, 2000; ISBN 3-527-29972-6.
33. Sōmiya, S. Hydrothermal corrosion of nitride and carbide of silicon. *Mater. Chem. Phys.* **2001**, *67*, 157–164. [CrossRef]
34. Linck, C.; Ionescu, E.; Papendorf, B.; Galuskova, D.; Galusek, D.; Sajgalik, P.; Riedel, R. Corrosion behavior of silicon oxycarbide-based ceramic nanocomposites under hydrothermal conditions. *Int. J. Mater. Res.* **2012**, *103*, 31–39. [CrossRef]
35. Yuan, J.; Luan, X.; Riedel, R.; Ionescu, E. Preparation and hydrothermal corrosion behavior of C f/SiCN and C f/SiHfBCN ceramic matrix composites. *J. Eur. Ceram. Soc.* **2015**, *35*, 3329–3337. [CrossRef]
36. Herrmann, M. Corrosion of Silicon Nitride Materials in Aqueous Solutions. *J. Am. Ceram. Soc.* **2013**, *96*, 3009–3022. [CrossRef]
37. Stellwag, B. The mechanism of oxide film formation on austenitic stainless steels in high temperature water. *Corros. Sci.* **1998**, *40*, 337–370. [CrossRef]
38. Ziemniak, S.E.; Hanson, M.; Sander, P.C. Electropolishing effects on corrosion behavior of 304 stainless steel in high temperature, hydrogenated water. *Corros. Sci.* **2008**, *50*, 2465–2477. [CrossRef]
39. Duan, Z.; Arjmand, F.; Zhang, L.; Abe, H. Investigation of the corrosion behavior of 304L and 316L stainless steels at high-temperature borated and lithiated water. *J. Nucl. Sci. Technol.* **2016**, *53*, 1435–1446. [CrossRef]

40. Badin, V.; Diamanti, E.; Forêt, P.; Darque-Ceretti, E. Characterization of Oxide Scales Formed on Ferritic Stainless Steel 441 at 1100 °C under water vapor. *Oxid. Met.* **2014**, *82*, 347–357. [CrossRef]
41. Chen, Z.; Wang, L.; Li, F.; Chou, K.C.; Sun, Z. The effects of temperature and oxygen pressure on the initial oxidation of stainless steel 441. *Int. J. Hydrog. Energy* **2014**, *39*, 10303–10312. [CrossRef]
42. Srisrual, A.; Coindeau, S.; Galerie, A.; Petit, J.P.; Wouters, Y. Identification by photoelectrochemistry of oxide phases grown during the initial stages of thermal oxidation of AISI 441 ferritic stainless steel in air or in water vapour. *Corros. Sci.* **2009**, *51*, 562–568. [CrossRef]
43. Badin, V.; Diamanti, E.; Forêt, P.; Darque-Ceretti, E. Water Vapor Oxidation of Ferritic 441 and Austenitic 316L Stainless Steels at 1100 °C for Short Duration. *Procedia Mater. Sci.* **2015**, *9*, 48–53. [CrossRef]
44. Bawane, K.; Lu, K.; Li, Q.; Bordia, R. High temperature oxidation behaviors of SiON coated AISI 441 in Ar + O_2, Ar + H_2O, and Ar + CO_2 atmospheres. *Corros. Sci.* **2020**, *166*, 108429. [CrossRef]
45. Parchovianský, M.; Parchovianská, I.; Švančárek, P.; Motz, G.; Galusek, D. PDC glass/ceramic coatings applied to differently pretreated AISI441 stainless steel substrates. *Materials* **2020**, *13*, 629. [CrossRef] [PubMed]
46. Padture, N.P.; Gell, M.; Jordan, E.H. Thermal barrier coatings for gas-turbine engine applications. *Science* **2002**, *296*, 280–284. [CrossRef]

 *materials*

Article

Preparation and Properties of Mo Coating on H13 Steel by Electro Spark Deposition Process

Wenquan Wang [1], Ming Du [1], Xinge Zhang [1,*], Chengqun Luan [1] and Yingtao Tian [2]

[1] Key Laboratory of Automobile Materials, School of Materials Science and Engineering, Jilin University, Changchun 130025, China; wwq@jlu.edu.cn (W.W.); mingdu19@mails.jlu.edu.cn (M.D.); chengqun16@mails.jlu.edu.cn (C.L.)

[2] Department of Engineering, Lancaster University, Bailring, Lancaster LA1 4YW, UK; yingtao.tian@manchester.ac.uk

* Correspondence: zhangxinge@jlu.edu.cn; Tel.: +86-431-8509-4687

Abstract: H13 steel is often damaged by wear, erosion, and thermal fatigue. It is one of the essential methods to improve the service life of H13 steel by preparing a coating on it. Due to the advantages of high melting point, good wear, and corrosion resistance of Mo, Mo coating was fabricated on H13 steel by electro spark deposition (ESD) process in this study. The influences of the depositing parameters (deposition power, discharge frequency, and specific deposition time) on the roughness of the coating, thickness, and properties were investigated in detail. The optimized depositing parameters were obtained by comparing roughness, thickness, and crack performance of the coating. The results show that the cross-section of the coating mainly consisted of strengthening zone and transition zone. Metallurgical bonding was formed between the coating and substrate. The Mo coating mainly consisted of $Fe_{9.7}Mo_{0.3}$, Fe-Cr, FeMo, and Fe_2Mo cemented carbide phases, and an amorphous phase. The Mo coating had better microhardness, wear, and corrosion resistance than substrate, which could significantly improve the service life of the H13 steel.

Keywords: electro spark deposition (ESD); Mo coating; H13 steel; microhardness; wear resistance; corrosion resistance

Citation: Wang, W.; Du, M.; Zhang, X.; Luan, C.; Tian, Y. Preparation and Properties of Mo Coating on H13 Steel by Electro Spark Deposition Process. *Materials* **2021**, *14*, 3700. https://doi.org/10.3390/ma14133700

Academic Editor: Daniel de la Fuente

Received: 31 May 2021
Accepted: 29 June 2021
Published: 1 July 2021

1. Introduction

H13 mold steel is mainly used for casting, extrusion, hot forming, and plastic molding applications due to its excellent hot-cold manufacturing properties, dimensional stability, and impact toughness resistance [1–4]. H13 mold steel is usually manufactured by casting and forging processes [5], which result in coarse carbides formation because of its low cooling and solidification rate [6,7]. The surface performance will be significantly reduced if H13 mold steel is exposed to high temperature for a long time [8–10]. Nowadays, some researchers have proposed to refine carbides by forging and rolling [11]. Although some achievements have been made, the problem has not been fundamentally solved. Moreover, mold steel can be easily damaged or fail due to wear and electrochemical corrosion, which vastly shortens its service life. To further improve the mold's service life, it is necessary to strengthen mold's wear resistance and corrosion resistance [12,13]. The appropriate strengthening methods of improving wear and corrosion performance have become a hotspot on the surface modification of hot work molds [14]. Therefore, many scholars from various countries have conducted extensive research on mold's surface modification. Kong et al. [15] prepared WC-12Co coatings to improve H13 mold steel's wear resistance and electrochemical corrosion performance. Żórawski et al. [16] revealed that the nanostructured WC-12Co composite coating has a denser structure and higher abrasive resistance than conventionally sprayed coatings. Meng et al. [17] investigated the mechanical properties and microstructures of H13 steel processed by a bionic laser surface alloying with different amount of TiC addition. With the increase of TiC fraction, the laser alloying zone's microstructure was refined and the microhardness was increased. Salmaliyan et al. [18] studied the effect of ESD process parameters on WC-Co coating with substrate of H13 steel. The results indicate that the crack propagation properties of the coating at low spark energy are different from those at high spark energy. Compared with other coating preparation technologies, ESD has attracted more interest as a promising surface technique for engineering materials [19–21]. ESD technology has unique advantages in preparing coatings such as WC-Co, FeNiCrBSiC-MeB2, Ti-Al intermetallic, CoCrFeNiMo High-Entropy Alloy, Ti6Al4V, and Cr coating due to its simple equipment, flexible operation, low cost, ecological greenness, wide adaptation range, and low heat input [22–28].

Mo is important to the industry due to its high-melting-point, high thermal conductivity, and low thermal expansion coefficient [29]. Moreover, Mo is an excellent coating metal in surface modification due to its superior metallurgical bonding with various metals and alloys. Existing researchers mostly use thermal spraying technology to prepare Mo coating. The only problem of thermal spraying technology is to have a high degree of oxidation [30]. In the present study, Mo coating was fabricated on the H13 steel using ESD technology to obtain better properties (microhardness, wear resistance, and corrosion resistance) than substrate. Properties of Mo coating with different ESD process parameters were investigated in detail.

2. Materials and Experimental Methods

2.1. Materials

In this study, the H13 steel was chosen as the substrate with sizes of 10 mm × 10 mm × 5 mm. The chemical compositions of H13 steel are shown in Table 1. The specimens were washed with acetone to remove oil and then ground with 600# sandpaper to remove oxides.

Table 1. Chemical compositions of H13 steel. (wt%).

Cr	Mo	Si	V	C	Mn	S	P	Fe
5.0	1.30	0.95	0.92	0.40	0.35	0.05	0.03	Bal.

2.2. Coating Preparation

The DZS-1400 ESD system (Institute of Surface Engineering Technology, Chinese Academy of Agricultural Mechanization Sciences, Beijing, China) was used to prepare the Mo coating on the substrate. Figure 1 shows the schematic diagram of preparing Mo coating on H13 steel by ESD process. Mo was used as the electrode. The electrode was cut into a cylinder with diameter of 3 mm. The inclination angle between the electrode and the substrate was 45 degrees. The rotation speed and moving speed of the processing gun were 1500 r/min during ESD. The shielding gas was argon with a flow rate of 8 L/min. The main processing parameters for ESD are presented in Table 2.

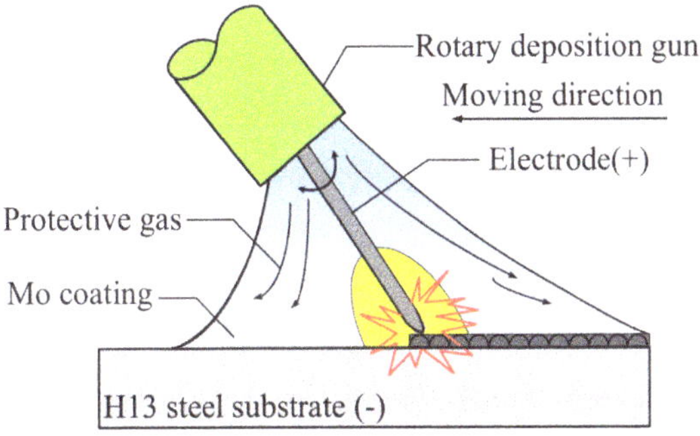

Figure 1. Schematic diagram of preparing Mo coating on H13 steel by ESD.

Table 2. ESD process parameters.

Deposition Power	Output Voltage	Discharge Frequency	Specific Deposition Time
100~1400 W	100 V	50~650 Hz	1~5 min/cm^2

2.3. Microstructure and Morphology Testing

The microstructure and wear scar morphology and the chemical compositions were observed and analyzed by scanning electron microscope (SEM, TESCAN VEGA3, Czech Republic) equipped with an energy dispersive spectrometer (EDS). The roughness and three-dimensional morphology of the coating were tested by the OLS3000 (China) laser confocal microscope. Roughness was evaluated quantitatively by arithmetical mean deviation of profile Ra. The phase constitutions were determined via the X-ray diffractometer (XRD) machine (D/Max2500Pc, Japan) using Cu Kα radiation.

2.4. Microhardnes and Wear Resistance Testing

The microhardness of the coating was tested by the HVS-1000 microhardness machine (Beijing Times Guangnan Testing Technology Co., Ltd, China) with a load of 100 g and dwell time of 10 s. The wear resistance of the coating was investigated for the low-speed rotation by the ML-100 abrasive wear test machine (Jinan Jingcheng Test Technology Co., Ltd, Jinan, China). Figure 2 shows the schematic of friction and wear. The silicon carbide sandpaper of 2000# was used as friction material. The travel period of a clockwise rotation and an anticlockwise of the testing machine is taken as the wear time. The load of abrasive wear test machine is 2~10 N. The mass of samples before and after wear is measured by electronic balance (precision 0.0001). The samples with the same parameters were subjected to three wear tests. The wear weight loss was averaged and recorded.

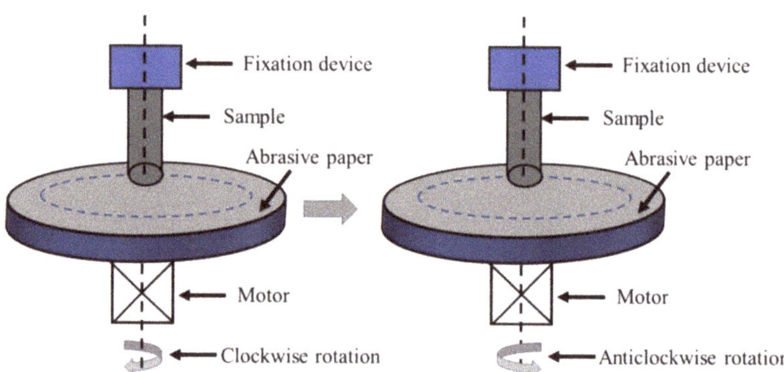

Figure 2. Schematic of friction and wear.

2.5. Electrochemical Corrosion Testing

The electrochemical corrosion performance and corrosion mechanism were studied by the VersaSTAAT3 electrochemical corrosion workstation (Guangzhou Beituo Science and Technology Co., Ltd, Guangzhou, China) with a three-electrode system. The saturated calomel electrode was used as the reference electrode and the platinum black electrode as the auxiliary electrode. The potential scanning range was −1.5 V~1 V with the scanning speed of 45 mV/min, and the corrosion medium was 3.5 wt.% NaCl solution. Because the sample was a metal material with electrical conductivity, the non-corrosive area had to be coated with a layer of insulating material to prevent other areas from conducting and affecting the test results.

3. Results and Discussion

3.1. Coating Appearance

The Mo coating appearance is shown in Figure 3. The arrowed area is an enlargement view of the area in the red frame. Figure 3 displays the orange peel-like morphology of Mo coating, which is a typical morphology of ESD coating. The coating is formed by fusion and superposition of numerous deposition points during ESD process. As a result, the Mo coating has a rough surface due to the uneven coverage of deposition points.

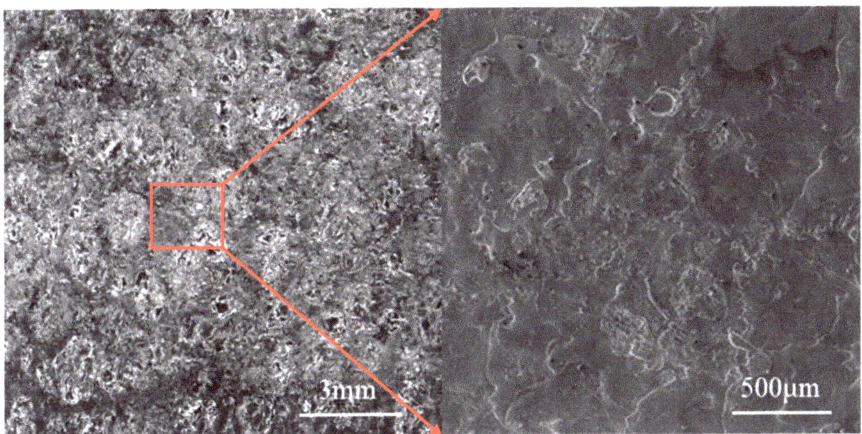

Figure 3. Typical Mo coating appearance.

Figure 4 shows the three-dimensional morphology of the coating with different deposition powers. Figure 5 indicates the effect of the deposition power on the roughness of the coating. It can be seen that the roughness of the coating is different. When the deposition power is less than 800 W, the roughness of the coating increases with the deposition power. However, when the deposition power exceeds 800 W, the roughness decreases. The roughness is mainly related to the discharge energy per unit time. During the ESD process, the pulse discharge energy per unit time can be expressed by the following formula:

$$E = P \times f \times t_{on} \tag{1}$$

where E is the pulse electric spark discharge energy per unit time, P is the deposition power; f is the pulse frequency, and t_{on} is the single pulse discharge time. According to Equation (1), the greater the deposition power P is, the greater the pulse spark discharge energy E per unit time is. When the deposition power is less than 800 W, the melting amount of the electrode increases with the deposition power. It is difficult to uniformly connect and stack of the deposition points during the deposition process due to the rapid cooling rate of the liquid metal, resulting in the increase of roughness. The melting amount of the electrode increases significantly when the deposition power exceeds 800 W. The flow of liquid metal could fill the concave of the deposited layer, so the roughness of the coating can be reduced.

Figure 6 shows the effect of the discharge frequency on the roughness of the coating. The results indicate that the roughness of the coating increases with the increase of discharge frequency at first and then decreases slowly. However, the discharge energy used to melt the Mo electrode increases with the increase of discharge frequency. The maximum roughness of the coating is obtained when the discharge frequency is 500 Hz. When the discharge frequency increases from 50 Hz to 500 Hz, the more quickly the Mo deposition points superimpose and the larger of surface height difference is, which results in an increase of roughness. When the discharge frequency increases from 500 Hz to 650 Hz, the amount of liquid metal increases significantly enough to make the liquid metal flow to fill the concave part of the deposition layer. Therefore, the roughness of the coating is reduced.

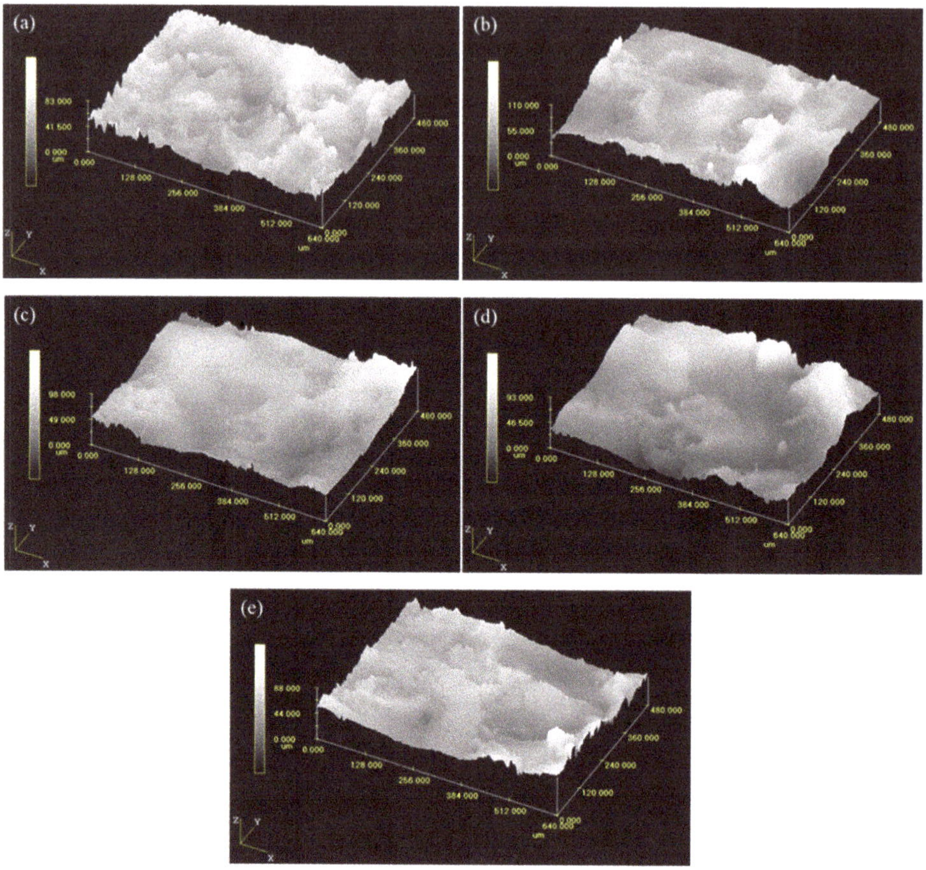

Figure 4. The three-dimensional morphology of coating deposited with different deposition power: (**a**) 100 W (Ra = 2.6 μm), (**b**) 500 W (Ra = 2.9 μm), (**c**) 800 W (Ra = 4.75 μm), (**d**) 1000 W (Ra = 4.35 μm), and (**e**) 1400 W (Ra = 3.8 μm).

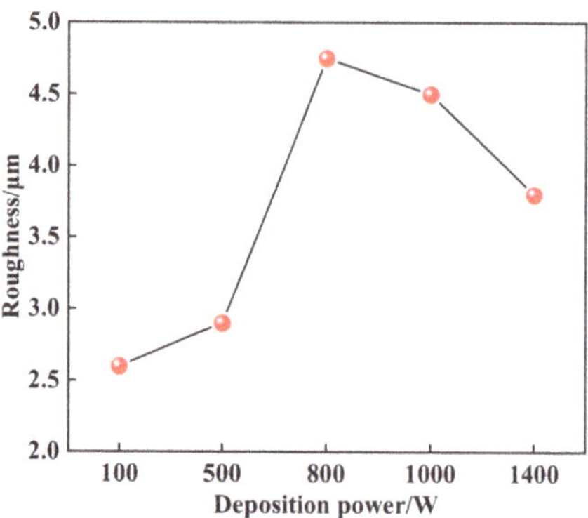

Figure 5. Effect of deposition power on the roughness of coating.

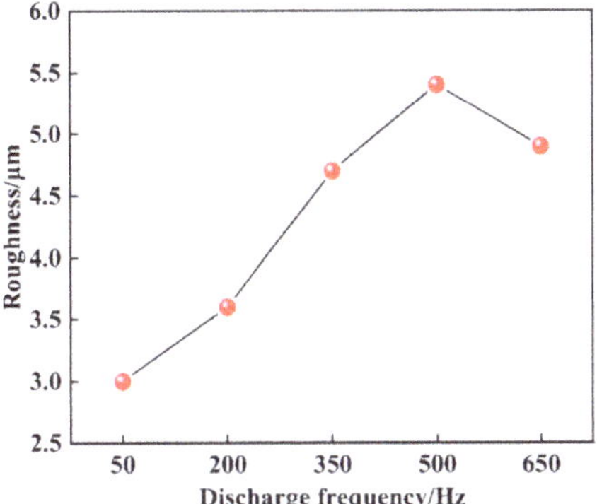

Figure 6. Effect of discharge frequency on the roughness of coating.

The substrate of the 10 mm ×10 mm area is carried out by ESD to prepared Mo coating with different specific deposition time. The specific deposition time is 1~5 min/cm^2. Figure 7 shows the effect of the specific deposition time on the roughness of the coating. It can be seen from the figure that the roughness of the coating increases with the increase of specific deposition time. The longer the specific deposition time is, the more difficult it is to stack and connect the deposition points evenly and the larger the roughness is. As the specific deposition time increases, the heat input of the coating increases. The thermal stress increases due to multiple rapid melting and rapid cooling. The surface defects of the coating such as shedding, cracks and pores, and the roughness of the deposition layer increase sharply when the specific deposition time exceeds 4 min/cm^2.

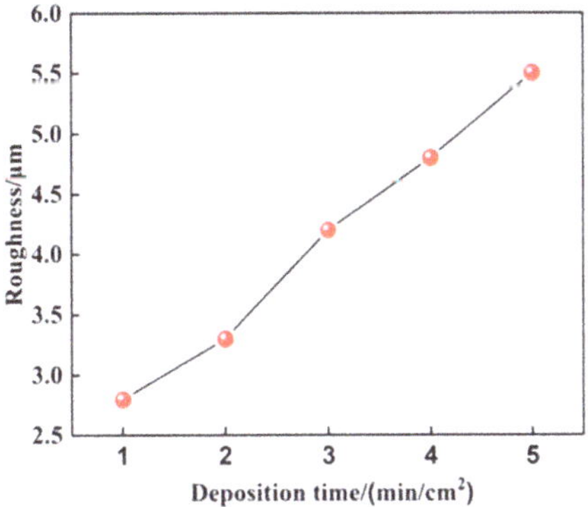

Figure 7. Effect of specific deposition time on the roughness of coating.

3.2. Coating Thickness

The thickness seriously affects the properties of the coating. The specific deposition time is the crucial parameter that affects the thickness of the coating during ESD. The cross-section images of the coating obtained with different specific deposition time are shown in Figure 8. The thickness of the coating is relatively uniform when the specific deposition time is less than 3 min/cm^2. The thickness of the coating is uneven when the specific deposition time exceeds 3 min/cm^2. At the same time, there are microcracks that appear in the coating. There are many reasons for the occurrence of microcracks. On the one hand, the ESD process is a process of instantaneous heating and rapid cooling. If the liquid electrode material that has transitioned to the substrate surface has not completely solidified, the next pulse discharge will arrive. Thereby, thermal stress is generated in the material. On the other hand, the ESD process is accompanied by thermal diffusion. The heat from the coating surface was diffused along the depth direction of the substrate. Appearance of microcracks in the coating is due to the large temperature gradient during the ESD process.

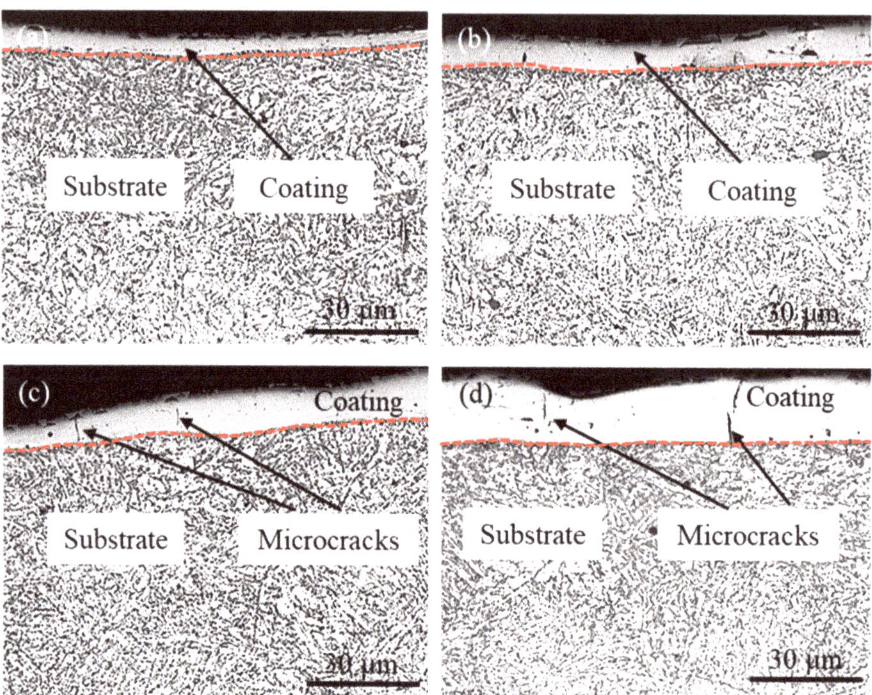

Figure 8. Cross-sectional images of coating: (**a**) 2 min/cm^2, (**b**) 3 min/cm^2, (**c**) 4 min/cm^2, and (**d**) 5 min/cm^2.

The frontal images of coating fabricated by different processing parameters are shown in Figure 9. The coatings display a splash pattern. With the increasing of the specific deposition time, the coating surface has more microcracks.

Figure 10 shows the effect of specific deposition time on the thickness of the coating. The thickness of the coating increases with specific deposition time when the specific deposition time is less than 4 min/cm^2. However, the thickness of the coating decreases when the specific deposition time increases to 5 min/cm^2. Due to the long specific deposition time, the coating undergoes many times rapid melting and cooling processes. Therefore, the surface of the coating produces thermal stress. It is shown in Figure 8c,d that the microcracks defects are formed in the coating. These defects cause the coating to fall off, and then the thickness of the coating decreases significantly. The plasticity and toughness of the coating can be decreased.

The effect of deposition power on the thickness of the coating is shown in Figure 11. The thickness of the coating first increases and then decreases with the increase of the deposition power. The maximum thickness of the coating is obtained when the deposition powers is 1000 W. The energy of electro spark discharge and melting rate of the electrode increase with the increase of the deposition power when the deposition powers are 100~1000 W. Therefore, the coating thickness increases gradually with the increase of deposition power. However, the melting rate of the electrode increases considerably when the deposition power exceeds 1000 W, resulting in the increase of cooling time of molten Mo, and the splatters of molten Mo increase with the rotation of the electrode. In addition, due to the increase of high thermal stress, it is easy for the coating to fall off and crack, so the coating thickness decreases with the improvement deposition power.

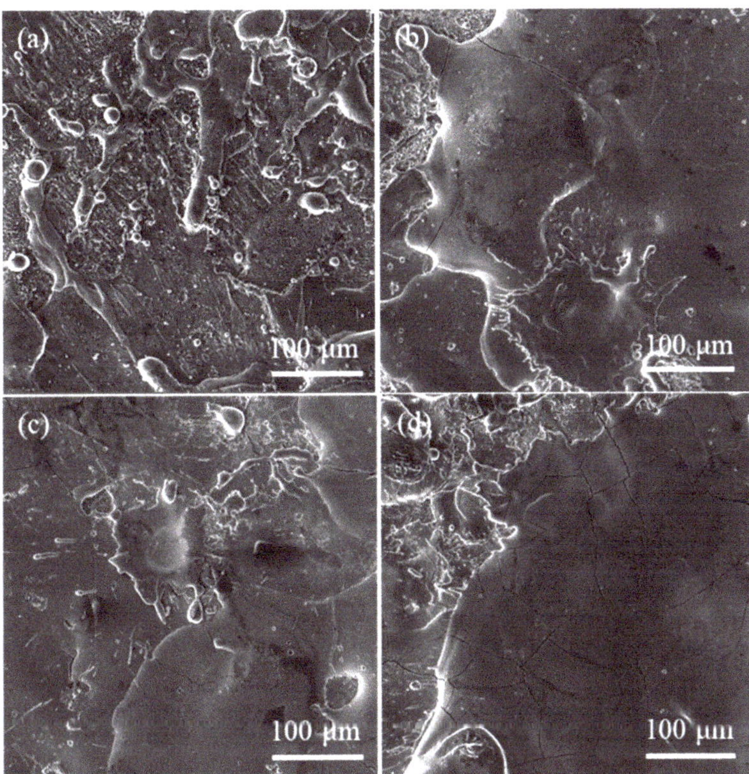

Figure 9. Frontal images of coating: (**a**) 2 min/cm^2, (**b**) 3 min/cm^2, (**c**) 4 min/cm^2, and (**d**) 5 min/cm^2.

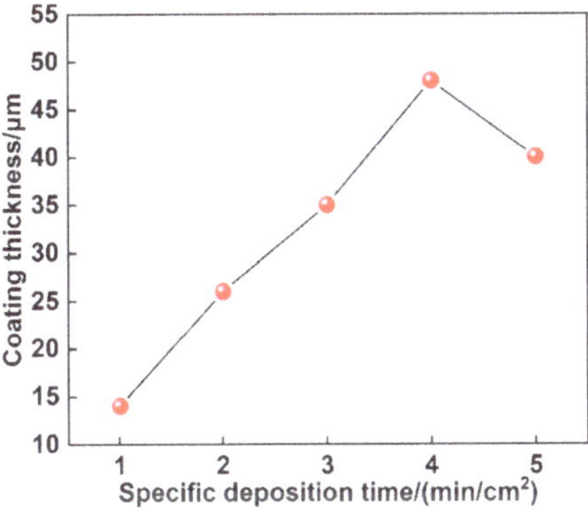

Figure 10. Effect of specific deposition time on the thickness of coating.

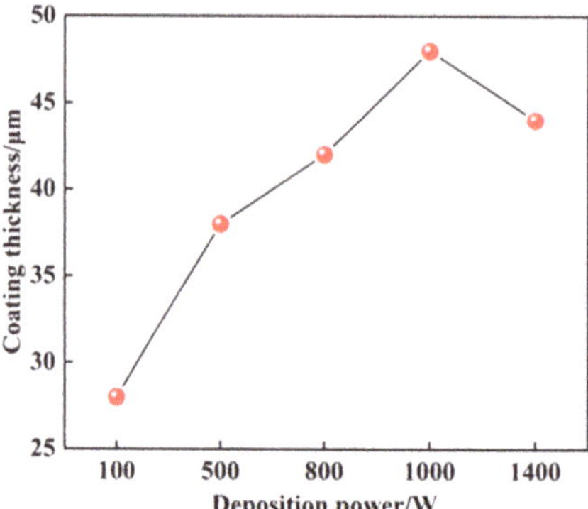

Figure 11. Effect of deposition power on the thickness of coating.

Figure 12 shows the relationship between the discharge frequency and the coating thickness. The discharge frequency was set to 50 Hz, 200 Hz, 350 Hz, 500 Hz, and 650 Hz as variable parameters to prepare the coating. The results show that the thickness of the coating increases with increasing discharge frequency. According to Equation (1), the discharge energy used to melt the electrode increased with the increase of discharge frequency, and then the thickness of the coating increased with the same specific deposition time. The electrode melting rate increases with the increase of discharge frequency when the discharge frequency exceeds 350 Hz. However, the thickness of the coating growth rate also slows down due to the slow growth of melting rate and melting amount of the electrode.

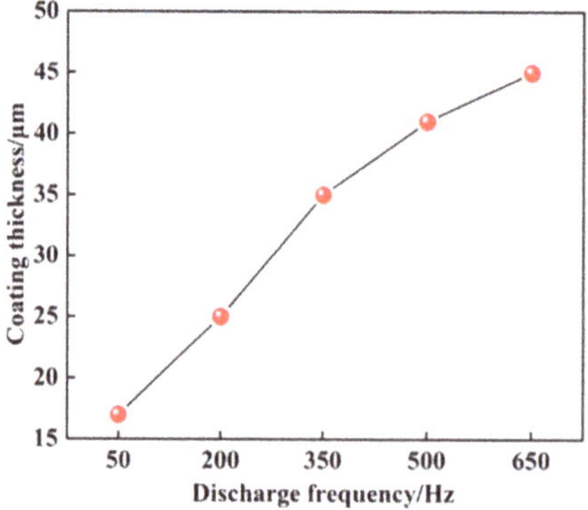

Figure 12. Effect of discharge frequency on the thickness of coating.

3.3. XRD and Chemical Compositions Analysis of the Typical Coating

The thickness and roughness of the coating are important indexes to evaluate the quality of the coating. The effects of the deposition power, discharge frequency, and specific deposition time on the coating quality were studied, and the optimal parameters were obtained as follows: the deposition power, discharge frequency, and specific deposition time were 1000 W, 350 Hz, and 3 min/cm^2, respectively. Finally, the Mo coating with 35 μm thickness and about 5.0 μm roughness was prepared without severe cracks.

The XRD pattern of the coating fabricated under 1000 W/350 Hz/(3 min/cm^2) by the ESD process is shown in Figure 13. The results show that the coating was composed of $Fe_{9.7}Mo_{0.3}$, Fe-Cr, FeMo, and Fe_2Mo cemented carbide phases. Jia Delong et al. obtained similar XRD analysis results at the plasmas sprayed Mo coating on stainless steel substrate [31]. The $Fe_{9.7}Mo_{0.3}$, FeMo, and Fe_2Mo phases were formed by the metallurgical reaction between the Mo electrode material and the H13 steel substrate during the ESD process. It is beneficial for good mechanical properties that the coating was formed by the metallurgical bond between the Mo coating and the H13 steel substrate [32]. In addition, the diffraction peaks of the Mo coating were scattered and miscellaneous, indicating that the coating contained a large number of amorphous phase structure.

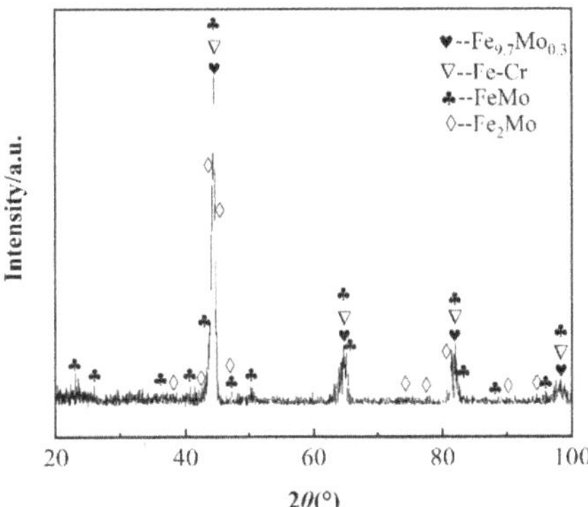

Figure 13. XRD pattern of coating fabricated under 1000 W/350 Hz/(3 min/cm^2).

Figure 14 shows the cross-section image and line scanning results of elements distribution of the coating. The elemental mapping of Figure 14a is shown in Figure 15. The cross-section of the coating mainly consists of the strengthening zone and transition zone. The thickness of the coating is about 35 μm, and that of the transition zone is about 10 μm. There are continuous and close bonds between the coating and the substrate, and there are no cracks and pores at the interface. The line and map scanning results of elements distribution indicate that the coating is mainly composed of Mo, Fe, and Cr elements. The contents of Fe and Cr elements gradually increase from the coating top surface to the substrate, while the content of Mo element gradually decreases. As shown in Figure 15, the enrichment of Mo element in the upper part of the coating is pronounced, while there are also a few Mo element in the substrate. The results also indicate that coating and substrate elements diffuse each other, and the metallurgical bond is formed between the coating and substrate. During the ESD process, there would be a small amount of Fe and Cr elements in the coating due to a small amount of substrate melted by electric erosion. Besides, the coating temperature is higher than the substrate, which is beneficial to diffusing the Fe and Cr elements into the coating. Simultaneously, this is also the reason for the formation of the transition zone.

3.4. Properties Analysis of the Typical Coating

3.4.1. Microhardness

Microhardness tests are conducted on the coating cross-section, and Figure 16 shows the hardness distribution of the coating. The results indicate that the microhardness of the coating gradually decreased from the coating top surface to the substrate. The highest microhardness of the coating value is 1369.5 HV, which is about 6.7 times the substrate. Due to the thin thickness of the coating, the element content in the micro zone of the coating changes greatly, so the microhardness gradient of the coating changes greatly. The analysis shows that during the ESD process, the Mo electrode has a metallurgical reaction with an H13 steel substrate, and the cemented carbide phase is generated on the surface of the H13 steel substrate. At the same time, the amorphous structure is formed inside the coating. The amorphous phase and the cemented carbide phase improve the microhardness and strengthen the bearing capacity of the coating. However, the microhardness distribution of the coating shows obvious gradient, which is beneficial to improve the friction and wear capacity of the coating.

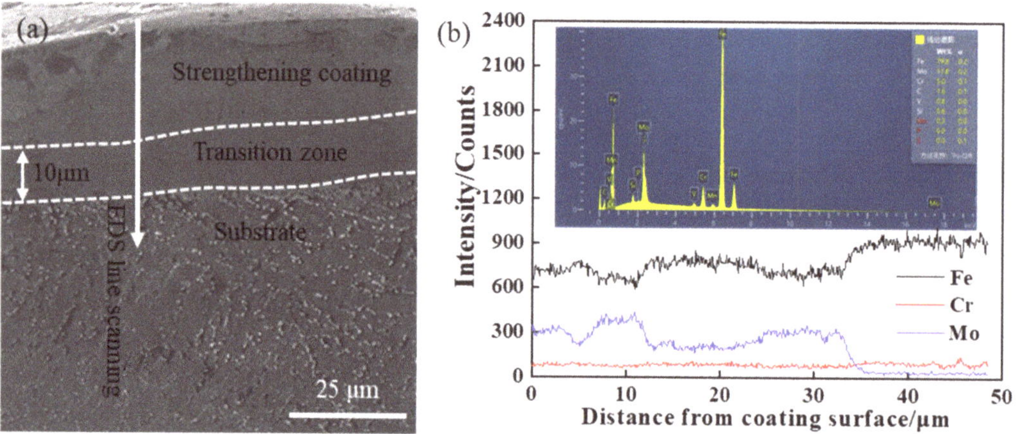

Figure 14. Morphology cross-section and elements distribution of coating fabricated under 1000 W/350 Hz/(3 min/cm^2): (**a**) the cross-section of coating and (**b**) line scanning results of elements distribution.

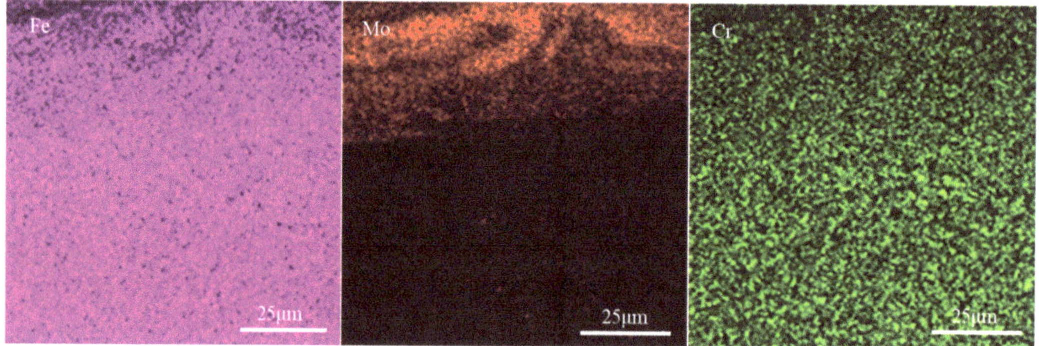

Figure 15. Map scanning results of elemental distribution of coating fabricated under 1000 W/350 Hz/(3 min/cm^2).

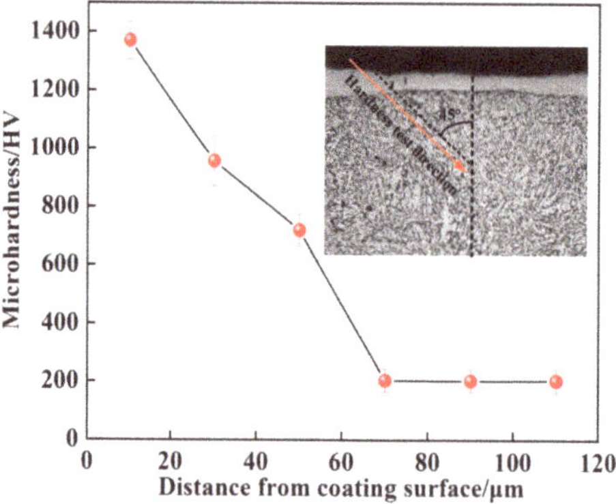

Figure 16. Hardness distribution of coating fabricated under 1000 W/350 Hz/(3 min/cm^2).

3.4.2. Wear Resistance

The coating and the substrate samples are conducted at room temperature by the abrasive wear test, and the weight losses of the samples under different loads are measured to evaluate the wear resistance. The wear resistance curves of substrate and Mo coating samples are shown in Figure 17. Under the same load, the weight loss of the coating is less than that of the substrate. With the increase of the load, the weight loss of the coating increases slightly and gently, but the weight loss of substrate increases sharply. According to the weight loss under the same load condition, the wear resistance of the coating is about seven times higher than that of the substrate. The good wear resistance of the coating corresponds to the defect-free and high microhardness of the coating and the good metallurgical bonding between the substrate and coating. In addition, there are a large number of cemented carbides in the coating, which are dispersed in the coating, thus significantly improving the wear resistance of the coating. The wear scar morphology of substrate and coating samples is shown in Figure 18. For the wear sample of the substrate, the furrow-shaped wear scars are deep and wide, and there are many pits formed by the material peeling off locally. However, the furrow-shaped wear scars on the coating surface are smoother and shallower, and there are no pits, which is consistent with the microhardness and weight loss results.

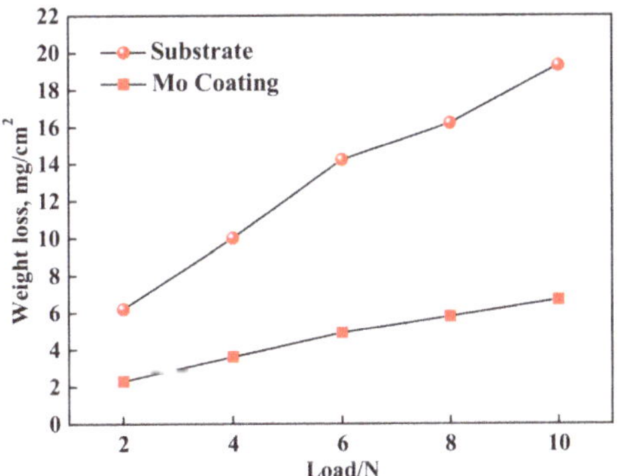

Figure 17. Curves of wear resistance for substrate and coating.

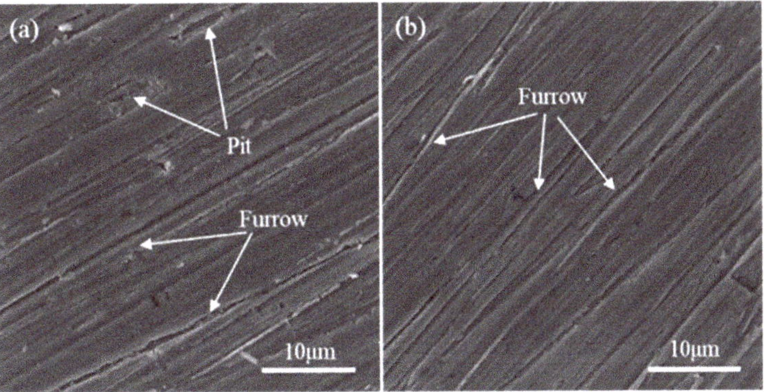

Figure 18. Wear scar morphology: (**a**) substrate and (**b**) coating.

3.4.3. Corrosion Resistance

The polarization curves of substrate and coating are obtained by electrochemical etching as shown in Figure 19. It can be seen that the sample of the substrate self-corrosion potential is −682 mV, and the self-corrosion current density is 2.01×10^{-3} A/m^2. The coating's self-corrosion potential was −621 mV, and the self-corrosion current density is 4.89×10^{-4} A/m^2. Compared with the substrate, the self-corrosion of the coating potential increased by 9%, and the self-corrosion current density was 24% of the substrate, which meant that the corrosion rate of the coating was significantly reduced. The corrosion resistance of the coating was much better than that of the substrate due to Mo's excellent corrosion resistance.

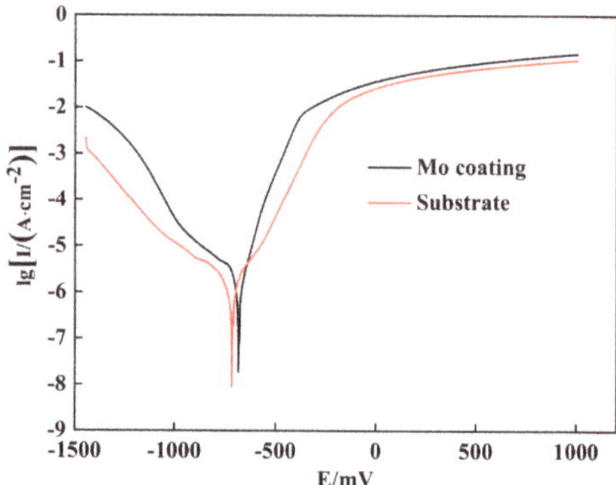

Figure 19. Polarization curves of substrate and coating.

4. Conclusions

In this study, the ESD process was used to prepare Mo coating on the H13 steel substrate, and the properties of the coating were investigated. The main conclusions are as follows.

(1) The Mo coating without defects was successively prepared on the H13 steel substrate by ESD process. The effect of main parameters on the thickness and roughness of the coating was studied. The well main parameters were obtained as follows: the deposition power, discharge frequency, and specific deposition time were 1000 W, 350 Hz, and 3 min/cm^2, respectively.

(2) The Mo coating was mainly composed of $Fe_{9.7}Mo_{0.3}$, Fe-Cr, and FeMo, Fe_2Mo cemented carbide phases and amorphous phase. The coating cross-section mainly consisted of the strengthening zone and transition zone. The metallurgical bonding was formed between the coating and substrate.

(3) The microhardness of the coating distribution showed a noticeable gradient, and the microhardness gradually decreased from the coating top surface to the substrate. The wear resistance of the coating was about seven times higher than that of the substrate. The amorphous phase and cemented carbide phase improved the microhardness and strengthened the bearing capacity of the coating.

(4) The self-corrosion of the coating potential increased by 9%, and the self-corrosion current density was 24% of the substrate. The corrosion resistance of the coating was much better than that of the substrate due to Mo's excellent corrosion resistance by ESD technology preparation.

Author Contributions: Conceptual and experimental design, W.W. and X.Z.; data collection and analysis, drafting, and writing, M.D.; literature research and data investigation, C.L.; supervision and validation, Y.T.; editing and revising, X.Z. All authors have read and agreed to the published version of the manuscript.

Funding: This work was financially supported by the Jilin Province Science and Technology Development Plan Project (No. 20200401034GX), Jilin Province Development and Reform Commission

Industrial Technology Research and Development Project (grant no.2020C029-1), and the Fundamental Research Funds for the Central Universities (grant no.45120031B004).

Institutional Review Board Statement: Not applicable.

Informed Consent Statement: Not applicable.

Data Availability Statement: The data presented in this study are available on request from the corresponding author.

Conflicts of Interest: The authors declare no conflict of interest.

References

1. Lu, H.; Xue, K.; Xu, X.; Luo, K.; Xing, F.; Yao, J.; Lu, J. Effects of laser shock peening on microstructural evolution and wear property of laser hybrid remanufactured Ni25/Fe104 coating on H13 tool steel. *J. Mater. Process. Technol.* **2021**, *291*, 117016. [CrossRef]
2. Besler, R.; Bauer, M.; Furlan, K.P.; Klein, A.N.; Janssen, R. Effect of Processing Route on the Microstructure and Mechanical Properties of Hot Work Tool Steel. *Mater. Res.* **2017**, *20*, 1518–1524. [CrossRef]
3. Pagounis, E.; Talvitie, M.; Lindroos, V.K. Microstructure and mechanical properties of hot work tool steel matrix composites produced by hot isostatic pressing. *Powder Metall.* **1997**, *40*, 55–61. [CrossRef]
4. Bahrami, A.; Anijdan, S.M.; Golozar, M.; Shamanian, M.; Varahram, N. Effects of conventional heat treatment on wear resistance of AISI H13 tool steel. *Wear* **2005**, *258*, 846–851. [CrossRef]
5. Zhang, B.; Yang, F.; Qin, Q.; Lu, T.; Sun, H.; Lin, S.; Chen, C.; Guo, Z. Characterisation of powder metallurgy H13 steels prepared from water atomised powders. *Powder Metall.* **2019**, *63*, 9–18. [CrossRef]
6. Randelius, M.; Sandström, R.; Melander, A. Fatigue strength of conventionally cast tool steels and its dependence of carbide microstructure. *Steel Res. Int.* **2012**, *83*, 83–90. [CrossRef]
7. Fischmeister, H.F.; Riedl, R.; Karagöz, S. Solidification of high-speed tool steels. *Met. Mater. Trans. A* **1989**, *20*, 2133–2148. [CrossRef]
8. Yao, F.; Fang, L. Thermal Stress Cycle Simulation in Laser Cladding Process of Ni-Based Coating on H13 Steel. *Coatings* **2021**, *11*, 203. [CrossRef]
9. Chen, H.; Lu, Y.; Sun, Y.; Wei, Y.; Wang, X.; Liu, D. Coarse TiC particles reinforced H13 steel matrix composites produced by laser cladding. *Surf. Coat. Technol.* **2020**, *395*, 125867. [CrossRef]
10. Li, G.S.; Li, J.H.; Feng, W.L. Effects of specific powder and specific energy on the characteristics of NiWC25 by laser cladding. *Surf. Technol.* **2019**, *48*, 253–258.
11. Boccalini, M.; Goldenstein, H. Solidification of high-speed steels. *Int. Mater. Rev.* **2001**, *46*, 92–115. [CrossRef]
12. Kudryashov, A.E.; Potanin, A.Y.; Lebedev, D.N.; Sukhorukova, I.V.; Shtansky, D.V.; Levashov, E.A. Structure and properties of Cr-Al-Si-B coatings produced by pulsed electrospark deposition on a nickel alloy. *Surf. Coat. Technol.* **2016**, *285*, 278–288. [CrossRef]
13. Cadney, S.; Brochu, M. Formation of amorphous Zr41. 2Ti13. 8Ni10 Cu12. 5Be22. 5 coatings via the ElectroSpark Deposition process. *Intermetallics* **2008**, *16*, 518–523. [CrossRef]
14. Kong, W.C.; Shen, H.; Gao, F.X.; Wu, F.; Lu, Y.L. Electrochemical corrosion behavior of HVOF sprayed WC-12Co coating on H13 hot work mould steel. *Anti Corros. Methods Mater.* **2019**, *66*, 827–834.
15. Kong, D.J.; Sheng, T.Y. Wear behaviors of HVOF sprayed WC-12Co coatings by laser remelting under lubricated condition. *Opt. Laser Technol.* **2017**, *89*, 86–91.
16. Zórawski, W. The microstructure and tribological properties of liquid-fuel HVOF sprayed nanostructured WC-12Co coatings. *Surf. Coat. Technol.* **2013**, *220*, 276–281. [CrossRef]
17. Meng, C.; Cao, R.; Li, J.Y.; Geng, F.Y.; Zhang, Y.W.; Wu, C.; Wang, X.L.; Zhuang, W.B. Mechanical properties of TiC-reinforced H13 steel by bionic laser treatment. *Opt. Laser Technol.* **2021**, *136*, 106815–106823. [CrossRef]
18. Salmaliyan, M.; Malek Ghaeni, F.; Ebrahimnia, M. Effect of electro spark deposition process parameters on WC-Co coating on H13 steel. *Surf. Coat. Technol.* **2017**, *321*, 81–89. [CrossRef]
19. Frangini, S.; Masci, A. A study on the effect of a dynamic contact force control for improving electrospark coating properties. *Surf. Coat. Technol.* **2010**, *204*, 2613–2623. [CrossRef]
20. Alexander, V.R.; Sahin, O.; Korkmaz, K. A modified electrospark alloying method for low surface roughness. *Surf. Coat. Technol.* **2009**, *203*, 3509–3515.
21. Luo, C.; Dong, S.J.; Xiang, X. Microstructure and properties of TiC coating by vibrating electrospark deposition. *Key Eng. Mater.* **2008**, *27*, 180–183. [CrossRef]
22. Li, Z.M.; Zhu, Y.L.; Sun, X.F.; Huang, Y.L.; Zhang, H.W. Development in Research and Application of Electro Spark Deposition Technology. *Hot Work. Technol.* **2013**, *42*, 32–37.
23. Alexander, A.B.; Sergey, A.P. Formation of WC-Co coating by a novel technique of electrospark granules deposition. *Mater. Des.* **2015**, *80*, 109–115.

24. Umanskyi, O.P.; Storozhenko, M.S.; Tarelnyk, V.B.; Koval, O.Y.; Gubin, Y.V.; Tarelnyk, N.V. Protective and Functional Powder Coatings Electrospark Deposition of FeNiCrBSiC-MeB2 Coatings on Steel. *Powder Metall. Met. Ceram.* **2020**, *59*, 330–341. [CrossRef]
25. Alexander, A.B.; Chigrin, P.G. Synthesis of Ti-Al intermetallic coatings via electrospark deposition in a mixture of Ti and Al granules technique. *Surf. Coat. Technol.* **2020**, *387*, 125550–125558.
26. Karlsdottir, S.N.; Geambazu, L.E.; Csaki, L.; Thorhallsson, A.I.; Stefanoiu, R.; Magnus, F.; Cotrut, C. Phase Evolution and Microstructure Analysis of CoCrFeNiMo High-Entropy Alloy for Electro-Spark-Deposited Coatings for Geothermal Environment. *Coatings* **2019**, *9*, 406. [CrossRef]
27. Wang, W.F.; Han, C. Microstructure and Wear Resistance of Ti6Al4V Coating Fabricated by Electro-Spark Deposition. *Metals* **2018**, *9*, 23. [CrossRef]
28. Cao, G.J.; Zhang, X.; Tang, G.Z.; Ma, X.X. Microstructure and Corrosion Behavior of Cr Coating on M50 Steel Fabricated by Electrospark Deposition. *J. Mater. Eng. Perform.* **2019**, *28*, 4086–4094. [CrossRef]
29. Wang, Y.W.; Wang, X.Y.; Wang, X.L.; Yang, Y.; Cui, Y.H.; Ma, Y.D. Microstructure and properties of in-situ composite coatings prepared by plasma spraying MoO3-Al composite powders. *Ceram. Int.* **2021**, *47*, 1109–1120. [CrossRef]
30. Satish, T.; Ankur, M.; Modi, S.C. High-Performance Molybdenum Coating by Wire–HVOF Thermal Spray Process. *J. Therm. Spray Technol.* **2018**, *27*, 757–768.
31. Jia, D.L.; Yi, P.; Liu, Y.C.; Sun, J.W.; Yue, S.B.; Zhao, Q. Effect of laser textured groove wall interface on molybdenum coating diffusion and metallurgical bonding. *Surf. Coat. Technol.* **2021**, *405*, 126561–126569. [CrossRef]
32. Chen, Z.; Zhou, Y. Surface modification of resistance welding electrode by electro-spark deposited composite coatings: Part I. Coating characterization. *Surf. Coat. Technol.* **2006**, *201*, 1503–1510. [CrossRef]

 materials

Article

Optimization of Pickling Solution for Improving the Phosphatability of Advanced High-Strength Steels

Sangwon Cho [1], Sang-Jin Ko [1], Jin-Seok Yoo [1], Joong-Chul Park [2], Yun-Ha Yoo [2] and Jung-Gu Kim [1,*]

[1] Department of Materials Science and Engineering, Sungkyunkwan University, Suwon-Si 16419, Korea; ppkh@skku.edu (S.C.); tkdwls121@skku.edu (S.-J.K.); wlstjr5619@skku.edu (J.-S.Y.)

[2] Steel Solution Marketing Dept., POSCO Global R&D Center, Incheon 21985, Korea; heavymetal@posco.com (J.-C.P.); yunha778@posco.com (Y.-H.Y.)

* Correspondence: kimjg@skku.edu; Tel.: +82-31-290-7360

Abstract: This study investigated the optimum pickling conditions for improving the phosphatability of advanced high-strength steel (AHSS) using surface analysis and electrochemical measurements. To remove the SiO_2 that forms on the surface of AHSS, 30 wt.% NH_4HF_2 was added to the pickling solution, resulting in a significant reduction in the amount of SiO_2 remaining on the surface of the AHSS. The phosphatability was improved remarkably using HNO_3 concentrations higher than 13% in the pickling solution. Furthermore, phosphate crystals became finer after pickling with a HNO_3-based solution rather than a HCl-based solution. Electrochemical impedance spectroscopy (EIS) data indicated that the corrosion resistance of AHSS subjected to HNO_3-based pickling was higher than that of AHSS subjected to HCl-based pickling. Fluorine compounds, which were involved in the phosphate treatment process, were only formed on the surface of steel in HNO_3-based solutions. The F compounds reacted with the phosphate solution to increase the pH of the bulk solution, which greatly improved the phosphatability. The phosphatability was better under HNO_3-based conditions than a HCl-based condition due to the fineness of the phosphate structure and the increased surface roughness.

Keywords: advanced high-strength steel; phosphate coating; acid cleaning; electrochemical impedance spectroscopy

Citation: Cho, S.; Ko, S.-J.; Yoo, J.-S.; Park, J.-C.; Yoo, Y.-H.; Kim, J.-G. Optimization of Pickling Solution for Improving the Phosphatability of Advanced High-Strength Steels. *Materials* **2021**, *14*, 233. https://doi.org/10.3390/ma14010233

Received: 2 December 2020
Accepted: 30 December 2020
Published: 5 January 2021

Publisher's Note: MDPI stays neutral with regard to jurisdictional claims in published maps and institutional affiliations.

1. Introduction

Consumer demand for automobiles has been increasing over the last decade, with particular regard for features such as appearance, driving performance, comfort, safety, fuel efficiency, and environmental friendliness. In response, the automotive industry has spurred the development of new and advanced technologies to improve safety while reducing vehicle weight. However, the technology used to manufacture lightweight automobiles is associated with environmental problems [1]. Research into automotive weight reduction can be divided into the fields of lightweight materials and high-strength steel sheets [2]. Studies on the use of lightweight materials for automobiles have mainly focused on aluminum and fiber-reinforced plastics [3–5], but the practical use of these materials is difficult due to product design limitations and manufacturing problems. However, high-strength materials can be used without making significant changes to conventional manufacturing methods, so they have the advantage of minimal additional investment costs. In addition, multi-material applications can be used with high-strength materials, depending on the function of the component; thus, a variety of designs are possible. Another advantage of using high-strength materials is that reductions in the thickness and weight of the material are obtained at the same time. Therefore, various grades of steel, from mild steel to advanced high-strength steel (AHSS), have been used in recent automotive bodies. In addition, steel manufacturers are actively researching applications for more advanced steel sheets to further reduce the weight.

Generally, a steel sheet for automobiles is subjected to a phosphate treatment to improve corrosion resistance and ensure coating film adhesion before painting [6,7]. Phosphate crystals form on the steel sheet surface from the phosphate treatment, which greatly enhances the adhesion of electro-painting [7]. The size of the phosphate crystals and coating weight are very important factors: relatively small crystals result in the formation of denser phosphate crystals, resulting in better adhesion to the coating.

During phosphate treatment, an acidic zinc phosphate solution reacts with the surface of the metal and consumes hydrogen ions in the solution. This reaction increases the pH near the metal surface. Then, the phosphate solution is saturated with zinc phosphate. As a result of the increase in pH, the formation of phosphate crystals is promoted. Therefore, phosphatability can be improved by an accelerator that increases the rate of hydrogen ion consumption on the metal surface [8].

Alloy elements such as Si and Mn are indispensable elements for obtaining high strength and high ductility. Therefore, AHSS has a higher content of these alloying elements compared to general steels. However, during steel manufacturing and use, Si and Mn oxides are easily formed on the steel surface. When these composite oxides densely form on the surface of a steel sheet, they have a significant effect on the phosphate treatment. In particular, a Si oxide film that forms on the AHSS surface acts as a barrier to phosphating, thereby decreasing the phosphatability [9]. Therefore, to improve the phosphatability of AHSSs that contain a large amount of Si, the formation of these oxides on the surface of the steel must be suppressed.

Currently, two main methods are used to remove the Si oxide that forms on the surface of AHSS. The first method uses a high-temperature, high-concentration inorganic acid (e.g., hydrochloric acid, sulfuric acid, nitric acid, or formic acid) solution, which removes the Si oxide by dissolving the base metal under the Si oxide [10]. In this method, if the concentration of the inorganic acid is low, the Si oxide on the surface is not removed; conversely, if the concentration is too high, over-etching occurs and the material cannot be used in automotive manufacturing. Therefore, it is important to appropriately control the concentration of the inorganic acid.

The second method utilizes hydrofluoric acid (HF) in a pickling solution to remove the Si oxide, using the following reaction:

$$SiO_2 + 4HF \rightarrow SiF_4 + 2H_2O \quad (1)$$

This method does not use HF alone; rather, HF is mixed with an inorganic acid to activate its reaction [11]. However, because HF is very dangerous, it is challenging to use it in a pickling solution. To react with silicate, which is a basic form of Si oxide, F^- is required. Therefore, many researchers have attempted to replace HF with ammonium fluoride (NH_4F) and ammonium hydrogen fluoride (NH_4HF_2) as a new source of F^- because they are eco-friendly and less hazardous than HF. NH_4F and NH_4HF_2 have been used in a mixture with an inorganic acid solution; the etching rate can be adjusted by controlling the type and concentration of the inorganic acid [12]. Because NH_4HF_2 has more F than NH_4F, its etching efficiency is better and its cost is lower; thus, it is suitable as a HF substitute.

Although many studies on the phosphatability of AHSS have been carried out, some problems still exist, such as a decrease in phosphatability due to Si oxides. The phosphatability of AHSS under the conventional pickling conditions used in industry are not optimal; as shown in Figure 1, the coating coverage and the coating weight are insufficient. Therefore, in this study, we investigated the optimum pickling conditions to improve the phosphatability of AHSS while minimizing environmental hazard.

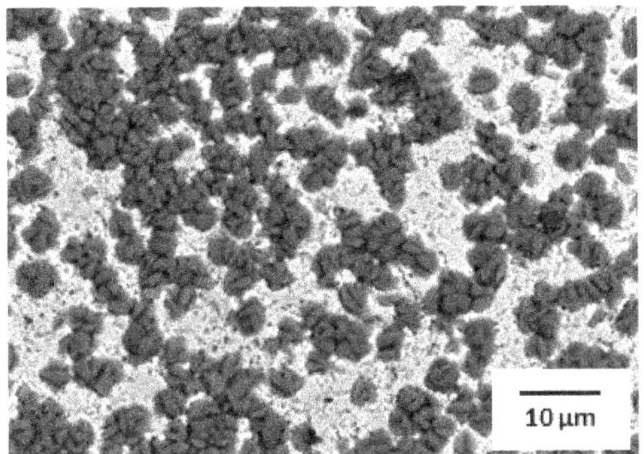

Figure 1. Surface SEM image of advanced high-strength steel (AHSS) after phosphate treatment under the conventional pickling condition.

2. Materials and Methods

The specimens used in the experiment were AHSSs with Si contents of 0.4 and 1.0, as described in Table 1. The specimens were dipped in an alkaline solution at 45 °C for 2 min for degreasing before testing. The phosphate treatment process was divided into two steps: pickling and phosphating. As listed in Table 2, the pickling solutions used in the experiment were prepared by adding 30 wt.% NH_4HF_2 to HCl and HNO_3 at various concentrations (5.5, 8, 10.5, 13, 15.5, and 18 wt.%). Pickling tests were performed at 55 °C for 7 s. Then, after a degreasing and surface conditioning process, phosphate treatment was performed in a Zn–phosphate solution according to the common automotive process.

Table 1. Chemical compositions of the steels. (Unit: wt.%).

Grade	C	Mn	Si	P	S
0.4Si steel	0.17	2.0	0.25	0.02	0.005
1.0Si steel	0.1	2.8	1.2	0.03	0.003

Table 2. Pickling conditions for the steels.

Parameter	Conventional Pickling Condition	HCl-Based Pickling Condition	HNO_3-Based Pickling Condition
Inorganic acid	HCl	HCl	HNO_3
Acid concentration (wt.%)	5.5, 18	5.5, 8, 10.5, 13, 15.5, 18	5.5, 8, 10.5, 13, 15.5, 18
Additives	–	NH_4HF_2 (30 wt.%)	NH_4HF_2 (30 wt.%)
Temperature (°C)	55	55	55
Time (s)	Within 7	7	7

Phosphatability was evaluated by coating coverage and coating weight. In an image obtained using the back scattered electron (BSE) mode in scanning electron microscopy (SEM; JSM-6700F, JEOL Ltd., Akishima, Japan), a difference in brightness existed between the areas where phosphate crystals were formed and where they did not form. The difference between the light and dark was measured to determine coating coverage using Image J S/W (Laboratory for Optical and Computational Instrumentation, University of Wisconsin). The coating weight was analyzed using energy dispersive X-ray fluorescence (EDXRF; EDX-8000, SHIMADZU, Kyoto, Japan). After calibrating the P signal detected in a standard specimen of known phosphate coating thickness, the coating weight can be calculated by multiplying the measured P signal by density. The oxides that remained on the steel surface after pickling were analyzed using secondary ion mass spectrometry

(SIMS; TOF-SIMS-5, Ion-ToF company, Munster, Germany), X-ray diffraction (XRD; D/max-2500V/PC, Rigaku, Tokyo, Japan), energy dispersive spectroscopy (EDS; JEOL JSM-6700F, JEOL Ltd., Akishima, Japan), and an electron probe micro-analyzer (EPMA; JEOL, JXA-8530F, JEOL Ltd., Akishima, Japan).

Electrochemical impedance spectroscopy (EIS; VSP300, Neoscience, Seoul, Korea) tests were performed with an amplitude of 10 mV in the frequency range of 100 kHz to 1 mHz. The electrochemical cell consisted of a three-electrode system. The counter electrode was a graphite rod, and a saturated calomel electrode was used as the reference electrode. The area exposed to the electrolyte was 1 cm^2. The test solution was a cyclic corrosion test solution from the Society of Automotive Engineering (SAE solution), and its chemical composition is listed in Table 3. EIS data were fit to the form of a Bode plot, and Bode spectra were fit to equivalent circuit models by using ZSimpWin software (Ver. 3.21).

Table 3. Chemical composition of Society of Automotive Engineering (SAE) solution.

Chemicals	NaCl	$CaCl_2$	$NaHCO_3$	$(NH_4)_2SO_4$
SAE solution (wt.%)	0.05	0.1	0.075	0.35

To analyze the pH change of the phosphate solution due to the oxides, 1 M of the iron oxides and the F compounds generated during pickling were dissolved in the phosphate solution. The change in pH was measured using a pH meter. To analyze the correlation between surface roughness and phosphatability, the surface roughness of the specimen after pickling was measured using an α-step (Alpha-Step IQ, KLA-Tencor, Milpitas, CA, USA).

3. Results and Discussion

3.1. Surface Analysis

Figures 2 and 3 show the amount of SiO_2 remaining on the 1.0Si steel surface layer as analyzed by TOF-SIMS and XPS after pickling with various solutions. Figure 2a shows a conventional pickling condition, in which the amount of SiO_2 remaining on the surface is larger and the distribution of SiO_2 is more uneven than under the other pickling conditions. In Figure 2b,c, however, in the pickling solutions with added NH_4HF_2, the SiO_2 on the steel surface was noticeably reduced, even though it was not completely removed. Figure 3 shows that the intensity of the SiO_2 peak in the pickling solution with added NH_4HF_2 was reduced compared with the intensity of the SiO_2 peak under conventional pickling conditions. Thus, the amount of SiO_2 remaining on the surface was significantly reduced by NH_4HF_2-added pickling.

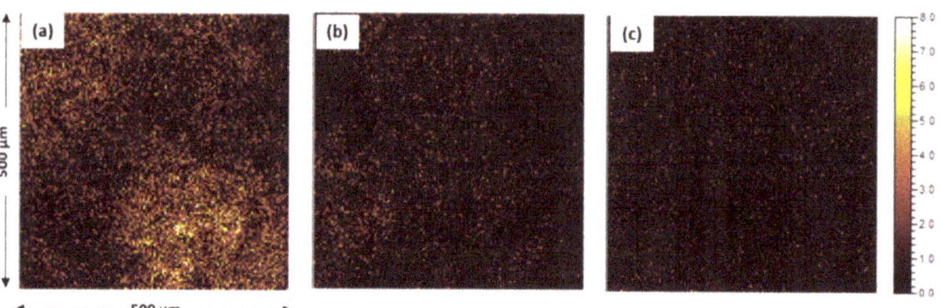

Figure 2. TOF-secondary ion mass spectrometry (SIMS) images of SiO_2 remaining on the 1.0Si steel surface after pickling under (**a**) conventional condition, (**b**) a HCl-based condition, and (**c**) a HNO_3-based condition.

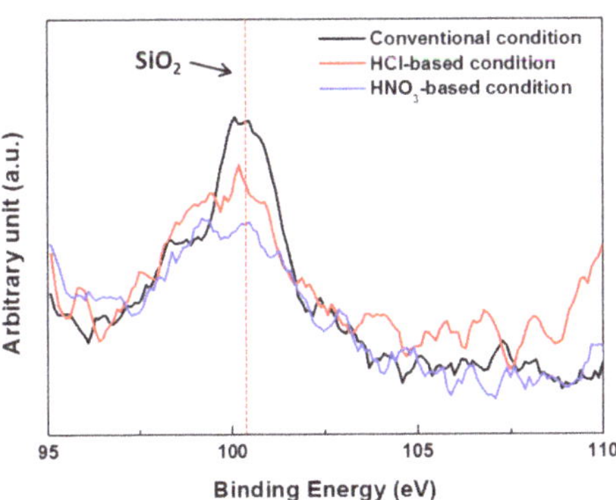

Figure 3. XPS spectra of Si on the 1.0Si steel surface after pickling.

After the pickling of 1.0Si steel under various conditions, the surface was analyzed with optical microscopy (OM), as shown in Figure 4. The surface of the specimen before and after HCl-based pickling was the same; however, after HNO_3-based pickling, the surface of the specimen was uniformly covered with a new gray product. Figures 5–7 and Table 4 reveal the oxides that remained on the surface after pickling, as analyzed by EDS, EPMA, and XRD.

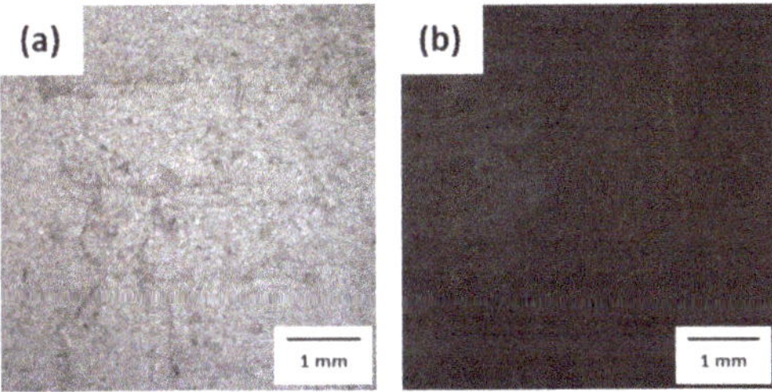

Figure 4. Optical microscopy (OM) images of the 1.0Si steel surface after pickling under (**a**) a HCl-based condition and (**b**) a HNO_3-based condition.

As shown in Figure 5, the EDS peaks of O, Si, Cr, and Fe can be observed from 1.0Si steel under all pickling conditions, indicating that the remaining oxides on the surface were mainly Fe, Cr, and Si oxides. However, the F peak was observed only under HNO_3-based pickling conditions. The results of EPMA analysis (Figure 6) were similar to those of EDS: an F component was observed throughout the surface only under HNO_3-based pickling conditions. This was largely due to NH_4HF_2, which means that the F compounds were formed on the surface only after HNO_3-based pickling. XRD analysis (Figure 7) indicated that the Fe of the base material was only detected in the specimens that underwent conventional pickling and HCl-based pickling. However, various components were detected in the specimen that underwent HNO_3-based pickling. The products formed under HNO_3-based pickling were $(NH_4)FeF_5{\cdot}H_2O$, $FeSiF_6{\cdot}6H_2O$, FeF_2, and/or FeF_3. The products were produced by the following reactions [13,14]:

$$Fe_2O_3 + 5NH_4HF_2 \rightarrow 2(NH_4)_2FeF_5 + 3H_2O + NH_3\uparrow \quad (2)$$

$$Si + 2H^+ + 2F^- + 4HF \rightarrow H_2SiF_6{\cdot}6H_2O + 2H_2\uparrow \quad (3)$$

$$Fe^{2+} + 2F^- \rightarrow FeF_2 \quad (4)$$

$$FeF_2 + H_2SiF_6 \rightarrow FeSiF_6 + 2HF \quad (5)$$

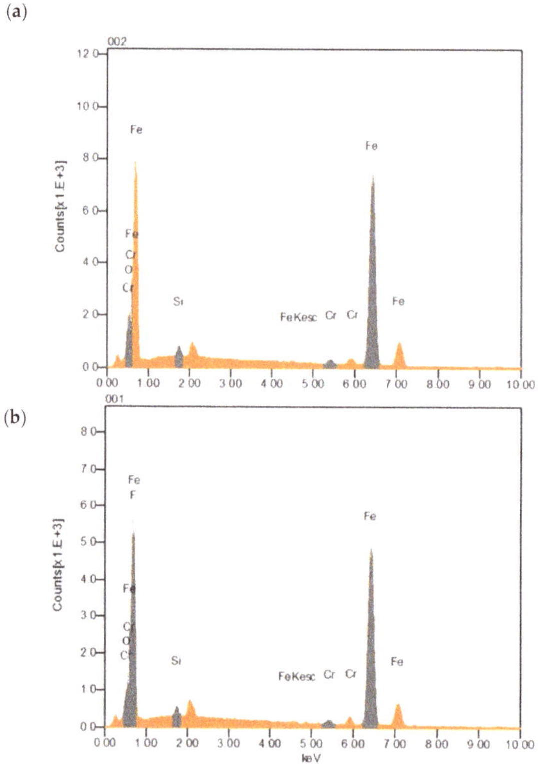

Figure 5. EDS analysis of the 1.0Si steel surface after pickling under (**a**) a HCl-based condition and (**b**) a HNO_3-based condition.

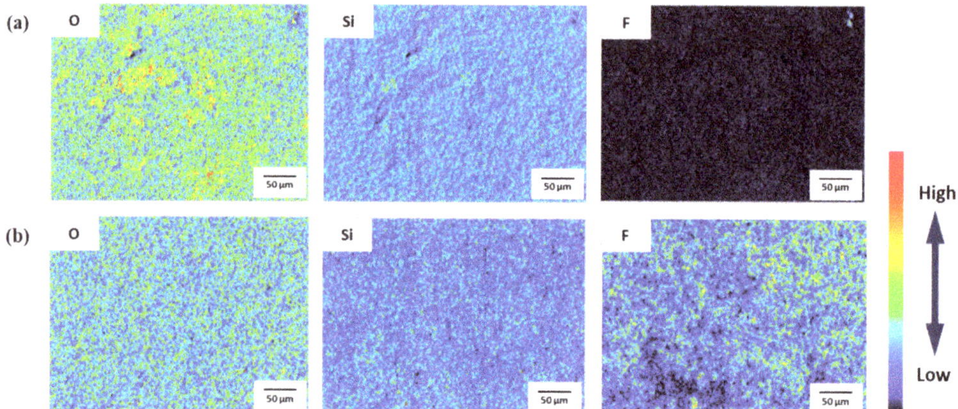

Figure 6. EPMA analysis of the 1.0Si steel surface after pickling under (**a**) a HCl-based condition and (**b**) a HNO_3-based condition.

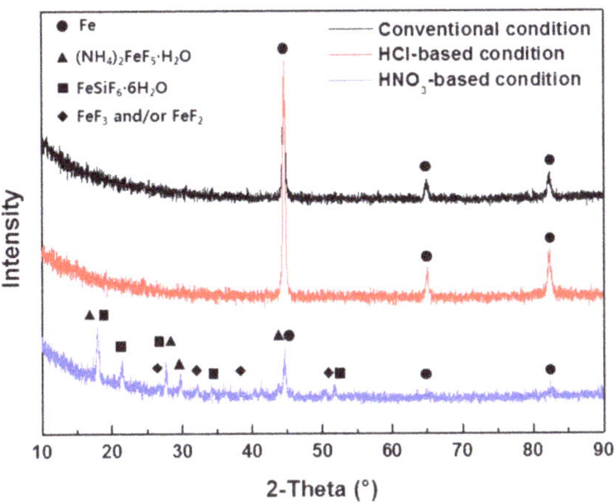

Figure 7. XRD analysis of the 1.0Si steel surface after pickling.

Table 4. EDS results of the 1.0Si steel surface after pickling.

Parameter	Fe	O	Si	F	Cr
HCl-based pickling condition (%)	94.13	3.77	0.85	–	1.25
HNO_3-based pickling condition (%)	92.25	1.98	0.85	3.56	1.36

In the HNO_3/HF system, $FeF_3 \cdot 3H_2O$ was in a stable phase at low pH (pH < 3.16) and then transformed into FeF_2 and FeF_3 at higher pH [15,16]. However, in the HCl/HF system, HCl made insoluble ferric fluoride, preventing the formation of fluoride precipitates [17].

3.2. *Phosphatability*

Figure 8 shows EPMA images of a cross section of 1.0Si steel on which a phosphate coating was formed. Under conventional pickling conditions, phosphate crystals were rarely formed; under HCl-based pickling, the areas where phosphate crystals did not form were sparse. However, under HNO_3-based pickling, the phosphate crystals were uniformly formed. As shown in Figures 5 and 6, the F compounds found after pickling were not observed after the phosphate treatment. Therefore, the F compounds were involved in the formation of phosphate crystals. In other words, the F compounds introduced by pickling had a positive effect on the phosphatability of the steel.

The phosphate treatment of AHSS was performed with various concentrations of hydrochloric acid (HCl) and nitric acid (HNO_3). SEM images of the surface after phosphate treatment are shown in Figures 9–12. When pickling was performed with NH_4HF_2, more phosphate crystals were formed compared to with the conventional pickling. However, as shown in Figures 9 and 10, phosphate crystals were not formed in many regions. Figures 11 and 12 show significantly fewer areas where phosphate crystals were not formed compared to Figures 9 and 10. Furthermore, the phosphate crystals became finer after pickling with a HNO_3-based solution than with a HCl-based solution. In addition, phosphate crystals were formed more uniformly on the surface of 1.0Si steel in both the HCl- and HNO_3-based solutions compared to those formed on 0.4Si steel.

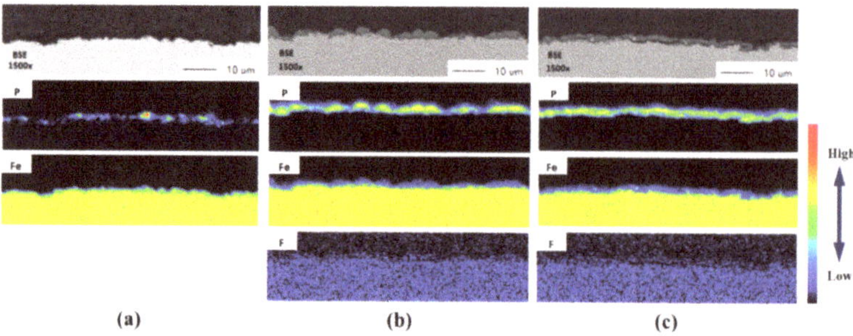

Figure 8. EPMA images of cross sections of 1.0Si steel formed with phosphate coating using (**a**) conventional conditions, (**b**) a HCl-based condition, and (**c**) a HNO_3-based condition.

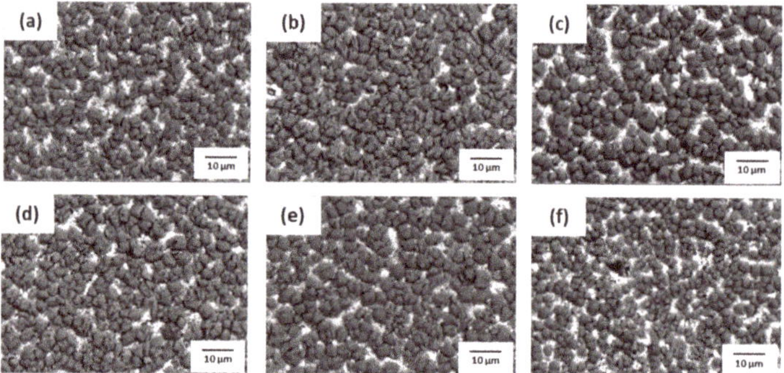

Figure 9. Surface SEM images of 1.0Si steel after phosphate treatment under HCl-based pickling condition with HCl concentration of (**a**) 5.5%, (**b**) 8%, (**c**) 10.5%, (**d**) 13%, (**e**) 15.5%, and (**f**) 18%.

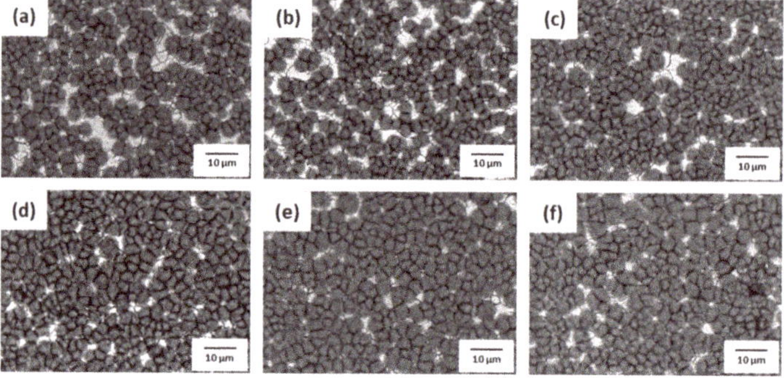

Figure 10. Surface SEM image of 0.4Si steel after phosphate treatment under HCl-based pickling condition with HCl concentration of (**a**) 5.5%, (**b**) 8%, (**c**) 10.5%, (**d**) 13%, (**e**) 15.5%, and (**f**) 18%.

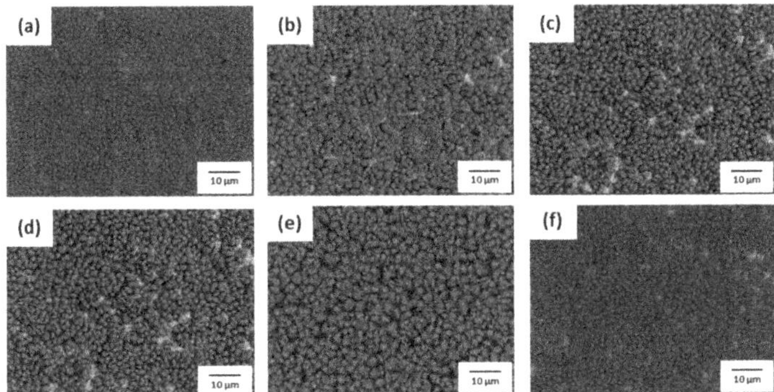

Figure 11. Surface SEM images of 1.0Si steel after phosphate treatment under HNO_3-based pickling condition with HNO_3 concentration of (**a**) 5.5%, (**b**) 8%, (**c**) 10.5%, (**d**) 13%, (**e**) 15.5%, and (**f**) 18%.

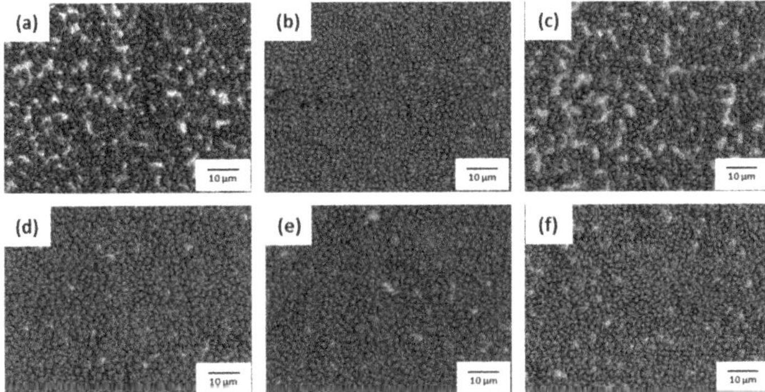

Figure 12. Surface SEM images of 0.4Si steel after phosphate treatment under HNO_3-based pickling condition with HNO_3 concentration of (**a**) 5.5%, (**b**) 8%, (**c**) 10.5%, (**d**) 13%, (**e**) 15.5%, and (**f**) 18%.

Figure 13 shows the coating coverage and coating weight as a function of the HCl and HNO_3 concentrations. Figure 13a shows that the coating coverage of 1.0Si steel decreased as the concentration of hydrochloric acid increased, whereas the coating coverage of 0.4Si steel increased as the concentration of hydrochloric acid increased. In addition, the coating coverage of both steels did not meet the phosphate quality requirements (95% or more) for automobiles. In the case of the coating weight, as the concentration of HCl increased, the coating weight of both 1.0Si and 0.4Si steel decreased. However, the phosphate quality requirement (2 g/m^2 or more) for automobiles was satisfied at all HCl concentrations except for 18%. As shown in Figure 13b, the coating coverage was not uniform at HNO_3 concentration below 10.5%, whereas HNO_3 concentration of 13% or more showed excellent coating coverage of at least 98%. A HNO_3 concentration of 8% was sufficient to achieve the desired coating coverage for 0.4Si steel. However, 1.0Si steel does not satisfy the desired coating coverage. Therefore, at least 13% of HNO_3 was required to meet the desired coating coverage. Furthermore, the coating weight was excellent at all concentrations.

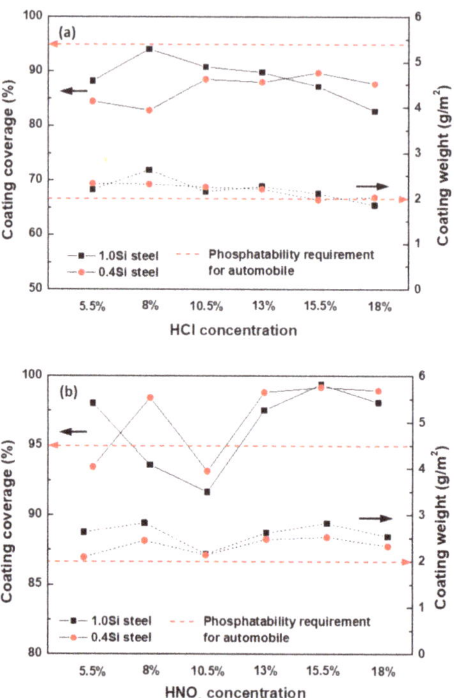

Figure 13. Coating coverage and coating weight vs. the inorganic acid concentration under (**a**) HCl-based condition and (**b**) HNO_3-based condition.

3.3. Electrochemical Imedance Spectroscopy

The results of EIS in the form of a Bode plot are shown in Figure 14. In the Bode plot, the low-frequency region is related to the surface film and the metal/surface interface, whereas the high-frequency region is attributed to the defects on the metal surface [18–20]. In the low-frequency region, the impedance (|Z|) values of 0.4Si steel were higher than those of 1.0Si steel. However, there was no significant difference in |Z| value with the use of different types of inorganic acids in both steels. The phase angle maxima and shoulder widths of the phase angles also showed the same trends as the |Z| values. The results of |Z| values and phase angle data indicate that the corrosion resistance of 0.4Si steel was superior to that of 1.0Si steel. However, there was no relationship between the concentration of inorganic acid and the |Z| values. Furthermore, improvement in corrosion resistance from the phosphate coating was not significant due to the short phosphating time used in the experiment. In general, the phosphate coating improves the corrosion resistance of a substrate. However, noticeable improvement in corrosion resistance is not observed during the stage of induction and commencement of film growth, which are the stages from the start of the phosphate treatment to 3 min [21,22]. Since the phosphating time in this study was within 2 min, the EIS results indicate that phosphate crystals had not sufficiently formed to be able to improve corrosion resistance for both types of inorganic acid. The phosphating industry generally uses coating weight and coverage as part of quality control, but coating weight and coverage do not have a direct relationship to corrosion resistance [21]. Therefore, the low correlation between coating weight, coverage, and the |Z| values was due to various factors, such as thickness and structure homogeneity.

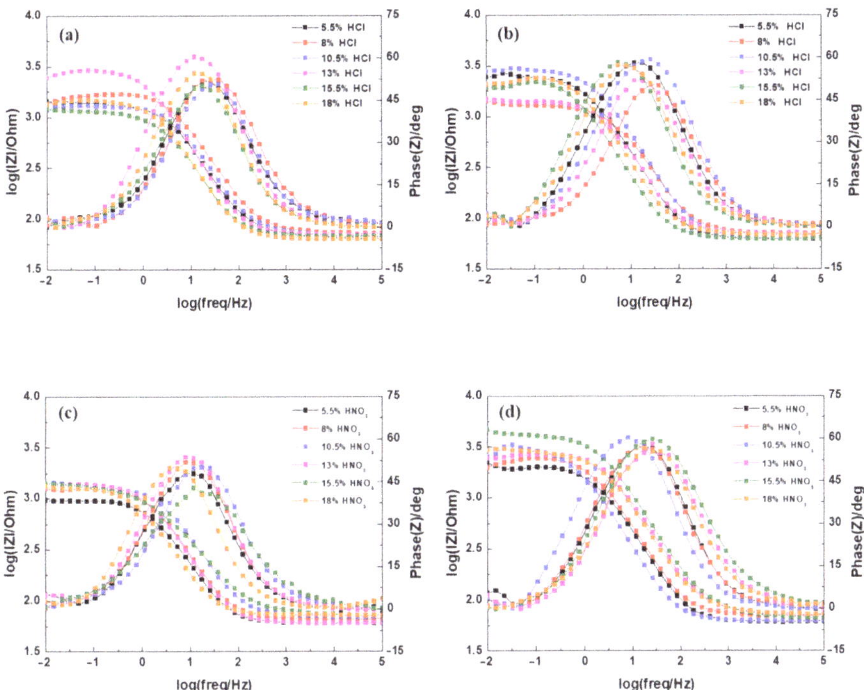

Figure 14. Bode impedance plots of electrochemical impedance spectroscopy (EIS) data of (**a**,**c**) 1.0Si steel and (**b**,**d**) 0.4Si steel in SAE solution.

The equivalent electric circuit model used to determine the optimized value for resistance and capacitance parameters is shown in Figure 15, and the corresponding EIS results are listed in Table 5. R_s is the solution resistance, CPE1 and R_{coat} are the capacitance and resistance of the phosphate coating layer, CPE2 is the double layer capacitance, R_{ct} is the charge transfer resistance, Q_{coat} and Q_{ct} are coating's capacitance and charge transfer capacitance, and n is an empirical exponent ($0 \leq n \leq 1$) measuring the deviation from the behavior of the ideal electric capacity. Figure 16 shows the variation of coating resistance (R_{coat}) from EIS measurement results as a function of inorganic acid concentration.

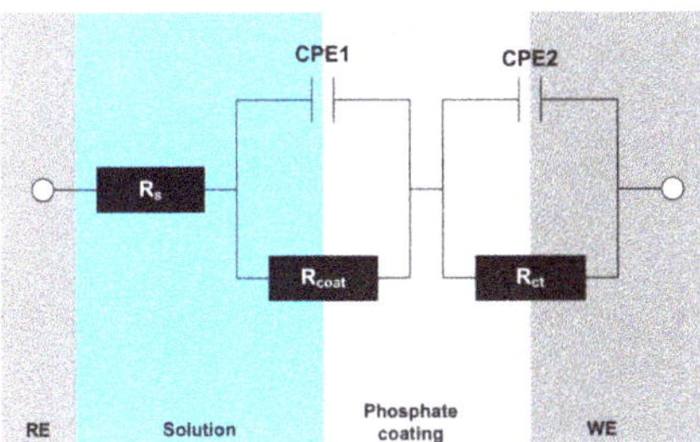

Figure 15. Equivalent circuit for phosphate coated steel in SAE solution. R_s is the solution resistance; CPE1 and R_{coat} are the capacitance and resistance of the phosphate coating layer; CPE2 is the double layer capacitance; R_{ct} is the charge transfer resistance; WE and RE are working electrode and reference electrode.

Table 5. Parameters from electrochemical impedance spectroscopy measurements.

Acid	Steel	Concentration	R_s ($\Omega\cdot cm^2$)	CPE1		R_{coat} ($\Omega\cdot cm^2$)	CPE2		R_{ct} ($\Omega\cdot cm^2$)
				Q_{coat} ($\mu F/cm^2$)	n_1		Q_{ct} ($\mu F/cm^2$)	n_2	
HCl	1.0Si	5.5%	62.98	56	0.8	1183	45.5	0.8	279.8
		8%	72.35	149	0.7301	358.2	42.4	0.9004	1285
		10.5%	63.58	178.3	0.5432	89.6	46.6	0.9395	1223
		13%	70.29	189.3	1	1083	64.7	0.9535	1777
		15.5%	67.04	108	0.8995	430.8	243	0.7957	673.2
		18%	63.41	208	0.975	140.6	104	0.8421	1,287
	0.4Si	5.5%	62.37	75	0.8419	1504	265	1	918.9
		8%	70.87	8990	0.6406	145.9	49.1	0.8305	1168
		10.5%	63.16	207.9	0.9996	940.4	45.8	0.8414	1878
		13%	70.72	981	1	179.4	70.6	0.8408	1186
		15.5%	62.67	179	0.8284	775.1	220	1	1320
		18%	69.29	134	0.8182	819.5	153	1	1413
HNO_3	1.0Si	5.5%	60.91	476	1	367.8	157	0.8468	528.2
		8%	66.97	120	0.8452	967.2	910	0.98	235.9
		10.5%	70.41	74.2	0.8295	1227	14,200	0.9539	138.1
		13%	61.01	186	0.8669	1059	250	0.8649	301.2
		15.5%	72.76	266	0.6565	1066	204	0.8387	272.2
		18%	73.47	300.1	0.8455	4300	397.7	0.9307	766.8
	0.4Si	5.5%	61.46	72.2	0.8527	1015	169	0.982	945.6
		8%	70.7	209	1	892.6	65	0.8414	1525
		10.5%	62.75	345	1	1332	108	0.8473	1680
		13%	72.12	48.7	0.9522	1564	66.2	0.7682	1036
		15.5%	65.98	29.4	0.8391	3812	2085	0.4284	690.9
		18%	73.74	38.9	0.8746	2000	426	0.635	932.9

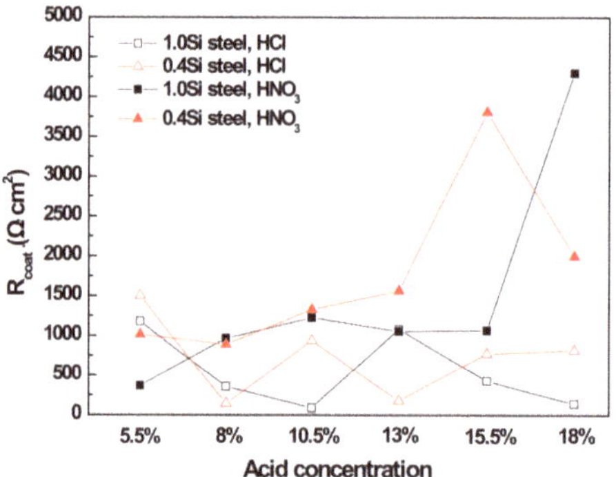

Figure 16. Variation of coating resistance (R_{coat}) from EIS analysis vs. inorganic acid concentration.

Unlike the |Z| values, there was no significant difference between the R_{coat} of 1.0Si and 0.4Si steel. However, the R_{coat} under HNO_3-based pickling conditions was higher than that under HCl-based pickling condition in both steels, and R_{coat} increased significantly as the concentration increased under HNO_3-based pickling conditions. This result is consistent with the coating coverage data shown in Figure 6. R_{coat} was affected more by coating coverage than by coating weight.

3.4. *pH Measurement*

In general, during the early stages of phosphate treatment, the reaction of iron or iron oxide with hydrogen ions increases locally with the pH at the metal surface, which promotes the nucleation of phosphate crystals [8]. The effects of iron oxides and F compounds generated under HNO_3-based pickling on the pH of the phosphate solution, are shown in Figure 17. The pH of the bulk solution did not change significantly when iron and iron oxides reacted with the phosphate solution. However, with F compounds, the pH of the bulk solution increased from 3.3 to 3.6 after reacting with the phosphate solution. Among F compounds, FeF_3 has a pH of 3.5–4.0 when dissolved in water (Equation (6)) [23]. Therefore, this F compound could have increased the pH of the phosphate solution by reacting with water. This increase in the pH of the phosphate solution indicates that the F compounds had a positive effect on the phosphatability of the steel because F compounds act as an accelerator for consumption of hydrogen ions.

$$FeF_3 + 3H_2O \rightarrow Fe(OH)_3 + 3HF \quad (6)$$

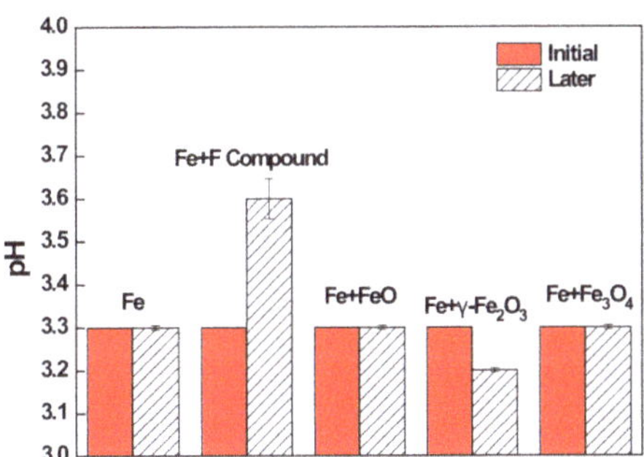

Figure 17. pH variation of phosphate solution before and after the addition of various iron oxides and F compounds.

3.5. *Surface Roughness Measurement*

The correlation between the surface roughness after pickling and the phosphatability of the steel is shown in Figure 18. The surface roughness was higher under a HNO_3-based pickling condition than under a HCl-based pickling condition. This result is due to NO_3^- ions acting as oxidizing agents; thus, the corrosiveness of HNO_3 is higher than that of HCl [24]. Generally, a higher surface roughness is associated with a higher coating weight and the formation of finer crystals; a level of 0.76–1.77 μm was reported to be the most suitable average roughness (Ra) [21,25]. In other words, the phosphatability was better with HNO_3-based pickling than with HCl-based pickling due to the surface roughness. However, the phosphatability did not have a significant correlation with surface roughness.

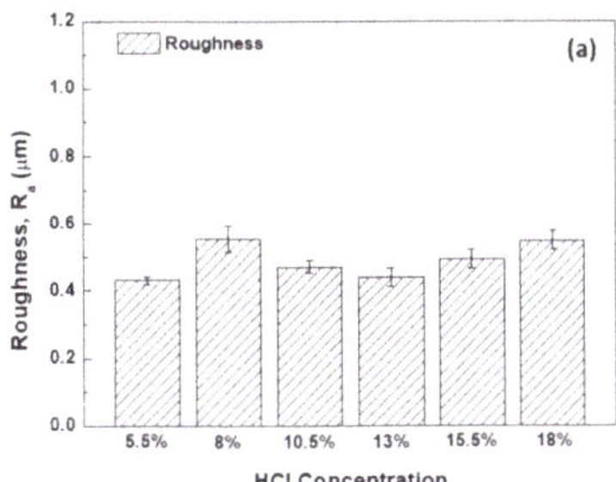

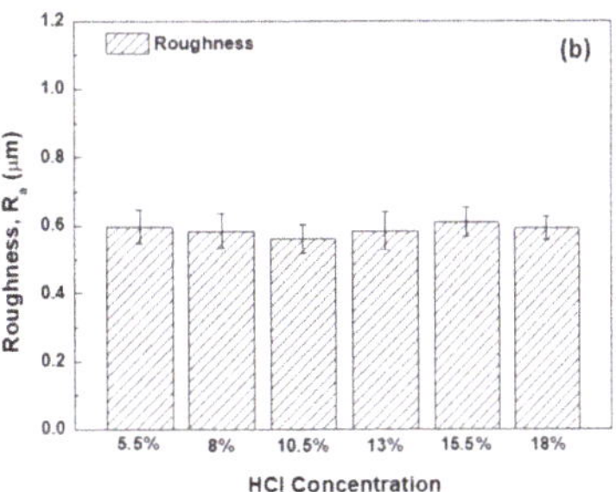

Figure 18. *Cont.*

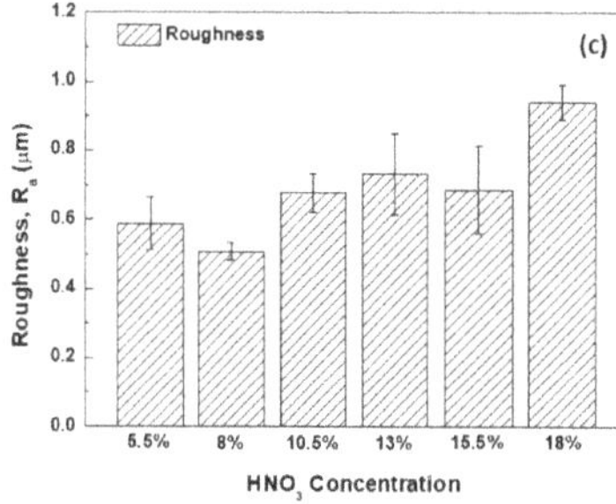

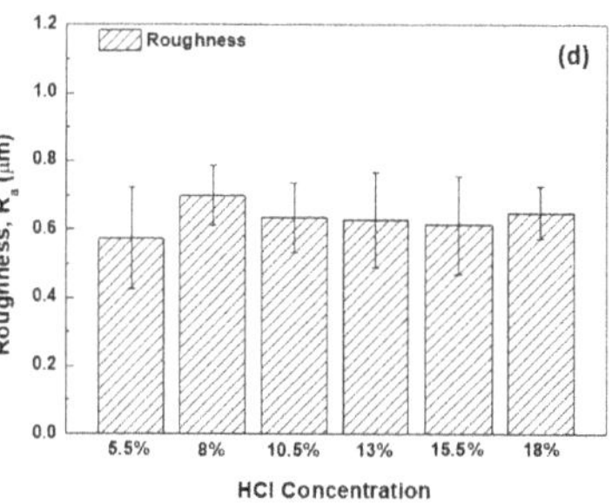

Figure 18. Correlation between surface roughness and phosphatability; (**a**,**c**) 1.0Si steel and (**b**,**d**) 0.4Si steel.

4. Conclusions

In this study, the optimum pickling conditions to improve the phosphatability of AHSS were investigated using various types of surface analysis and EIS. The conclusions based on our investigation are as follows:

- With HNO_3-based pickling solutions, the phosphatability improved remarkably at HNO_3 concentrations higher than 13% for both steels. Furthermore, the phosphate crystals became finer after pickling with a HNO_3-based solution compared to those with a HCl-based solution.
- SiO_2 was noticeably removed by a pickling solution with NH_4HF_2.
- The corrosion resistance of phosphate-treated AHSS was higher using a HNO_3-based pickling condition compared to a HCl-based pickling condition.
- With HNO_3-based pickling solutions, F compounds, which are involved in the phosphate treatment process, formed on the surface of the AHSS. The F compounds reacted with the phosphate solution to increase the pH of the phosphate solution, thereby greatly improving the phosphatability of AHSS.
- The phosphatability was better under HNO_3-based pickling conditions than under HCl-based pickling conditions due to the increased surface roughness.

Author Contributions: Conceptualization, S.C. and Y.-H.Y.; methodology, J.-G.K.; software, S.-J.K. and J.-S.Y.; validation, Y.-H.Y., J.-C.P. and J.-G.K.; formal analysis, S.C.; investigation, S.C. and S.-J.K.; resources, J.-C.P. and Y.-H.Y.; data curation, S.C.; writing—original draft preparation, S.C.; writing—review and editing, S.C., S.-J.K., J.-S.Y., Y.-H.Y. and J.-G.K.; visualization, S.C.; supervision, J.-G.K.; project administration, Y.-H.Y. and J.-G.K.; funding acquisition, J.-G.K. All authors have read and agreed to the published version of the manuscript.

Funding: This research was funded by POSCO, grant number 2017Z068.

Institutional Review Board Statement: Not applicable.

Informed Consent Statement: Not applicable.

Data Availability Statement: Data sharing is not applicable to this article.

Conflicts of Interest: The authors declare no conflict of interest.

References

1. Kulekei, M.M. Magnesium and its alloys applications in automotive industry. *Int. J. Adv. Manuf. Technol.* **2008**, *39*, 851–865. [CrossRef]
2. Tisza, M.; Czinege, I. Comparative study of the application of steels and aluminum in lightweight production of automotive parts. *Int. J. Lightweight Mater. Manuf.* **2018**, *1*, 229–238.
3. Machado, J.J.M.; Nunes, P.D.P.; Marques, E.A.S.; Da Silva, L.F.M. Adhesive joints using aluminium and CFRP substrates tested at low and high temperature under quasi-static and impact conditions for the automotive industry. *Compos. Part B* **2019**, *158*, 102–116. [CrossRef]

4. Kang, J.; Rao, H.; Zhang, R.; Avery, K.; Su, X. Tensile and fatigue behavior of self-piercing revets of CFRP to aluminium for automotive application. *IOP Conf. Ser. Mater. Sci. Eng.* **2016**, *137*, 12–25. [CrossRef]
5. Matheis, R.; Eckstein, L. Aluminium-carbon fibre-reinforced polymer hybrid crash management system incorporating braided tubes. *Int. J. Automot. Compos.* **2016**, *2*, 330–355. [CrossRef]
6. Fathyunes, L.; Azadbeh, M.; Tanhaei, M.; Sheykholeslami, S.O.R. Study on an elaborated method to improve corrosion resistance of zinc phosphate coating. *J. Coat. Technol. Res.* **2017**, *14*, 709–720. [CrossRef]
7. Kathavate, V.S.; Pawar, D.N.; Bagal, N.S.; Deshpande, P.P. Role of nano ZnO particles in the electrodeposition and growth mechanism of phosphate coatings for enhancing the anti-corrosive performance of low carbon steel in 3.5% NaCl aqueous solution. *J. Alloys Compd.* **2020**, *823*, 153812. [CrossRef]
8. Tegehall, P.-E.; Vannerberg, N.G. Nucleation and formation of zinc phosphate conversion coating on cold-rolled steel. *Corros. Sci.* **1991**, *32*, 635–652. [CrossRef]
9. Nomura, M.; Hashimoto, I.; Kamura, M.; Kozuma, S. Development of high strength cold-rolled steel-sheets with excellent phosphatability. *Kobeco Technol. Rev.* **2007**, *28*, 44–48.
10. Jung, B.H.; Lee, K. Method and Device for Pickling Ultra-High Strength Steel Sheet. Kr. Patent KR101696117B1, 13 January 2017.
11. Park, H.K.; Nho, H.S.; Kwak, S.J.; Park, S.H. Method for Pickling Hot Rolled Steel Sheet Having Advanced High Strength. Kr. Patent KR20120074135A, 5 July 2012.
12. Lee, C.T. Non-HF type etching solution for slimming of flat panel display glass. *Appl. Chem. Eng.* **2016**, *27*, 101–109. [CrossRef]
13. Kraydenko, R.I.; Dyachenko, A.N.; Malyutin, L.N.; Petlin, I.V. The mechanism for production of beryllium fluoride from the product of ammonium fluoride processing of beryllium-containing raw material. *IOP Conf. Ser. Mater. Sci. Eng.* **2016**, *135*, 12–21. [CrossRef]
14. Pastushenko, A.; Lysenko, V. Electrochemical synthesis of luminescent ferrous fluorosilicate hexahydrate ($FeSiF_6{\cdot}6H_2O$) nanopowders. *RSC Adv.* **2016**, *6*, 8093–8095. [CrossRef]
15. Reddy, R.G.; Wang, S.; Chen, B. Solubility of iron in spent pickling solutions. *Min. Met. Explor.* **1993**, *10*, 102–107. [CrossRef]
16. Sartor, M.; Buchloh, D.; Rogener, F.; Reichardt, T. Removal of iron fluorides from spent mixed acid pickling solutions by colling precipitation at extreme temperatures. *Chem. Eng. J.* **2009**, *153*, 50–55. [CrossRef]
17. Park, H.S.; Cho, J.H.; Jung, J.H.; Duy, P.P.; Le, A.H.T.; Yi, J. A review of wet chemical etching of glass in hydrofluoric acid based solution for thin film silicon solar cell application. *Curr. Phot. Res.* **2017**, *5*, 75–82.
18. Kim, K.H.; Lee, S.H.; Nguyen, D.N.; Kim, J.G. Effect of cobalt on the corrosion resistance of low alloy steel in sulfuric acid solution. *Corros. Sci.* **2011**, *53*, 3576–3587. [CrossRef]
19. Hong, M.S.; Kim, S.H.; Im, S.Y.; Kim, J.G. Effect of Ascorbic acid on the pitting resistance of 316L stainless steel in synthetic tap water. *Met. Mater. Int.* **2016**, *22*, 621–629. [CrossRef]
20. Kissi, M.; Bouklah, M.; Hammouti, B.; Benkaddour, M. Establishment of equivalent circuits from electrochemical impedance spectroscopy study of corrosion inhibition of steel by pyrazine in sulphuric acidic solution. *Appl. Surf. Sci.* **2006**, *252*, 4190–4197. [CrossRef]
21. Sankara Narayanan, T.S.N. Surface pretreatment by phosphate conversion coatings—A review. *Rev. Adv. Mater. Sci.* **2005**, *9*, 130–177.
22. Asadi, V.; Danaee, I.; Eskandari, H. The effect of immersion time and immersion temperature on the corrosion behavior of zinc phosphate conversion coatings on carbon steel. *Mater. Res.* **2015**, *18*, 706–713. [CrossRef]
23. Jones, D.A. *Principles and Prevention of Corrosion*, 2nd ed.; Prentice Hall, Inc.: Upper Saddle River, NJ, USA, 1996.
24. Kim, H.J. Variation of phosphatability with chemical composition and surface roughness of steel sheet. *Surf. Eng.* **1998**, *14*, 265–267. [CrossRef]
25. Meshri, D.T. *Kirk-Othmer Encyclopedia of Chemical Technology; Fluorine Compounds, Inorganic, Iron*, 5th ed.; John Wiley & Sons, Inc.: Hoboken, NJ, USA, 2000.

MDPI
St. Alban-Anlage 66
4052 Basel
Switzerland
www.mdpi.com

Materials Editorial Office
E-mail: materials@mdpi.com
www.mdpi.com/journal/materials

www.ingramcontent.com/pod-product-compliance
Lightning Source LLC
LaVergne TN
LVHW071010170726
843515LV00006B/1126

* 9 7 8 3 0 3 6 5 8 9 8 5 5 *